The World of
Niagara Wine

The World of
Niagara Wine

Michael Ripmeester, Phillip Gordon Mackintosh,
and Christopher Fullerton, editors

**WILFRID LAURIER
UNIVERSITY PRESS**

Wilfrid Laurier University Press acknowledges the financial support of the Government of Canada through the Canada Book Fund for our publishing activities.

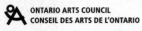
ONTARIO ARTS COUNCIL
CONSEIL DES ARTS DE L'ONTARIO

Library and Archives Canada Cataloguing in Publication

The world of Niagara wine / Michael Ripmeester, Phillip Gordon Mackintosh, and Christopher Fullerton, editors.
Includes bibliographical references and index.
Issued also in electronic formats.

ISBN 978-1-55458-360-7

1. Wine industry—Ontario—Niagara Peninsula. 2. Wine and wine making—Ontario—Niagara Peninsula. 3. Wineries—Ontario—Niagara Peninsula. I. Ripmeester, Michael R. (Michael Robert), 1962– II. Mackintosh, Phillip Gordon, 1960– III. Fullerton, Christopher Adam, 1970–

TP559.C3W67 2013   663'.200971338   C2012-907188-9

Electronic monograph.
Issued also in print format.
ISBN 978-1-55458-406-2 (EPUB).—ISBN 978-1-55458-405-5 (PDF)

1. Wine industry—Ontario—Niagara Peninsula. 2. Wine and wine making—Ontario—Niagara Peninsula. 3. Wineries—Ontario—Niagara Peninsula. I. Ripmeester, Michael R. (Michael Robert), 1962– II. Mackintosh, Phillip Gordon, 1960– III. Fullerton, Christopher Adam, 1970–

TP559.C3W67 2013   663'.200971338   C2012-907189-7

Cover design by Angela Booth Malleau. Cover photo by Arthur Kwiatkowski / iStockphoto. Text design by Angela Booth Malleau.

© 2013 Wilfrid Laurier University Press
Waterloo, Ontario, Canada
www.wlupress.wlu.ca

This book is printed on FSC recycled paper and is certified Ecologo. It is made from 100% post-consumer fibre, processed chlorine free, and manufactured using biogas energy.

Printed in Canada

For our families

Contents

Section Three: The Vineyard to the Bottle

Section Four: A Cultural Perspective on Niagara Wines

Map of the Niagara Region and Its Wineries

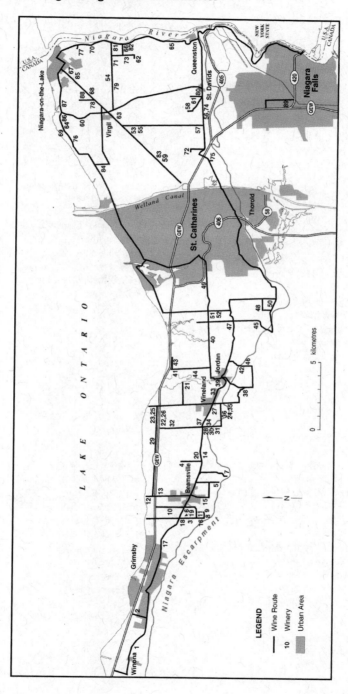

Niagara Wineries – December 2012

WINONA

1 Puddicombe Farms & Winery
1468 Hwy. #8

GRIMSBY

2 Kittling Ridge Estate Wines & Spirits
297 South Service Rd.

BEAMSVILLE

3 Angels Gate Winery
4260 Mountainview Rd.

4 Cornerstone Estate Winery
4390 Tufford Rd.

5 Crown Bench Estates Winery
3850 Aberdeen Rd.

6 Daniel Lenko Estate Winery
4246 King St.

7 De Sousa Wine Cellars
3753 Quarry Rd.

8 EastDell Estates Winery
4041 Locust Lane

9 Fielding Estate Winery
4020 Locust Lane

10 Good Earth Vineyard and Winery (The)
4556 Lincoln Ave.

11 Hidden Bench Vineyards & Winery
4152 Locust Lane

12 Legends Estates Winery
4888 Ontario St. North

13 Magnotta Winery
4701 Ontario St.

14 Malivoire Wine Co.
4260 King St.

15 Mountain Road Wine Company
4016 Mountain Rd.

16 Organized Crime Winery (The)
4043 Mountainview Rd.

17 Peninsula Ridge Estates Winery
5600 King St.

18 Rosewood Estates Winery & Meadery
4352 Mountainview Rd.

19 Thirty Bench Wines
4281 Mountainview Rd.

20 Thomas & Vaughan Estate Winery
4245 King St.

VINELAND

21 Alvento Winery
3048 Second Ave.

22 Birchwood Estate Wines
4679 Cherry Ave.

23 Crispino Wines & Crispino Estate Winery
4890 Victoria Ave.

24 Featherstone Winery & Vineyard
3678 Victoria Ave.

25 Foreign Affair Winery (The)
4890 Victoria Ave. North

26 Green Lane Estate Winery
4679 Cherry Ave.

27 Kacaba Vineyards
3550 King St.

28 Lakeview Cellars Estate Winery
4037 Cherry Ave.

29 Maple Grove Estate Winery
4063 North Service Rd.

30 Megalomaniac
(John Howard Cellars of Distinction)
3930 Cherry Ave.

31 Ridgepoint Wines
3900 Cherry Ave.

32 Royal DeMaria Wines
4551 Cherry Ave.

33 Stoney Ridge Estate Winery
3201 King St.

34 Tawse Winery
3944 Cherry Ave.

35 Twenty Twenty-Seven Cellars
3678 Victoria Ave.

36 Vineland Estates Winery
3620 Moyer Rd.

37 Wayne Gretzky Estates Winery
3751 King St.

JORDAN

38 Calamus Estate Winery
3100 Glen Rd.

39 Cave Spring Cellars
3836 Main St.

40 Creekside Estate Winery
2170 Fourth Ave.

41 Harbour Estates Winery
4362 Jordan Rd.

42 Flat Rock Cellars
2727 Seventh Ave.

43 Le Clos Jordanne
2540 South Service Rd.

44 Nyarai Cellars
4055 Nineteenth St.

45 Rockway Glen Estate Winery Inc.
3290 Ninth St.

Niagara Wineries – December 2012

46 Sue-Ann Staff Estate Winery
3210 Staff Ave.

47 Trillium Hill Estate Winery
3420 Ninth St.

ST. CATHARINES
48 13th Street Winery
1776 Fourth Ave.

49 Harvest Estate Wines
1179 Fourth Ave.

50 Henry of Pelham Winery
1469 Pelham Rd.

51 Hernder Estate Wines
1607 Eighth Ave

52 Kings Court Estate Winery
2083 Seventh St

NIAGARA-ON-THE-LAKE
53 Diamond Estates
1067 Niagara Stone Rd.

54 Caroline Cellars
1029 Line 2

55 Cattail Creek Estate Winery
R.R. 4, 1156 Concession 6

56 Chateau Des Charmes Wines
1025 York Rd.

57 Colaneri Estate Winery
348 Concession 6

58 Coyote's Run Estate Winery
485 Concession 5

59 De Moura Winery Way (The)
545 Niagara Stone Rd.

60 Domaine Vagners
1973 Four Mile Creek Rd.

61 Five Rows Craft Wine
361 Tanbark Rd.

62 Frogpond Farm Organic Winery
1385 Larkin Rd.

63 Hillebrand Estates Winery
1249 Niagara Stone Rd.

64 Hinterbrook Estate Wines
1181 Lakeshore Rd., R.R. 3

65 Ice House (The)
14778 Niagara Parkway Rd.

66 Inniskillin Wines
1499 Line 3

67 Jackson-Triggs Niagara Estate Winery
2145 Niagara Stone Rd.

68 Joseph's Estate Wines
1811 Niagara Stone Rd.

69 Konzelmann
1098 Lakeshore Rd.

70 Lailey Vineyard Wines
15940 Niagara River Pkwy.

71 Reimer Vineyards
1289 Line 3

72 Maleta Winery
450 Queenston Rd., R.R. 4

73 Marynissen Estates Winery
1208 Line One

74 Mike Weir Estate Winery Limited
1025 York Rd.

75 Niagara College Teaching Winery
135 Taylor Road

76 Palatine Hills Estate
911 Lakeshore Rd., R.R. 3

77 Peller Estates Winery
290 John Street East, R.R. 1

78 Pillitteri Estates Winery
1696 Niagara Stone Rd.

79 PondView Estate Winery
925 Line 2, 905-468-0777

80 Ravine Vineyard Estate Winery
1366 York Rd.

81 Reif Estate Winery
15608 Niagara River Pkwy.

82 Riverview Cellars Estate Winery
15376 Niagara Pkwy.

83 Southbrook Vineyards
581 Niagara Stone Rd.

84 Stonechurch Vineyards Inc.
1270 Irvine Rd., R.R. 5

85 Stratus Vineyards
2059 Niagara Stone Rd.

86 Strewn Winery
1339 Lakeshore Rd., R.R. 3

87 Sunnybrook Farm Estate Winery
1425 Lakeshore Rd.

88 Vignoble Rancourt Winery
1829 Concession 4

NIAGARA FALLS
89 Vincor Canada
4887 Dorchester Rd.

Foreword

Konrad Ejbich

Back in the days when I was a novice "student" of wine—there were no actual wine courses back then—I read every book on the subject I could find in the Ottawa public library or in the used bookstores I scrutinized for a discarded gem. My treasure trove uncovered a few classic but dusty recollections from writers like George Saintsbury, *Notes on a Cellar-Book*; Harry Waugh, *Harry Waugh's Wine Diary*; and Frank Schoonmaker, *Frank Schoonmaker's Encyclopedia of Wine*. On the reference side, veteran wine merchant Alexis Lichine had just issued a first edition of *The Encyclopedia of Wines & Spirits*, and a new fellow on the scene, Hugh Johnson, had released his first *Pocket Wine Book*. The only Canadian-authored wine book I was able to dig up at the time was *Toronto Star* editorial writer George Bain's wonderfully chatty *Champagne Is for Breakfast*.

All of these writers focused on Old World wines, the main reason being that this was all that counted in the mid-1970s. Canada produced ports, sherries, and pop wines with cute "duckie" names.

Fast forward a few decades and we are awash in books by international and Canadian authors about wines made in every corner of the world, with many titles appearing about our homegrown industry. This presents a challenge to every writer who is looking for something new to say. I speak from experience.

What sets *The World of Niagara Wine* apart is its voice and its perspective. It is authored by a group of deeply curious academics who have an abiding passion for wine and for all that happens in the Niagara Peninsula. This book is the fruit, so to speak, of their labour of love.

Now, there isn't much I enjoy more than tasting wine, except perhaps talking with folks who know more about it than I do, a privilege I enjoyed thoroughly during a working visit to Brock University in the fall of 2010, where I met with many of the contributors to this collection. In the hierarchy of wine pleasures, reading books, essays, and articles written by thoughtful people is as much a source of joy as the others. That's where this book comes in.

Here, many of the writers, but certainly not all, are geography instructors in Brock's Department of Geography. I admit that before our first meeting and editorial consultation, I had imagined a limited study of weather patterns, terrain, subsoils, and geological origins of the rocks and The Falls. I discovered much more than a predictable collection of academic papers.

Leafing once again through the manuscript to write these words, I feel as if I'm sitting in on a roundtable discussion with a group of like-minded folks, all talking about wines while sipping some. In fact, I recommend the same for you. Get yourself a glass of something local before curling up with this book to listen in on the conversation.

Acknowledgements

We would like to acknowledge several layers of support. First, we thank all of the contributors to this volume. Their hard work and insightful research has really made this book what it is. Sadly, we note that Alun Hughes, a long time colleague in the Department of Geography, passed away just as the volume neared publication. He will be missed.

Second, we would like to extend gratitude to the following Brock units for their financial and institutional support:

The Office of the Dean of Social Sciences
The Office of the Dean of Applied Health Studies
The Office of the Dean of Math and Sciences
The Brock University Advancement Fund
The Cool Climate and Viticulture Institute at Brock University.

We also gratefully acknowledge the cartographic skills of Loris Gasparotto. His maps and diagrams always make good work better.

Finally, we are deeply indebted to Lisa Quinn and Rob Kohlmeier at Wilfrid Laurier University Press. We are very grateful for your initial interest in this project and for your continued support during the long writing and publication process.

Introduction

The World of Niagara Wine

Michael Ripmeester, Phillip Gordon Mackintosh, and Christopher Fullerton

U ntil it was recently replaced, a billboard proclaiming "Niagara Wine Country: More to discover" welcomed visitors and residents crossing into Niagara. Such a welcome was not always so specific. A scant forty years ago, this sign could have singled out the region's celebrated and thriving tender-fruit industry, or even alluded to its strong manufacturing base. At that time, wine was hardly the symbol of Niagara, whose grape growers raised native *Vitis labrusca* grapes that, according to at least one observer, were suited only for making "jug wine" for a "pop-drinking public."[1] By the early 1990s, however, grape growers and winemakers in Niagara had adopted new varietals and imposed new standards for winemaking. It was not long before people started to notice the improvements. In 1993, for instance, Ken MacQueen wrote in the *Kitchener-Waterloo Record*: "[C]learly something wonderful is happening here in Niagara. It's like returning to find that the snot-nosed brat next door has grown into the young Kate Hepburn. That the Baby Duck of our youth is now a swan, or at least a graceful Gewürztraminer—a wine much easier to pronounce after the second glass."[2] Now, as we move into the second decade of this new century, grapes and wine busily consume the region's social, economic, cultural, and ideational resources, and preoccupy the imaginations of local governments, planners, and entrepreneurs.

Our purpose in this book is to explore the combination of nature and human effort that has shaped Niagara as one of Canada's leading grape-growing and wine-producing regions. If this statement is true, it is because of both policy and people. In a policy context, the Government of Ontario's wine strategy has "focus[ed] on securing a strong, prosperous future for the Ontario grape and wine industry" in order to fulfill a government commitment to make "the Ontario wine industry a thriving $1.5 billion business [by] 2020."[3] With political support, Niagara wines are flourishing; the status of grape growing is less certain. All the while, Niagara-centred oenologists and oenophiles have scrutinized the Niagara wine industry and its burgeoning geography, publishing numerous Niagara wine books that include, but are hardly limited to, the following: Linda Bramble's *Niagara's Wine Visionaries: Profiles of the Pioneering Winemakers*; Rod Phillips's *Ontario Wine Country*; Walter Sendzik's *Insider Guide to the Niagara Wine Region*; Konrad Ejbich's *A Pocket Guide to Ontario Wines, Wineries, Vineyards, & Vines*; and Donald Ziraldo's *Anatomy of a Winery: The Art of Wine at Inniskillin*.[4] These, and books like them, play a crucial part in the construction not just of a wine industry, but of a wine imaginary in Niagara, at once social and geographical, scientific and cultural, economic and aesthetic.

The World of Niagara Wine is supported by the research efforts of local academics. Researchers affiliated with Brock University, Niagara College, and numerous other institutions conduct inquiries as varied as precision viticulture, the psychology of taste, sub-appellation, soil composition, and Icewine climatology, among others. In particular, Brock's multi-million-dollar Cool Climate Oenology and Viticulture Institute (CCOVI) houses much of this scholarship. Since its founding in 1996, CCOVI's mandate has expanded to include scientific and social scientific research, with representation from Brock's Faculties of Applied Health Sciences, Business, Math and Science, and Social Sciences, as well as from independent researchers, authors, and consultants.[5] Yet despite Niagara's growing reputation as a centre of wine research excellence, the often breathtaking human and physical geographies of the region merely hint at the historical, economic, scientific, psychological, political, social, and cultural underpinnings of Niagara wine.

While acknowledging the worth of the efforts that have preceded it, *The World of Niagara Wine* fills a gap in the thinking about, and the ever-growing literature on, Niagara wines. It is the first to offer both an introductory and transdisciplinary accounting of the Niagara grape and wine industry, approaching the subject academically but for a wide and thoughtful readership. Our insistence on transdisciplinary approaches and our hopes of an audience both within and beyond the academy provides opportunities as well as challenges.

One of the goals of this book is to demonstrate that while the maturation of Niagara's grape and wine industry has done much to boost the region's economic fortunes and image, some uneasy tensions have emerged in this process. Few would suggest that the jobs created, the visitors attracted, and the libations concocted by Niagara wineries are in and of themselves negative: in fact, many in the region regard the continued growth of the industry as critical to the region's economic renaissance. As wine-lured visitors to the region are often unaware, the wine economy in Niagara has boomed as manufacturing has busted, with decades of factory closings and thousands of job losses. Similarly, producers and processors associated with the region's once-thriving tender-fruit industry—grapes, peaches, cherries, pears, berries, nectarines, plums, and so on—have borne the brunt of a changing global economy, and they see themselves as having little to gain from Niagara's wine economy. Tender-fruit farmers, for example, who no longer have a market for their produce, consider switching to grape growing an expensive and risky proposition, and the skill sets of former factory workers are not those sought by wineries.

These tensions have certainly arisen and been on public display in the context of the ostensibly celebratory Niagara Wine Festival. The festival organizers' recent decision to change not only the event's name, from the original "Niagara Grape and Wine Festival," but also to use it as a premier tourist attraction, has not been well received by all local residents. Many see these shifts as a not-so-minor form of "treason" against "Grape and Wine," the nickname for the sixty-year-old fall grape-and-wine bash. For these and many other reasons, the story of Niagara's transformation into a world-class wine region, as it is thus far written, is somewhat bittersweet. It is this bittersweetness that has prompted in the editors and authors of this volume a commitment to presenting a wide-ranging discussion—both critical and celebratory—of the world of Niagara wine.

In this context, *The World of Niagara Wine* contributes to the debate on the sustainable and inclusive economic, social, and cultural development of Niagara. However, the diverse group of authors participating in this project, the specific languages in which they speak and write, and the range of topics—from history to marketing to science to social and cultural critique—may give the book an uneven feel, albeit one necessary for representing this diversity of experience. To offset that perception, we offer a guide to what follows. To provide the reader with a general sense of the cultural, emotional, and institutional shape of grapes and wine in Niagara, and to make the transitions between styles and content less disruptive, we have organized the book into four subsections. The first, *Niagara Wines Decanted*, puts the region's grape and wine industry into historical and contemporary context. The second, *Business and Bottles*, builds a scaffolding around the business aspects of the grape and wine industry. The third section, *The Vineyard to the Bottle*, is the book's largest; it introduces readers to a combination of nature and science that makes Niagara wines unique. Finally, Section 4, *A Cultural Perspective on Niagara Wines*, peers into the Niagara grape and wine milieu with the critical gaze of the social sciences.

Section One: Niagara Wines Decanted

The three authors in this section provide an account of an influential period in the development of the Niagara grape and wine story. First, Alun Hughes explores the putative origins and history of grape growing and wine production in Niagara, roughly from the late 1700s to the late 1800s. This is no arcane matter, as evidenced by recent attempts to declare 2011 the bicentennial of Ontario wine production. Hughes's account challenges the widely accepted Johan Schiller and Porter Adams wine history narratives. More specifically, Hughes argues that the former was not the father of commercial winemaking in Ontario and the latter not the first to systematically plant grape vines. Wisely, Hughes rejects any broad claims. Rather, he concludes the impossibility of identifying any individual pioneer of grape growing or commercial wine production in Ontario.

Dan Malleck's research of the history of alcohol consumption in Niagara brings the book into the early and mid-twentieth century. Malleck offers a curious analysis of the regulation of wine production and consumption. Wine, he contends, was treated differently than beer and spirits early on. Temperance activists worried less about a refined class of wine drinker than about saloon and tavern dwellers and their harder potations. Indeed, it was this broad acceptability of wine that urged the Ontario government to begin regulating grape and wine production in the interwar years.

In closing the first section, Christopher Fullerton brings a modern-day perspective to the cumulative history of Niagara grape and wine production. He suggests that "wine" is becoming a lynchpin for a great deal of economic thinking in Niagara, ranging from the development of local tourism and downtown revitalization plans through to the promotion of the region's two major post-secondary institutions, as well as its home-building industry. As a prelude to several chapters found later in the book, however, Fullerton contends that with so much riding on the success of Niagara's grape and wine industry, a whole series of concerted efforts to protect Niagara's agricultural landscape, its farmers, and the wine industry itself will have to be made. Without these initiatives, he argues, several years of successful economic revitalization may well have been wasted, and communities throughout the Niagara wine region could see their long-term development aspirations shattered.

Section Two: Business and Bottles

Niagara is a relatively small grape-growing region compared to many of its competitors in the Old World and the New. Yet the profile of Niagara wines has evolved from humble beginnings, characterized by international insignificance and poor quality, to increasing status as an international producer of award-winning wine, which is especially true of its iconic Icewines. It is widely agreed that the Niagara wine industry is poised for further growth. However, as the authors in this section contend, this will require a combination of skillful marketing, continued quality assurance that includes strident regulation, and the creation of consistent identity and branding.

Linda Bramble's long engagement with the documentation of the turn-of-the-twenty-first-century evolution of Niagara wine production is evident in her brief history of the VQA (Vintners Quality Alliance)—she also provides a primer on the often misunderstood differences between the VQA and the International-Canadian Blends designations.[6] This is a unique history that demonstrates a curious irony in a world of corporate deregulation: Niagara winemakers, in order to warrant the provenance of their grapes and to develop and maintain the reputation they now enjoy, demanded provincial regulation from a Conservative government of a 1990s Ontario bent on deregulation. Regulation, it seems, created the curious policy circumstance for VQA success.

Next, Astrid Brummer, as a marketing analyst for the Liquor Control Board of Ontario (LCBO), uses her "insider" perspective to illustrate the sales robustness of Ontario wines (with Niagara wines comprising a large proportion) in an extraordinarily competitive market. Remarkably, she points out that despite the bleakness of the 2007–8 recession, Niagara's VQA wines saw overall net sales and volume growth higher than European wines, which in fact experienced sales declines.

As Niagara's wine industry grows and matures, the search continues for a distinctive place in the world. With this in mind, Maxim Voronov, Dirk De Clercq, and Narongsak Thongpapanl investigate the marketing evolution of a Niagara wine identity. They confirm that aesthetics sit at the heart of the wine business and insist that winemakers in Niagara should not hesitate "to cultivate images of artisanal, personal winemaking and distance itself from the perception" of industrial or mass production. Furthermore, they argue that commercial and artistic strategies must occur simultaneously. The authors conclude that Niagara's best strategy would be emulation of the European model—an idea challenged in some ways by Brummer's earlier observations.

Finally, Janet McLaughlin investigates the indispensability of temporary migrant labourers to the Niagara grape and wine industry. Despite the need for temporary labourers, they are largely invisible and remain socially and culturally marginalized in Niagara. What is more, McLaughlin contends that though the industry characterizes these workers as reliable, flexible, and productive, these traits better reflect the expectations, and especially the requirements, of the Temporary Foreign Workers Program, the Seasonal Agricultural Workers Program, and the farmers themselves. She contends that while this portrayal holds some truth, the world of migrant workers is, in actuality, a much messier mixture of economics and politics on both global and local scales.

Section Three: The Vineyard to the Bottle

In their chapter on the development and application of "precision viticulture," Marilyne Jollineau and Victoria Fast outline what we can only describe as a leading edge of wine science. Old World vintners enjoy millennia of practical grape-growing experience; this is not the case in New World wine regions. Thus,

Jollineau and Fast carefully explain how Niagara wineries, and Stratus Vineyards in particular, use various geospatial technologies, including digital mapping, to overcome this deficit. This includes the development of vineyard management practices and techniques. Geographic information systems (GIS) provide vineyard managers with microdata on virtually every square centimetre of their vineyards.

Anthony B. Shaw conducted the painstaking research that led to the designation of the Niagara Peninsula's sub-appellations;[7] he gives a powerfully descriptive accounting of the topographic and climatic attributes of what oenologists and viticulturalists call Niagara's cool climate region. Shaw is particularly keen to demonstrate the nexus of Niagara's physical geography and its climate, and how both affect—and effect—the region's viticulture.

Daryl Dagesse, in an instructive chapter on the soil science of Niagara wines, seeks to correct "many misconceptions regarding the soil as it pertains to oenology and viticulture." He argues that while prehistoric bedrock geology certainly guided the formation of Niagara's grape-producing soils, the more recent retreat of the last glacier in southern Ontario imbued Niagara with distinctive and anomalous fine-grained, clay-rich soils not traditional to other wine regions.

Andrew Reynolds, through a series of Niagara-based investigations into what is known as "the terroir effect," explores the correlation between viticultural practices and berry composition, which ultimately leads to a better quality of wine. Reynolds maintains that grape growers who are attentive to factors such as fruit exposure, canopy manipulation, pre-fermentation practices, and the actual site of the vineyard may improve monoterpene concentration (the compounds that establish aroma) of berries and juices of several *Vitis vinifera* cultivars.

The critical assessment of taste preoccupies Ronald S. Jackson's chapter. For Jackson, all discussions of vinous quality begin and end in the glass, although this is hardly a quest for objective wine-truths. Wine drinkers, he contends, should avoid suppressing personal tastes for those of authorities or the interested. Apropos to this book, Jackson also offers readers a tasting approach to Niagara vintages.

The frozen grape preoccupies Debbie Inglis, director of CCOVI, and Gary Pickering, who collaborate in a chapter devoted to one of Niagara's signature wines: Icewine. They offer a brief but thorough introduction to its history, science, and regulation that includes its origins and development, as well as the meticulous growing, harvesting, pressing, and vinification processes involved in its production. They explain that, under the authority of the VQA, strict standards have been established to control what can legally be marketed as Icewine. However, as Inglis and Pickering also note, Canadian Icewines face significant threats from the counterfeiting of Icewine in countries such as Taiwan and China.

Section Four: A Cultural Perspective on Niagara Wines

The authors of this section discover that Niagara wine, like most things in life, is often not what it appears to be. Michael Ripmeester and Russell Johnston examine the ways in which Niagarans incorporate grapes and wine into local identity. They began this research curious to uncover the heritage traditions of Niagara, and were surprised to find that grapes and wines are at least as important as the War of 1812 and its luminaries. In their most recent research, they have learned that people in St. Catharines most commonly answer "grapes and wine" in response to the question "What is the first thing that comes to mind when you hear the word 'Niagara'?" Despite this response, these participants have a rather ambiguous relationship with the wine industry.

In their chapter, Nick Baxter-Moore and Carolyn Charest investigate how Niagara vintners, using the actual design and construction of their wineries, attempt to create the genuineness believed necessary for branding and marketing their wines. This process of "constructing authenticity" includes the wineries' intentional association of the quality of the wine with the calibre of the winery and its environs. They conclude that wine consumers have learned to connect wine architecture with the wine itself.

David Telfer and Atsuko Hashimoto contend that Niagara wine and regional food are part of an extensive rebranding of Niagara to make it a centre of wine and culinary tourism in Ontario. They review both the benefits and challenges emerging from this tourism transformation, noting that the enthusiastic push for Niagara wine and food has done little to encourage locals to consume either. In the penultimate chapter, Hugh Gayler recounts the highly contentious and historical land-use planning process in Niagara. He specifically focuses on the local agricultural economy. As mid- to late-twentieth-century suburbanization of Niagara ate up great swaths of arable land, creative policy interventions were contrived to halt it. In particular, Gayler suggests that the introduction of Ontario's greenbelt legislation was a defining moment in the long-standing effort to protect Niagara's agricultural lands.

In the final chapter, Phillip Gordon Mackintosh considers the implications for public space in St. Catharines, given the current eagerness to invest wine with municipal and regional political-economic aspirations. His analysis of the Niagara Wine Festival's annual Grande Parade accentuates a persistent controversy: the supposed appropriation of a local grape and wine tradition by a wine industry more concerned about wine tourism than local nostalgia. Mackintosh concludes that sentimental Niagarans, correctly or incorrectly, regard the Wine Festival board's actions as an elite misinterpretation of both the parade and its historical-cultural purpose in the streets of St. Catharines.

❦ ❦ ❦

The World of Niagara Wine endeavours to reveal the complex relations between science, business, and social science that give form and substance to Niagara wines. There is also something serendipitous about the book and its research. It happily confirms the sentiments of that billboard: "Niagara Wine Country: More to discover." There is indeed more to discover, although we are struck by the irony of this simple roadside greeting. With so much enthusiasm for and focus on the science, culture, and economic potential of wine, the "more" is withering from neglect. And by "more," we mean Niagara's small family-run farms, its besieged urban blue-collar population, and the tradition of inclusive celebrations around grapes and wine. So, beware. Even if the "welcome" is genuine and heartfelt, it twinges from time to time with ambivalence. Niagarans love to share their warm hospitality, the pride they feel in their ongoing tradition of "Grape and Wine," and the growing international reputation of their wine industry. Yet they may not, alas, always like the wine tourist's single-minded motivation for visiting. Such is the world of Niagara wine.

Ultimately, our intent here is not to provide a comprehensive introduction to Niagara wines: such an undertaking exceeds our purpose and could only end in disappointment. Nevertheless, we hold that Niagara, its wineries, and its wines provide a fascinating opportunity to study how geography, history, science, and business all come together in the real world of everyday living in the region. If *The World of Niagara Wine* raises as many questions as it answers, we will have accomplished our goal. As editors and as residents of Niagara, we look forward to the dialogue this volume may provoke.

Notes

1　David Bruser, "From High Finance to the Wine Cellar: Winery Business 'Not a Playground.' Executives Gravitating to Grape Trade a Sign Niagara's Industry Is Maturing," *Toronto Star*, September 29, 2005, A1.

2　Ken MacQueen, "Ontario Wineries Are Improving with Age," *The Record* (Kitchener-Waterloo, Ontario), August 14, 1993, A6.

3　Ontario Wine Strategy Steering Committee. *Poised for Greatness*. Ministry of Consumer and Business Services: 2, 1. 2001.

4　Linda Bramble, *Niagara's Wine Visionaries: Profiles of the Pioneering Winemakers* (Halifax: Lorimer, 2009); R. Phillips, *Ontario Wine Country* (Vancouver and Toronto: Whitecap Books, 2006); W. Sendzik, *Insider Guide to the Niagara Wine Region* (Toronto: HarperCollins, 2006); K. Ejbich, *A Pocket Guide to Ontario Wines, Wineries, Vineyards, & Vines* (Toronto: McClelland and Stewart, 2005); D. Ziraldo, *Anatomy of a Winery: The Art of Wine at Inniskillin* (Toronto: Key Porter Books, 2000).

5　CCOVI fellows and affiliates also include researchers from the University of Guelph; Emory University, Georgia; Agriculture and Agri-Food Canada; the Agriculture Pacific Agri-Food Research Centre (PARC); Foreign Affairs and International Trade Canada; and the Vineland Research and Innovation Centre.

6　Until recently, these wines were designated as "Cellared in Canada."

7　See the sub-appellations map on VQA Ontario's website, http://www.vqaontario.ca/Appellations/NiagaraPeninsula.

Niagara Wines Decanted

One

The Early History of Grapes and Wine in Niagara

Alun Hughes

The impact of the War of 1812 on the Niagara Peninsula was devastating. Between October 1812 and November 1814, the Peninsula was the scene of frequent action. Major battles were fought at Queenston Heights, Fort George, Stoney Creek, Beaverdams, Chippawa, Lundy's Lane, and Fort Erie, and at other times the Peninsula experienced severe disruption from ransacking soldiers and Native warriors. Not surprisingly, many farmers and other property owners suffered major losses as houses and barns were burned, fences torn down, crops trampled, livestock killed, and household items taken. This is very evident in the war loss claims submitted after the return of peace, not only for damage caused by the Americans but also by the British and their Native allies.

One of these claims was submitted by Thomas Merritt (Figure 1.1) of Grantham Township, father of William Hamilton Merritt of Welland Canal fame. The elder Merritt's farm was located east of Twelve Mile Creek at the southern tip of Martindale Pond, now a part of St. Catharines.[1] His losses were fairly typical—buildings torched, fences and crops destroyed, and items stolen (including a chestnut horse, two stoves, and two feather beds). He also lost fruit trees and grape vines.[2] The fact that Merritt claimed for the vines suggests that they were not just growing wild, which makes this one of the earliest references to grape culture in the Niagara Peninsula.

Elizabeth Simcoe, wife of John Graves Simcoe, first lieutenant-governor of Upper Canada, mentions grapes and vines several times in her diary. But although she spent almost two years at Newark (later Niagara) between 1792 and 1796, all her references come from outside the Peninsula—for example, across the Niagara River at the "Ferry House opposite Queenston," where she "breakfasted in the Arbour covered by wild Vines," and at York (later Toronto), where she "gathered wild grapes ... pleasant but not sweet." She even mentions one case of impromptu winemaking, when soldiers laying out Dundas Street

Figure 1.1 Thomas Merritt
Source: Wm. Hamilton Merritt, *Memoirs of Major Thomas Merritt*,
U.E.L., 1909.

"met with quantities of wild Grapes & put some of the Juice in Barrels to make vinegar … it turned out very tolerable Wine" (though the reference to vinegar makes one wonder what sort of wine this was).[3] Making wine from wild grapes was rare, however, and most of the wine consumed in early Upper Canada was imported. Thus when Scottish traveller Patrick Campbell was entertained by Mohawk leader Joseph Brant at his home on the Grand River in 1792, it was port and Madeira they savoured, not locally produced wine.[4]

Early references to grapes and winemaking in the Niagara Peninsula are few (Merritt's war claim is a rare exception), which may seem strange given the present-day importance of the Niagara wine industry. However, it is understandable given the area's history. The purpose of this chapter is to outline that history, and to consider agriculture and the grape and wine industry in particular in the broader context of the political and economic developments of the time. The history falls into three phases: the Pre-Loyalist Period prior to the end of the American Revolutionary War in 1783; the Pioneer Period from 1783 to the end of the War of 1812 in 1815; and the Postwar Period from 1815 to a few years beyond Confederation in the early 1870s.

The Pre-Loyalist Period

When Europeans first made contact with Native peoples in what is now Canada in the late sixteenth and early seventeenth centuries, the Niagara Peninsula was inhabited by the Neutral Nation. Their primary homeland was in the Hamilton area, but their territory extended west as far as the Thames River and east for a short distance across the Niagara River. They formed part of the broader Iroquoian family, which included the Huron and Petun tribes around Georgian Bay and the Iroquois of the Finger Lakes (the latter comprising the Seneca, Cayuga, Oneida, Onondaga, and Mohawk of the Five Nations or Iroquois Confederacy).[5] The Neutral were so named by French explorer Samuel de Champlain (Figure 1.2) because they refused to take sides in the long-standing hostilities between the Huron and the Iroquois (though they were far from peace-loving when it came to other tribes).[6] What they called themselves is not known.

Figure 1.2 Samuel de Champlain
Source: H.P. Biggar, ed., *The Works of Samuel de Champlain*, Vol. 1, 1922.

Unlike the tribes of the Algonquian family, which occupied land further north in what is now Ontario, the Neutral and other Iroquoians supported themselves by farming as well as hunting and fishing. They employed a form of shifting agriculture based mainly on the cultivation of Indian corn, squashes, and kidney beans, occupying a site for two or three decades until its productivity declined. This allowed them to live in semi-permanent villages, typically a group of longhouses surrounded by palisades. One of these, dating from the early seventeenth century and possibly a regional capital, was the so-called Thorold Site, located on the Escarpment brow near Brock University.[7] Important Neutral burial sites at Grimsby and St. Davids date from the same era. Archaeological evidence aside, our knowledge of the Neutral comes mainly from the writings of early French missionaries. Thus Recollect Joseph de La Roche Daillon visited Neutralia in 1626, as did Jesuits Jean de Brébeuf and Pierre Chaumonot in 1640–41, though whether they ventured into the Niagara Peninsula is debatable.[8]

Such missionaries, together with fellow French explorers and administrators, were responsible for the first references to grapes and wine in Canada (unless, of course, one accepts the claim by some that Norseman Leif Eriksson gave the name Vinland to Newfoundland in 1001 because of wild grapes growing there). Sailing up the St. Lawrence in 1535, Jacques Cartier saw "so many vines loaded with grapes that it seemed they could only have been planted by husbandmen,"[9] and in 1603 Samuel de Champlain reported "many vines, on which there were exceedingly fine berries," from which they "made some very good juice."[10] Champlain speculated about making wine, but Recollect Nicolas Viel, writing from Huronia in 1623, provided the first report of anyone doing so: "When the wine which we had brought from Quebec in a little barrel of twelve quarts failed, we made some of wild grapes which was very good."[11]

Father Viel and other missionaries needed wine, of course, to celebrate mass. Sacramental wine was routinely imported from Europe (Spanish wine apparently being a favourite among the Jesuits), but when it was not available, the only choice was to make one's own. This was especially common when travelling long distances. In 1669, Sulpician missionaries Francis Dollier de Casson and René Bréhant de Galinée made wine, "as good as vin de Grave," while wintering on the Lake Erie shore in the Port Dover area during a year-long circumnavigation of southern Ontario.[12] Recollect Louis Hennepin makes several references to making wine during his North American travels in the late 1670s, on one occasion availing himself of the "wild vines loaded with grapes" near the Detroit River. (Later, however, something went amiss, for in 1680 he complains of not having been able to celebrate mass for nine months for lack of wine!)[13]

But wine was not just made out of necessity. Isaac de Razilly, governor of Acadia, writing to Marc Lescarbot, a French writer, lawyer, and traveller, in 1635, says that "The vines grow wild here and from the wine that was made from them we said mass";[14] a year later, Jesuit Paul Le Jeune writes from Huronia, "In some places there are many wild vines loaded with grapes; some have made wine of them through curiosity; I tasted it, and it seemed to me very good."[15] In neither of these cases is there any suggestion that imported wine was unavailable. Indeed, in the case of Acadia, imported grape vines may have been available also, for Razilly goes on to say that "Bordeaux vines have been planted that are doing very well," which may be the very first mention of grape cultivation in Canada. Others saw no need for imported vines, as witness Jesuit Jacques Bruyas, writing from Iroquois country in 1668, reports: "There are also vines, which bear tolerably good grapes, from which our fathers formerly made wine for the mass. I believe that, if they were pruned two years in succession, the grapes would be as good as those of France."[16]

None of these early sources mention Native peoples being involved in wine-making, because, in eastern North America at least, they never seem to have made wine. To quote Gabriel Sagard, who was with Viel in Huronia in 1623, "The Savages do eat the grape, but they do not cultivate it and do not make wine from it"—this because they lacked "the imagination or the proper equipment."[17] We may assume that what Sagard said about the Huron applied also to the Neutral in the Niagara Peninsula. They no doubt ate the wild grapes growing here, and possibly even made juice, but they never went as far as making wine.[18]

But the days of the Neutral were numbered anyway. In the late 1640s, the continuing antagonism between the Huron and Iroquois erupted into full-scale warfare. The Iroquois, having the advantage of muskets acquired from the Dutch, easily defeated the Huron, and then turned on the Petun and Neutral, effect-ively destroying both by 1651.[19] The Iroquois did not occupy the conquered territories, however, opting instead to return to the Finger Lakes region, and the Niagara Peninsula served only as a hunting ground and routeway. Later in the seventeenth century, the Iroquois were themselves displaced by the Algonquian Mississauga tribe from further north, but they did not occupy the Peninsula either.[20] In 1721, the French established a presence across the Niagara River at Fort Niagara, and in the mid-1760s, after the defeat of the French in the Seven Years War, the British built Navy Hall and Fort Erie on the west bank. Otherwise the Peninsula remained uninhabited from 1651 until about 1780, during the American Revolutionary War. For a century and a quarter, therefore, there was obviously no grape growing or winemaking in the Niagara Peninsula.

Fort Niagara played an important role in the war, safeguarding the vital por-tage route around Niagara Falls and providing John Butler's corps of Rangers a base from which they conducted raids into rebel-held territory. But the fort was almost entirely dependent on imported supplies, and it was not long before the presence of troops, Rangers, Natives, and a growing number of refugees created a major provisioning problem. Since an agreement with the Seneca precluded cultivation of the land beyond the immediate vicinity of the fort, the only solution was to develop the west bank. In June 1779, Frederick Haldimand, Governor of Quebec (of which Niagara was then a part), gave authorization for a small number of "good husbandmen" to move across the river, though it was probably the next year that they did so.[21] The intention was clear—to help feed the troops and refugees at the fort, rather than to create a permanent settlement. The land was to remain the property of the Crown, and when the war was over (and won) the settlers would reclaim their former homes in New York or the other colonies.

Delays in providing seed and tools meant that progress was slow, but in August 1782 a census of the "Settlement at Niagara" recorded sixteen farmers, sixty-seven family members, one male slave, 236 acres of cleared land, an assortment of livestock, and what appear to be healthy crop yields, primarily of Indian corn and potatoes. The following year, the growing settlement was surveyed by Allan McDonell; his map showed most of the settlers occupying rectangular 100-acre lots arranged in a formal grid pattern along the Niagara River as far as the Escarpment (Figure 1.3). A second census in 1784 revealed considerable progress, listing forty-six settlers as well as the dimensions of their houses and barns.[22] The impression was now one of stability—appropriately so, for with the war having been lost, what had been a temporary expedient would now become a permanent settlement.[23]

The Pioneer Period

The immediate postwar need was to provide land for all those encamped around Fort Niagara or across the river—former Rangers and other refugees (together comprising Canada's renowned United Empire Loyalists, though their status as such was not formalized until 1789),[24] discharged soldiers, and Natives who had fought for the Crown. This required three things: policies for the allocation of land, the acquisition of land, and surveys to lay out the land in townships and lots.

The policies came in 1783, in the form of *Additional Instructions* from the Crown stipulating township size and structure and how much land was to be granted to different classes of settler. The amounts (later modified) ranged from 100 acres for civilian heads of families (plus 50 acres for each family member) to 1,000 acres for field officers.[25] Initially, everyone would hold land under a variant of the seigneurial system of tenure current in Quebec, which quickly became a cause of great discontent before the freehold system was introduced in 1788.[26]

Land acquisition involved the purchase in 1784 of a vast tract of about 4,500 square miles from the Mississauga extending west as far as the Thames River. This included the entire Niagara Peninsula, and was in addition to a four-mile-wide strip along the Niagara River that had been bought during wartime to provide

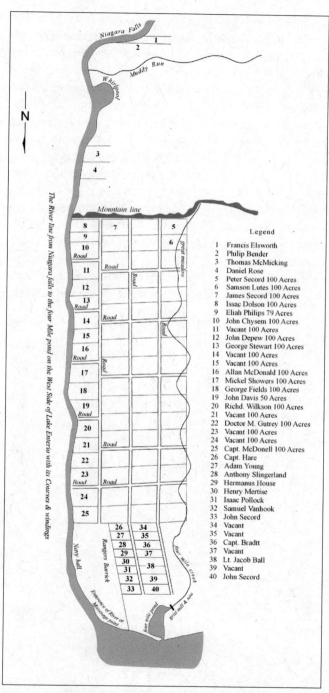

Legend

1 Francis Elsworth
2 Philip Bender
3 Thomas McMicking
4 Daniel Rose
5 Peter Secord 100 Acres
6 Samson Lutes 100 Acres
7 James Secord 100 Acres
8 Issac Dolson 100 Acres
9 Eliah Philips 79 Acres
10 John Chysem 100 Acres
11 Vacant 100 Acres
12 John Depew 100 Acres
13 George Stewart 100 Acres
14 Vacant 100 Acres
15 Vacant 100 Acres
16 Allan McDonald 100 Acres
17 Mickel Showers 100 Acres
18 George Fields 100 Acres
19 John Davis 50 Acres
20 Richd. Wilkson 100 Acres
21 Vacant 100 Acres
22 Doctor M. Gutrey 100 Acres
23 Vacant 100 Acres
24 Vacant 100 Acres
25 Capt. McDonell 100 Acres
26 Capt. Hare
27 Adam Young
28 Anthony Slingerland
29 Hermanus House
30 Henry Mertise
31 Isaac Pollock
32 Samuel Vanhook
33 John Secord
34 Vacant
35 Vacant
36 Capt. Bradtt
37 Vacant
38 Lt. Jacob Ball
39 Vacant
40 John Secord

Figure 1.3 McDonell's map of the Niagara Settlement, 1783
Source: *Haldimand Papers*, Add. Mss. 21829. Redrawn by Loris Gasparotto.

for settlement on the west bank. The cost came to one-tenth of a penny an acre, a "steal" by any standard, but the Mississauga had little understanding of the British concept of land ownership and were willing to accept what Haldimand himself called a "trifling consideration."[27] Later that year, a reservation 12 miles wide along the Grand River was set aside for the Iroquois who had fought for the Crown under Joseph Brant (it became known as the Six Nations Tract, the Five Nations having become Six with the addition of the Tuscarora in 1722).[28]

The third requirement was surveys (Figure 1.4). In other areas, such as Kingston, surveys began promptly in 1784, but in Niagara serious surveying was delayed until almost four years after the war ended, which created a huge problem. Rations were available from the government for a limited time only, leaving the would-be settlers with no option but to fend for themselves. The result was a rash of "promiscuous" squatter settlement, both above and below the Escarpment, which caused no end of difficulties for the surveyors after they eventually began work in June 1787.[29]

The ensuing twenty-month period saw the laying out, in whole or in part, of fourteen townships along the Lake Ontario shoreline and the Niagara River,

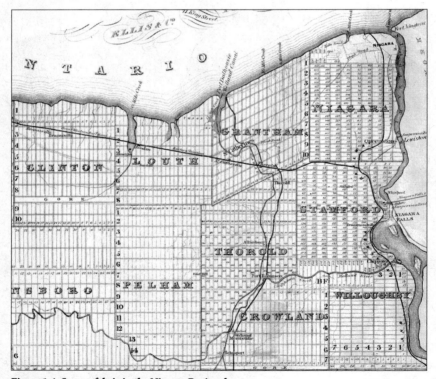

Figure 1.4 Survey fabric in the Niagara Peninsula
Source: *Ellis & Co, Map of the Counties of Wentworth, Brant, Lincoln, Welland, Haldimand,* c. 1859.

including a few inland. Originally numbered, they were not named until the early 1790s, at which time Township No. 1 became Newark (later Niagara), No. 2 became Stamford, No. 3 became Grantham, and so on. They varied somewhat in shape and size, but were all subdivided into 100-acre lots arranged in rows or concessions. Usually the lots were rectangles oriented north–south or east–west, but there were exceptions, as in Grantham, where the lots were parallelograms.[30] By the end of the century, the township fabric was complete throughout the Peninsula, and the original township, lot, and concession lines continue to impact our lives to this day, not least through their effect on farm (and vineyard) boundaries.

At first, the priority of every settler was survival. To this end, the authorities supplied not only emergency provisions and land, but also implements, seed, and other necessities. In 1783, a sawmill and gristmill were built just upstream from the Four Mile Pond; they were government owned and known as King's Mills. More mills followed after 1786, when restrictions on private ownership were relaxed, and even more after 1788, when freehold tenure was introduced.[31] The gristmills met a pressing need for flour, but flour was not produced for domestic consumption alone. A major market existed in the garrison stationed along the Niagara River from Fort Niagara (still in British hands until 1796) to Fort Erie, and over time both wheat and flour were exported to Montreal and even to the United States and Britain.[32] As a result, wheat became by far the dominant crop in the Peninsula by the turn of the century.

As a result also, Queenston merchant and mill owner Robert Hamilton became very rich, for in 1786 he engineered a deal that made him the principal supplier of flour to the garrison.[33] A prominent local figure for over two decades, Hamilton occupied many important positions, among them presidency of the Agricultural Society of Niagara, the first society of its kind in Upper Canada (Figure 1.5).[34] The society was established late in 1792, probably at the suggestion of Lieutenant-Governor Simcoe himself.[35] As patron of the society, Simcoe donated books on farming and an annual subscription of ten guineas "to be disposed of ... for the benefit of agriculture."[36] Few of the members were ordinary farmers, however, and the monthly meetings seem to have consisted mainly of merchants, politicians, clergymen, and gentlemen farmers dining together and chatting about rural affairs.[37] As early as 1796, Hamilton described the society as having been "rather neglected by its members,"[38] though it seems to have survived until about 1807.[39] There were some tangible achievements, however, including the importing of new varieties of fruit trees from Long Island in 1794,[40] the acquisition of four acres of land for use as an experimental farm in 1797,[41] the offer of "premiums" to farmers who produced the largest crops, also in 1797,[42] and (possibly) the sponsorship of Upper Canada's first agricultural fair at Queenston in 1799.[43]

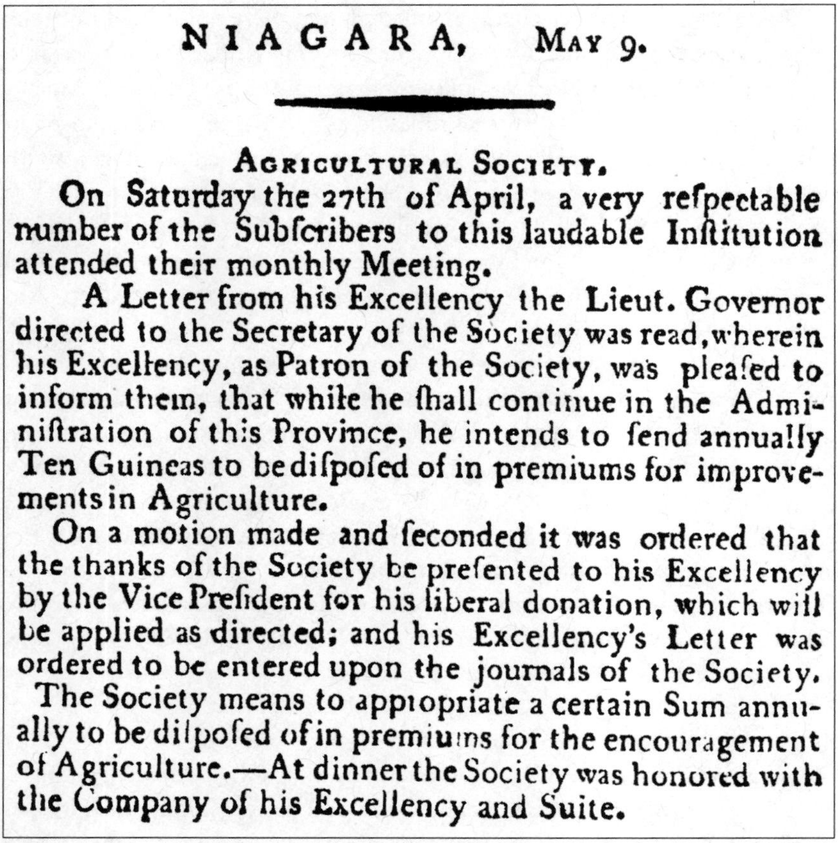

NIAGARA, MAY 9.

───────

AGRICULTURAL SOCIETY.

On Saturday the 27th of April, a very refpectable number of the Subfcribers to this laudable Inftitution attended their monthly Meeting.

A Letter from his Excellency the Lieut. Governor directed to the Secretary of the Society was read, wherein his Excellency, as Patron of the Society, was pleafed to inform them, that while he fhall continue in the Adminiftration of this Province, he intends to fend annually Ten Guineas to be difpofed of in premiums for improvements in Agriculture.

On a motion made and feconded it was ordered that the thanks of the Society be prefented to his Excellency by the Vice Prefident for his liberal donation, which will be applied as directed; and his Excellency's Letter was ordered to be entered upon the journals of the Society. The Society means to appropriate a certain Sum annually to be difpofed of in premiums for the encouragement of Agriculture.—At dinner the Society was honored with the Company of his Excellency and Suite.

Figure 1.5 Report of the meeting of the Agricultural Society, held April 27, 1793
Source: *Upper Canada Gazette*, May 9, 1793.

Throughout this time, the Peninsula remained sparsely populated. Settlers, most occupying lots of 100 acres or more, were widely scattered in clearings cut out from the wilderness, and there were very few villages. Over time, the early villages along the Niagara River, such as Newark (Niagara after 1798) and Queenston, were joined by others. St. Catharines, for example, emerged after a church, school, and tavern were built in 1796–97 where the main east–west route through the Peninsula crossed the Twelve Mile Creek.[44] But these communities were small, and the area as a whole remained a mixture of farms and uncleared land.

Grape growing and winemaking were hardly priorities in those early years, though there must have been settlers who made wine from wild grapes on an informal basis. One of these may have been Johann Schiller,[45] who lived in Niagara Township from 1798 to about 1808 and who had supposedly acquired winemaking skills in his native Germany. While there is no proof that he made wine in Niagara, it is claimed by many writers that by 1811, after he had moved to Toronto Township north of Lake Ontario (in what became Cooksville, now part of Mississauga), he was producing enough wine from local grapes to sell to his neighbours. Indeed, a number of books and articles about winemaking in Ontario and Canada have immortalized him as the "the acknowledged father of Canadian wine," or "the father of commercial winemaking in Canada," or some such accolade.[46]

But as it turns out, though Schiller may indeed have made wine in Toronto Township, he did nothing to merit any such distinction. Contemporary documents such as petitions and land grants provide basic details of his life (service in the American Revolutionary War, initial settlement in Lower Canada, move to Upper Canada, work as a shoemaker, and final move to Toronto Township), but say nothing about grapes or wine.[47] Indeed, the first mention of Schiller and winemaking does not appear until over a century later, this in a series of unattributed newspaper articles on grape growing and winemaking in the Cooksville area, published in 1929 and 1934.[48] These provide little by way of solid evidence and contain serious errors besides. The errors include the claim that Schiller was responsible for introducing the Clinton grape (Schiller died in 1816 and the Clinton grape was not recognized until 1830), and that he sent Clinton roots to France to combat the phylloxera disease, which at the time was devastating French vineyards (phylloxera was not an issue until the 1860s). These articles are very deficient, and seem to be the sole source of the Schiller story.[49]

At the same time, there is clear evidence from early-nineteenth-century sources linking others, if not to winemaking at least to grape growing, and this in the Niagara Peninsula. William Claus, Deputy Superintendent of Indian Affairs for Upper Canada, maintained an elaborate garden at his home in Niagara, and on April 22, 1806, wrote in his gardening notebook, "Sowed next the gate to the left 3 rows of grape seed & a row of Orange Seed."[50] While this is hardly grape growing on a major scale, Thomas Merritt's claim for the loss of grape vines during the War of 1812, quoted earlier, does suggest something significant. So does the claim of Robert Kerr of Niagara Township, who sought compensation for the destruction of "four vineyards."[51] Further evidence comes from Robert Gourlay's *Statistical Account of Upper Canada*, in the responses to an 1817 questionnaire survey of local townships, which include the telling observation from Louth Township that "Grapes have succeeded well in the Niagara district."[52] These examples are the first indications of something akin to systematic grape growing in the Niagara Peninsula. There remain many questions, of course, among them which varieties of grapes were grown, what they were used for, and the scale of the operations (for example, what exactly did Kerr mean by a "vineyard"?), but they do represent a beginning.

The Postwar Period

The first two phases in the history of the Niagara Peninsula have been considered in some detail, but the third phase, from the end of the War of 1812 to just beyond Confederation, saw so many major changes—political, economic, and social—that there is space to mention only one development of particular significance to the Niagara Peninsula: the building of the Welland Canal. (In a sense, this is just as well, for there is much to say about the emerging grape and wine industry during this period.)

The Peninsula had witnessed significant progress between 1783 and 1812, but the war years constituted a major setback. Recovery was slow, but steady, and by 1821, Scottish traveller John Howison was able to say, "Between Queenston and the head of Lake Ontario, the farms are in a high state of cultivation, and their possessors are comparatively wealthy."[53] However, their wealth was nothing compared to that generated a few years later by the construction of the First Welland Canal. Opened in 1829, it initially ran from Lake Ontario to the Welland River (at what became Port Robinson), but by 1833 it was extended to Lake Erie, and in 1845 the First Canal, with its forty tiny wooden locks, was replaced by the Second Canal, with twenty-seven larger stone locks.[54] The impact on the Niagara Peninsula, especially the eastern portion, was dramatic. The canal not only provided a lake-to-lake link to replace the portage around Niagara Falls, making for easier access to distant markets, but it also supplied a source of power for mills. As a result, a significant industrial component was added to the economy of the Peninsula. New communities—Port Colborne, Welland, Port Robinson, Allanburg, Thorold, Merritton, and Port Dalhousie (to use their present-day names)—were founded. St. Catharines, the one pre-existing settlement located on the canal, experienced rapid growth, and places away from the canal, like Niagara and St. Johns, went into decline.[55]

Despite increasing industrialization and urbanization, agriculture remained a staple activity, though now there were signs of diversification. An important aspect of this, with a direct bearing on the future grape and wine industry, was fruit growing. This already had something of a history in the area. In 1793, Elizabeth Simcoe commented on thirty large cherry trees behind her house (no doubt planted a few years earlier by military officers);[56] in 1794, an anonymous traveller wrote of orchards being "in great forwardness" in Upper Canada;[57] and in 1801, John Dun was advertising apples for sale in Niagara.[58] Only three years later, William Bond of York was seeking to dispose of a nursery-cum-orchard containing 10,000 apple seedlings and 41 mature apple trees.[59]

Further evidence comes from the 1812 war claims, where the destruction of orchards was a frequent complaint. Thus, in addition to his vines, Thomas Merritt lost thirty apple trees, twenty each of cherry and peach trees, plus apricot, plum, and quince trees,[60] while Robert Kerr lost "a large nursery of grafted and in-noculated [sic] fruit trees of all descriptions," among them six varieties of plums, eleven of peaches, and twelve of apples.[61] Similar losses are recorded in other claims. Ralfe Clench of Niagara sought compensation for seventy "large healthy bearing [apple] trees, [of which] almost the whole were the best of grafted fruit, and there was not an apple tree among them [that was not] choice."[62] People like Clench were clearly engaged in serious fruit growing, but it was on a small scale and probably intended mainly for local consumption. Though some may have shipped fruit to York, Kingston, and elsewhere, the distance to major centres of population and the rudimentary transportation facilities hindered serious commercial ventures.

But fruit growing continued to increase in importance, and with it the cul-tivation of grapes. There are many indications of this, among them the observa-tions of postwar travellers like Francis Hall, who spoke of "luxuriant orchards" along the Niagara River,[63] and also the emergence of professional nurserymen such as William Custead, whose land in Toronto Township lay just three miles east of what had been Johann Schiller's homestead. In 1827, Custead published an eighteen-page catalogue featuring many types of fruit and ornamental trees, plus eight varieties of grape vines—Early White, Boston Sweet Water, Bland's Virginia, Isabella, White's Sweet Water, Jersey, Black Frontenac, and French Chocolate.[64] To facilitate ordering, he engaged agents throughout Upper Canada, including ones at Queenston and Niagara. In 1839, acquiring nursery stock locally became even easier when Chauncey Beadle founded the St. Catharines Nursery; his first catalogue, dated 1841, featured grape vines selling for 25 cents each.[65] The nursery was continued by his son Delos Beadle, who became a major figure in the fruit-growing industry after the mid-century.[66]

The same period saw a number of important developments in support of agriculture. Provincial legislation was passed establishing regular local mar-kets; the first was at Kingston in 1801, with Niagara following in 1817 and St. Catharines in 1845.[67] Informal market fairs, intended mainly for the sale of livestock and grain, were held from time to time in various locations.[68] From the mid-1820s, a number of agricultural societies came into existence, especially after 1830, when the province made available a grant of £100 to any local society that could raise £50 through its own efforts. The earliest societies were district based, but were soon followed by county and township societies, most which held regular meetings and organized annual competitive exhibitions.[69] The Niagara District society was formed in 1831 and held its inaugural exhibition that year in Clinton Township.[70] By the early 1850s, there were separate Lincoln and Welland County societies, each with a number of township branches.[71]

At first the societies operated independently, but in 1846 a province-wide organization, The Agricultural Association of Upper Canada, was established, which both provided coordination and introduced its own annual exhibition; the first was held in Toronto that same year.[72] Niagara was the venue in 1850, attracting over 1,600 entries and about 14,000 visitors, the most to date.[73] Over time, specialized societies came into existence, such as the Fruit Growers' Association of Upper Canada, the St. Catharines Horticultural Society, and the County of Lincoln Grape Growers' Association.[74] Finally, there was the publication of agricultural periodicals, such as the *British American Cultivator*, launched in 1842. In 1849 it was superseded by *The Canadian Agriculturist*, which in turn gave way to the *Canada Farmer* in 1864.[75]

By mid-century, the importance of the fruit industry in Niagara and elsewhere was quite clear, as evidenced by lectures delivered at society meetings, articles in newspapers and periodicals, and competitions at agricultural exhibitions. In 1853–55, the industry received a major boost with the completion of the Great Western Railway from the new suspension bridge over the Niagara River at Clifton through Hamilton to Toronto and Windsor.[76] This provided rapid access to major markets, and by the late 1850s a number of Niagara Peninsula farmers were growing fruit for sale. The same decade also saw the beginnings of commercial grape growing and winemaking—and this brings us to one Porter Adams.

Adams is said to have been "the first to systematically plant and cultivate vines" in Ontario, this in 1857 on his farm located roughly midway between St. Davids and Queenston.[77] As already mentioned, others had planted grapes before this, but the key word here is "systematically," suggesting that his was a commercial operation. This, certainly, is what is implied, if not stated explicitly, in what many have written about Adams. Some add that by the 1860s, he was shipping grapes across Lake Ontario to the Toronto market.[78]

There is an obvious parallel here with Johann Schiller, the supposed father of commercial winemaking, and, as in the case of Schiller, it turns out that there is no basis for the claims made for Adams. As proof it is sufficient to point out that he was barely eleven years old in 1857 and did not buy the land in question until almost a quarter of a century later, in 1883.[79] Which invites the question, how could such a glaring misconception come about? Ironically, perhaps, it originates in an Ontario Department of Agriculture research bulletin published in 1912.

The author, T.B. Revett, mentions several grape-growing pioneers from the 1850s and '60s. The very first named is Adams, of whom Revett says, "About 1857 some grapes were planted on a farm belonging to Mr. Porter Adams, situated in Niagara Township."[80] What he should have said was "on a farm now belonging to Mr. Porter Adams," but by the time the error was corrected in a second bulletin written by F.M. Clement four years later, it was too late.[81] The claim that Adams was the first person in Ontario to plant grapes was in print, and has been accepted uncritically by many writers since.[82] Of course, Porter Adams may indeed have grown grapes on his land, which he held from 1883 till his death in 1921, but not in 1857. If anyone grew grapes in 1857, it would have been the owner at the time, Job Chubbuck, but this may be impossible to prove.[83]

If there is no basis for the Porter Adams claim, there is ample supporting evidence for some of the other grape growers mentioned by Revett and Clement, and for others besides. Indeed, the late 1850s and early '60s saw a significant increase in the numbers involved in the nascent grape and wine industry. Not surprisingly, there was considerable variation in what they were doing. Some operated on a limited scale, while others were more ambitious, though the work was rarely more than a sideline for anyone. Some were just growing table grapes (the main market at the time), while others were fermenting wine. There was also a great deal of experimentation with different grape varieties—the Concord, which was to dominate the industry well into the twentieth century, was introduced about 1850—and there was much debate on the relative merits of cultivating grapes under glass (with or without artificial heat) as opposed to growing them outdoors.[84]

One of the earliest commercial grape growers was William Haskins of Hamilton, who described himself as a self-taught amateur and whose full-time job was city engineer and waterworks manager. In 1880, he reported to the Ontario Agricultural Commission that he had been "engaged in the culture of grapes for twenty-one or twenty-two years," which could mean as early as 1858.[85] (It was Haskins who described the Concord as "the grape for the million, because it will grow anywhere.")[86] Four years later, according to Revett and Clement, William H. Read planted three acres of Concord, Delaware, and Hartford Prolific grapes on his farm in Louth Township. But he must have been growing grapes well before this, for he won three prizes at the 15th Exhibition of the Provincial Agricultural Association held in Hamilton in 1860.[87] Indeed, there are indications that he was a serious grape grower as early as 1850.[88]

Another prizewinner at Hamilton was John C. Kilborn[89] of Beamsville, for the best four clusters of black grapes "grown in open air." (Second was J.G. Keefer of Thorold—not normally considered part of grape and wine country.) Kilborn was a regular exhibitor at these events, and may have been growing grapes since the mid-1850s. He was also making wine, and in an 1860 letter to *The Canadian Agriculturist* he says, "The wine sells in this locality for one dollar and three quarters per gallon, and would probably bring more if we asked it, at all events it is worth four times as much as the miserable stuff generally sold by our merchants under the name of wine."[90] He adds that in 1857 "four or five barrels of wine were grown from a single vine in one season in the Township of Grimsby," but it is not clear if he was the vintner.

In these circumstances, it is impossible to identify conclusively any one person as the first to grow grapes or to make wine commercially in the Peninsula. Having said that, there is one name that does stand out: that of William Whitney

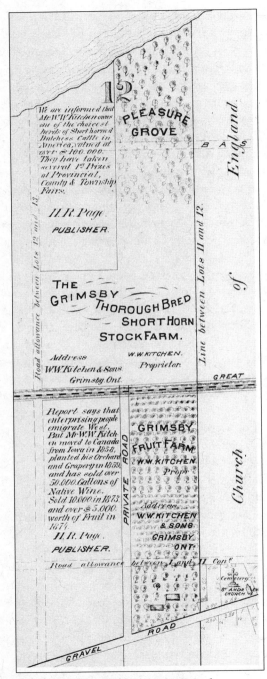

Figure 1.6 William Kitchen's land in Grimsby
Source: H.R. Page, *Illustrated Historical Atlas of the Counties of Lincoln & Welland, Ont.*, 1876.

Kitchen of Grimsby, who without question went much further than most (Figure 1.6). Born in 1824 in Dumfries Township, he spent time in the United States and came to Grimsby from Iowa in 1858. That same year, his father Charles bought about ninety acres of land just west of the village of Grimsby, and here William established his farm.[91] The Great Western Railway bisected the property; to the north was his stock farm, and to the south his fruit farm, which he planted with trees and vines in 1859. Just five years later he wrote of having 1,000 grape vines and "eight hundred gallons of beautiful, pure, and unadulterated wine" for sale in his cellar, all this in "Cold Canada."[92] In 1867 he bought the land from his father and evidently prospered. Perhaps the best indication of his success is the unusually detailed depiction of his property on the Grimsby village map of 1876 in H.R. Page's *Illustrated Historical Atlas of the Counties of Lincoln & Welland*.[93] Contrary to standard cartographic practice, the map is tantamount to an advertisement for Kitchen (and besides, it faces a page devoted solely to pictures extolling his cattle).

This would not have come cheap. Nor would the full-page advertisement he took out in the 1866 *Canadian Almanac* promoting his wine—winner of many awards at provincial exhibitions, "a pure article, good for Medicinal purposes ... sold by most of the principal Chemists in Canada East and West," and "in use by some Hundreds of Churches, for sacramental services" (Figure 1.7).[94] Curiously, there is no mention of consuming wine for its own sake, though local newspaper advertisements commencing in 1867 do add that "it is sent everywhere to private families and to hotel-keepers."[95] By this time he had for sale 20,000 gallons of wine, port, and sherry (at $2.50 a gallon, less 25 cents for orders of 10 gallons or more), as well as 80,000 grape vines. When Page's atlas came out in 1876, the publisher himself attested to the fact that Kitchen had by then sold over 50,000 gallons of "Native Wine."[96]

William Kitchen, who died in Rochester in 1909, was clearly a very significant figure in the early history of the grape and wine industry in the Niagara Peninsula, if not in Ontario and Canada. But mindful of the lessons learned about Schiller and Adams, we must resist the temptation to anoint him "father" of anything, for there were others in the province doing similar things about the same time. One in particular must be mentioned. This is Justin De Courtenay, who during the 1860s was manager of the Clair House vineyards in Cooksville, established by Englishman Henry Parker on land that had once belonged to Johann Schiller. De Courtenay, who despite his French surname was English born (his middle name was McCarthy), was without question an immensely important figure, in many respects much more so than Kitchen.[97]

De Courtenay wrote two pamphlets, *The Culture of the Vine and Emigration* and *The Canada Vine Grower*,[98] and was a tireless advocate for the grape and wine industry, even corresponding with John A. Macdonald on the subject (at

Figure 1.7 William Kitchen advertisement
Source: *Canadian Almanac*, 1866.

the time Macdonald was Attorney General of Canada West).[99] He was also a staunch proponent of North American varieties of grape (though not if they had a "*foxy* flavour"),[100] and a severe, though polite, critic of anyone who went "against nature" by cultivating hothouse grapes (among them Delos Beadle of St. Catharines).[101] In 1866, he co-founded an incorporated company known as the Canada Vine Growers' Association (cvga) based at his Clair House vineyards,[102] and in 1867 he won a medal for icewine at an international exposition in Paris.[103] De Courtenay has been described as a "visionary" who "more than any of his contemporaries ... brought the advantages of vine growing and winemaking to the Canadian farmer, public and government."[104]

De Courtenay seems to have had little direct involvement with the grape and wine industry in the Niagara Peninsula, though the cvga did advertise in local newspapers, and he is known to have attended a meeting of the Fruit

Growers' Association in Grimsby in 1866 and to have spoken to the County of Lincoln Grape Growers' Association in St. Catharines in February 1867.[105] Interestingly, just two weeks after the latter meeting, William Kitchen started advertising regularly in *The St. Catharines Constitutional*.[106] Whether it was anything to do with De Courtenay (a perceived threat from a competitor, perhaps), or no more than coincidence, is not known. But then in April, the CVGA offered for sale fifty company shares to which Kitchen had subscribed but never paid for—he was even singled out by name in a newspaper announcement[107]—so perhaps relations between Kitchen and De Courtenay were not so good after all. Whatever the truth of the matter, Kitchen still joined De Courtenay at a meeting of the Ontario Fruit Growers' Association held at the Clair House vineyards in October of 1867.[108]

But the prominence of De Courtenay and the CVGA must have challenged Kitchen, and he twice sought to incorporate companies of his own. The first, in about 1868, was the Grimsby Grape Growing, Wine Making and Fruit Canning Company,[109] and the second, in 1870, was the Grimsby Union Wine Company. For the latter, which was to be capitalized at $40,000, with each share selling for $40, he issued a detailed prospectus that was advertised prominently in local newspapers.[110] But though four local ministers (Anglican, Methodist, Presbyterian, and Baptist) and four doctors were listed as "references," this initiative, like the previous one, seems to have foundered. Almost certainly the explanation was lack of investors, partly, no doubt, because the CVGA had cornered the market. Indeed, an earlier attempt in 1866 by the County of Lincoln Grape Growers' Association to establish a company had failed for this very reason, though there were also suspicions of governmental bias toward the CVGA.[111] It was not long, however, before a Niagara-based company came into existence, for in 1873 George Barnes of St. Catharines founded the Ontario Grape Growing and Wine Manufacturing Company, which very soon became the largest company of its kind in the Dominion.[112] Later known simply as Barnes Wines, it continued until 1988.

In 1869, an unidentified correspondent for the Cincinnati *Enquirer*, who was staying at the Stephenson House overlooking the Twelve Mile Creek in St. Catharines, penned a tribute to the proprietor, Beverley Tucker, in which he said: "The view from the Stephenson is extremely beautiful, extending for miles over a very fertile country, richly diversified with hills and dale[s]. It is a splendid country for fruit. Cherries, currants, apples, pears and peaches grow in profusion. Grapes also, of the usual varieties, grow most luxuriantly, and I find that winemaking is becoming an important manufacture."[113] It took a long time for this to happen, but once it did, the Niagara Peninsula never looked back.[114]

Notes

1 Niagara Land Registry Office, Grantham Township Memorial Abstract Index, Concession 4, Lots 20 and 21.

2 Library and Archives Canada (hereafter LAC), Upper Canada: War of 1812 Losses Claims, RG 19 E 5 (a) Vol. 3741 File 2, Claim No. 68.

3 *Mrs. Simcoe's Diary*, ed. Mary Quayle Innis (Toronto: Macmillan of Canada, 1965), 99–100, 107, 109–10.

4 P. Campbell, *Travels in the Interior Inhabited Parts of North America. In the Years 1791 and 1792* (Edinburgh: John Guthrie, 1793), 192.

5 *Atlas of Great Lakes Indian History*, ed. Helen Hornbeck Tanner (Norman, OK: University of Oklahoma Press, 1987), 26–27, Map 5; Gordon K. Wright, *The Neutral Indians, A Source Book: Occasional Papers* of the New York State Archeological Association, No. 4 (Rochester, NY: New York State Archeological Association, 1963), 1–9.

6 *The Works of Samuel de Champlain,* Vol. 3, 1615–1618, ed. H.P. Biggar (Toronto: Champlain Society, *Publication,* Vol. 3, 1929), 99–100.

7 William C. Noble, "Thorold: An Early Historic Niagara Neutral Town," in John Burtniak and Wesley B. Turner eds., *Villages in the Niagara Peninsula, Proceedings, Second Annual Niagara Peninsula History Conference* (St. Catharines, ON: Brock University, 1980), 43–55.

8 For a discussion of this question, see Michael Power, *A History of the Roman Catholic Church in The Niagara Peninsula 1615–1815* (St. Catharines, ON: Roman Catholic Diocese of St. Catharines, 1983), 7–38.

9 *The Voyages of Jacques Cartier,* ed. H.P. Biggar, Publications of the Public Archives of Canada, No. 11 (Ottawa: F.A. Acland, King's Printer, 1924), 141.

10 *The Works of Samuel de Champlain,* Vol. 1, 1599–1607, ed. H.P. Biggar (Toronto: Champlain Society, *Publication,* Vol. 1, 1922), 329, 395.

11 *First Establishment of the Faith in New France by Father Christian Le Clercq,* Vol. 1, trans. John Gilmary Shea (New York: John G. Shea, 1881), 208.

12 "Galinée's Narrative and Map," trans. and ed. James H. Coyne, Ontario Historical Society, *Papers and Records* 4 (1903): 53.

13 *A Description of Louisiana, by Father Louis Hennepin, Recollect Missionary,* trans. John Gilmary Shea (New York: Shea, 1880), 92, 259.

14 Translated from Monique Hivert-Le Faucheux, "La vie quotidienne en Acadie au temps de Razilly: le témoignage d'un document manuscrit," *Les cahiers de la société historique Acadienne* 26, no. 2 (1995): 120.

15 *The Jesuit Relations and Allied Documents,* Vol. 9, Quebec, 1636, ed. Reuben Gold Thwaites (New York: Pageant Book Company, 1959), 155.

16 *The Jesuit Relations and Allied Documents,* Vol. 51, Ottawas, Lower Canada, Iroquois, 1666, Reuben Gold Thwaites (New York: Pageant Book Company, 1959), 121.

17 Translated from Gabriel Sagard-Théodat, *Histoire du Canada* (Paris: Claude Sonnius, 1636), 218.

18 Lyman Carrier, *The Beginnings of Agriculture in America* (New York: McGraw-Hill, 1923), 289.

19 Conrad E. Heidenreich, "History of the St. Lawrence-Great Lakes Area to A.D. 1650," in Chris J. Ellis and Neal Ferris, eds., *The Archaeology of Southern Ontario to A.D. 1650,* Occasional Publication of the London Chapter, Ontario Archaeological Society, No. 5 (London: London Chapter, O.A.S., 1990), 475–92.

20 Tanner, *Atlas of Great Lakes Indian History,* 32–33, Map 6; 40–41, Map 9; 58–59, Map 13.

21 "Extract of a Letter from General Haldimand to Lieut.-Colonel Bolton, Dated Quebec, June 7th, 1779," in *Records of Niagara, A Collection of Documents Relating to the First Settlement 1778–1783,* ed. E.A. Cruikshank (Niagara-on-the-Lake, ON: Niagara Historical Society, Publication No. 38, 1927), 12.

22 "Surveys, journals, plans, etc., relative to the settlements of Loyalists in Canada: 1782–1784," in British Library, *Haldimand Papers,* Add. Mss. 21829.

23 For a general account of this early phase of settlement, see Alun Hughes, "John Butler and Early Settlement on the West Bank of the Niagara River," in *The Butler Bicentenary* (Niagara-on-the-Lake: Colonel John Butler [Niagara] Branch, The United Empire Loyalists' Association of Canada, 1997), 64–82. This article also contains reproductions of the two censuses and the map referred to in this paragraph.

24 LAC, Quebec Land Book, RG 1 L 1, Vol. 18, Minutes of the Executive Council, Province of Quebec, Nov. 9, 1789.

25 "George R. Additional Instructions to … Frederick Haldimand … Given at Our Court at St James's the 16th Day of July 1783," in *Documents Relating to the Constitutional History of Canada 1759–1791,* Part 2, ed. Adam Shortt and Arthur G. Doughty (Ottawa: The Historical Documents Publication Board, 1918), 730.

26 "Lord Sydney to Lord Dorchester. Whitehall 3d Sepr 1788," in Shortt and Doughty, *Documents Relating to the Constitutional History of Canada 1759–1791,* Part 2, 957.

27 "Indian Council at Niagara," in *Records of Niagara, 1784–7,* ed. E.A. Cruikshank (Niagara-on-the-Lake: Niagara Historical Society, *Publication* No. 39, 1928), 28–32.

28 "Official Notice. Frederick Haldimand …," in Cruikshank, *Records of Niagara, 1784–7,* 49.

29 "P.R. Frey, D. Surveyor, to John Colins, D.S.G., Niagara, 18th October, 1788," in *Third Report of the Bureau of Archives for the Province of Ontario, 1905,* ed. Alexander Fraser (Toronto: Legislative Assembly of Ontario, 1906), 312–13.

30 For a general account of these surveys, see Alun Hughes, "The Early Surveys of Township No. 1 and the Niagara Peninsula," in *Niagara's Changing Landscapes,* ed. Hugh J. Gayler (Ottawa: Carleton University Press, 1994), 209–39.

31 "From the Quebec Gazette, Thursday, February 16, 1786," in Cruikshank, *Records of Niagara, 1784–7,* 81–84; For a general account of mills in Niagara, see Alun Hughes, "Secord, Servos and Niagara's First Mills," *Newsletter,* Historical Society of St. Catharines, December 2011, 9–10.

32 Robert Leslie Jones, *History of Agriculture in Ontario 1613–1880* (Toronto: University of Toronto Press, 1946), 27–30.

33 Bruce G. Wilson, *The Enterprises of Robert Hamilton: A Study of Wealth and Influence in Early Upper Canada 1776–1812* (Ottawa: Carleton University Press, 1983), 33, 77–78, 124–25.

34 In early accounts, the society is also referred to as the Agricultural Society of Upper Canada.

35 C.C. James, "The Pioneer Agricultural Society of Ontario," Fair Number, *The Farming World,* September 1902, 211–12.

36 "E.B. Littlehales to the Secretary of the Agricultural Society of Upper Canada, April 25, 1793," in *The Correspondence of Lieut. Governor John Graves Simcoe.* ed. E.A. Cruikshank, Vol. 1, 1784–1793 (Toronto: Ontario Historical Society, 1923), 318; *Upper Canada Gazette,* 9 May 1793.

37 Anonymous, "Canadian Letters. Description of a Tour thro' the Provinces of Lower and Upper Canada in the course of the year 1792 and '93," in *Early Travellers in the Canadas 1791–1867,* ed. Gerald M. Craig (Toronto: Macmillan Company of Canada, 1955), 9; *Upper Canada Gazette,* 4 July 1793.

38   "Robert Hamilton to John Graves Simcoe, Feb. 15, 1796," in *The Correspondence of Lieut. Governor John Graves Simcoe,* Vol. 4, 1795–1796, ed. E.A. Cruikshank (Toronto: Ontario Historical Society, 1926), 198.

39   *Upper Canada Gazette,* November 15, 1806; *The York Gazette,* 13 June 1807.

40   Janet Carnochan, *History of Niagara (In Part)* (Toronto: William Briggs, 1914), 266–67.

41   "The Petition of the Agricultural Society of Niagara," in E.A. Cruikshank, "Petitions for Grants of Land in Upper Canada, Second Series, 1796–99," Ontario Historical Society, *Papers and Records* 26 (1930): 102.

42   *Upper Canada Gazette,* 15 March 1797.

43   *Canada Constellation,* 8 November 1799.

44   John N. Jackson, *St. Catharines, Ontario: Its Early Years* (Belleville, ON: Mika Publishing Company, 1976), 124–30.

45   Schiller's name is spelled in numerous ways in old documents: Shealor, Shela, Sheler, Shieller, and Shilar. Usually his first name is given as John.

46   Percy Rowe, *The Wines of Canada* (Toronto: McGraw-Hill, 1970), 29; William F. Rannie, *Wines of Ontario: An Industry Comes of Age* (Lincoln, ON: W.F. Rannie, 1978), 17; Tony Aspler, *The Wine Atlas of Canada* (Toronto: Random House Canada, 2006), 129; Kathleen A. Hicks, *Cooksville: Country to City* (Mississauga, ON: Friends of the Mississauga Library System, 2007), 15–16; Linda Bramble, *Niagara's Wine Visionaries: Profiles of the Pioneering Winemakers* (Toronto: James Lorimer & Company, 2009), 14; One of the very few authors to express any sort of doubt about Schiller is Rod Phillips, *Ontario Wine Country* (Vancouver: Whitecap Books, 2006), 14.

47   LAC, Lower Canada Land Papers, RG 1 L 3L Vol. 15, 85105-85108; LAC, Upper Canada Land Petitions "S" Bundle 4 1797–1799, RG 1 L3, Vol. 451, No. 186; LAC, Civil Secretary's Correspondence, Upper Canada, Upper Canada Sundries, July–December 1808, RG 5 A 1, Vol. 8, 003532-003534; Archives of Ontario, RG 1 C IV, Box 485, 1145–1147; Region of Peel Archives, Peel County Land Registry Office, Toronto Township, Concession I North of Dundas Street, Lots 9 and 17; Johannes Helmut Merz, *The Hessians of Upper Canada* (Ameliasburg, ON: Seventh Town Historical Society, 2008), 64–66.

48   The sequence seems to be as follows. In 1929, the *Toronto Evening Telegram* published three articles on the Cooksville vineyards and winery, starting on August 30. The final article, on September 10, 1929, was all about Schiller and his family, and was reprinted in the *Brampton Conservator* on September 12. Then on 22 April 1934, *The Globe* (Toronto) published its own article, which was reprinted in the *Port Credit News* on 25 April 1934. A much-shortened version came out in the *Streetsville Review and Port Credit Herald* on April 26.

49   For a detailed analysis of the Schiller claim, see Alun Hughes, "Johann Schiller: Father of Canadian Wine?" *Newsletter,* Historical Society of St. Catharines, November 2010, 9–10.

50   "William Claus 1806. From William Claus's Garden Book for 1806–1810," in *Garden Voices: Two Centuries of Canadian Garden Writing,* ed. Edwinna von Baeyer and Pleasance Crawford (Toronto: Random House of Canada, 1995), 83–84.

51   LAC, Upper Canada: War of 1812 Losses Claims, RG 19 E 5 (a) Vol. 3741, File 3, Claim No. 125.

52   Robert Gourlay, *Statistical Account of Upper Canada,* Vol. 1 (London: Simpkin and Marshall, 1822), 426.

53   John Howison, *Sketches of Upper Canada, Domestic, Local and Characteristic* (Edinburgh: Oliver & Boyd, 1821), 135.

54 Much has been written about the Welland Canals, both popular and scholarly, over the years. See, for example, John N. Jackson, *The Four Welland Canals: A Journey* (St. Catharines, ON: Vanwell, 1988), and Roberta M. Styran and Robert R. Taylor, *The "Great Swivel Link": Canada's Welland Canal* (Toronto: Champlain Society, Publication No. 64, 2001). See also other works by these same authors.

55 Jackson, *The Welland Canals and Their Communities: Engineering, Industrial, and Urban Transportation* (Toronto: University of Toronto Press, 1997).

56 Innis, *Mrs. Simcoe's Diary*, 97.

57 Anonymous, "A Letter from a Gentleman to his Friend, Descriptive of the Different Settlements, in the Province of Upper Canada, New York, 20th Nov. 1794," in John Ogden, *A Tour through Upper and Lower Canada* (Litchfield, CT, 1799), 103–4.

58 *Niagara Herald*, November 7, 1801.

59 *Upper Canada Gazette*, August 4, 1804; Pleasance Crawford, "Some Early Ontario Nurserymen," *Canadian Horticultural History* 1, no. 1 (1985): 29–65.

60 LAC, Upper Canada: War of 1812 Losses Claims, RG 19 E 5 (a) Vol. 3741, File 2, Claim No. 68.

61 LAC, Upper Canada: War of 1812 Losses Claims, RG 19 E 5 (a) Vol. 3741, File 3, Claim No. 125.

62 LAC, Upper Canada: War of 1812 Losses Claims, RG 19 E 5 (a) Vol. 3745, File 1, Claim No. 324.

63 Francis Hall, *Travels in Canada, and the United States, in 1816 and 1817* (Boston: Wells and Lilly, 1818), 126.

64 William W. Custead, *Catalogue of Fruit & Ornamental Trees, Flowering Shrubs, Garden Seeds and Green-house Plants, Bulbous Roots & Flower Seeds, Cultivated and for Sale at the Toronto Nursery* (York [U. C.]: William Lyon Mackenzie, 1827).

65 Chauncey Beadle, *Catalogue of Fruit Trees, Cultivated and for Sale at the St. Catharines Nursery* (St. Catharines, ON: Hiram Leavenworth, 1841); Pleasance Crawford, "Some Early Niagara Peninsula Nurserymen," in *Agriculture and Farm Life in the Niagara Peninsula*, Proceedings, ed. John Burtniak and Wesley B. Turner, Fifth Annual Niagara Peninsula History Conference (St. Catharines, ON: Brock University, 1983), 63–90.

66 Pleasance Crawford, "Beadle, Delos White," *Dictionary of Canadian Biography*, Vol. 13, 1901–1910 (Toronto: University of Toronto Press, 1994), 46–48.

67 These markets were established by *Statutes of Upper Canada*, 41 Geo. III, c. 3 (1801), 57 Geo. III, c. 4 (1817), *Statutes of the Province of Canada*, 8 Vic., c. 63 (1844–45); Brian S. Osborne, "Trading on a Frontier: The Function of Peddlers, Markets, and Fairs in Nineteenth-Century Ontario," in *Canadian Papers in Rural History*, Vol. 2, ed. Donald H. Akenson (Gananoque, ON: Langdale Press, 1980), 69.

68 Jones, *History of Agriculture in Ontario 1613–1880*, 159–61.

69 "A History of the Agriculture and Arts Association," *Fiftieth Annual Report of the Agriculture and Arts Association of Ontario, 1895*, Appendix D (Toronto: Warwick Bro's & Rutter, 1896), 137–38; Robert W. Carbert, "Agricultural and Horticultural Societies and Fairs in the Niagara Peninsula," in *Agriculture and Farm Life in the Niagara Peninsula*, Proceedings, ed. John Burtniak and Wesley B. Turner, Fifth Annual Niagara Peninsula History Conference (St. Catharines, ON: Brock University, 1983), 49–53; Jones, 162–70.

70 Jones, *History of Agriculture in Ontario 1613–1880*, 164–65.

71 Carbert, "Agricultural and Horticultural Societies and Fairs in the Niagara Peninsula," 51–52.

72 "A History of the Agriculture and Arts Association," 138–42; Jones, *History of Agriculture in Ontario 1613–1880*, 171–73.

73   Philip Dodds, *Ontario Agricultural Fairs and Exhibitions 1792–1867* (Picton, ON: Ontario Association of Agricultural Societies, 1967), 30.

74   Carbert, "Agricultural and Horticultural Societies and Fairs in the Niagara Peninsula," 57; References to all three associations are to be found in the following issues of the *St. Catharines Constitutional:* 19 July, 2 August, 13 December 1866, and 17 January, 7 February 1867.

75   Fred Landon, "The Agricultural Journals of Upper Canada (Ontario)," *Agricultural History* 9, no. 4 (1935): 167–75.

76   John N. Jackson and John Burtniak, *Railways in the Niagara Peninsula* (Belleville, ON: Mika Publishing Company, 1978), 40–43.

77   Rowe, *The Wines of Canada*, 28.

78   Tony Aspler, *Vintage Canada* (Toronto: McGraw-Hill Ryerson, 1993), 11.

79   Niagara Land Registry Office, Niagara Township, Instrument No. 1841, Bargain and Sale, Lucius S. Oille et al. and Elias Porter Adams, June 11 1883; Norris Counsell Woodruff, *Twelve Generations from the Colony of Connecticut in New England and the Province of Upper Canada 1636–1959, A Woodruff Genealogy* (N.p.: 1959?), 96.

80   T.B. Revett, *The Grape Growing Industry in the Niagara Peninsula* (Toronto: Ontario Department of Agriculture, *Bulletin* 202, May 1912), 1.

81   F.M. Clement, *The Grape in Ontario* (Toronto: Ontario Department of Agriculture, *Bulletin* 237, March 1916), 4.

82   One of the few skeptics is *Wines of Ontario,* 21, 26–28.

83   Niagara Land Registry Office, Niagara Township Memorial Abstract Index, Concession 2, Lots 42, 43 and 44; For a detailed analysis of the Adams claim, see Alun Hughes, "Porter Adams and Grape-Growing in Niagara," *Newsletter,* Historical Society of St. Catharines, March 2011, 8–10.

84   J.M. De Courtenay, *The Canada Vine Grower: How Every Farmer in Canada May Plant a Vineyard and Make His Own Wine* (Toronto: James Campbell and Son, 1866), 22–23; Reproduced in the *St. Catharines Constitutional*, 31 January 1867.

85   Ontario Agricultural Commission, *Report of the Commissioners,* Vol. 3, Appendix D, *Evidence Relating to Grape Culture and Wine Making* (Toronto: C. Blackett Robinson, 1881), 3; *City of Hamilton Twelfth Annual Alphabetical, General, Street, Miscellaneous and Subscribers' Classified Business Directory for the Year March, 1885, to March, 1886* (Hamilton, ON: W.H. Irwin, 1886), 381.

86   Ontario Agricultural Commission, 4.

87   *Canadian Agriculturist* 12, no. 15 (1860): 590.

88   *Canadian Agriculturist* 12, no. 15 (1860): 466.

89   Kilborn's name is spelled inconsistently in the literature; sometimes it is Kilborn, and sometimes Kilborne.

90   *Canadian Agriculturist* 12, no. 15 (1860): 365.

91   Niagara Land Registry Office, Grimsby Township Memorial Abstract Index, Concession 2 and Broken Front, Lot 12.

92   *Moore's Rural New-Yorker* 14, no. 1 (1863): 7.

93   H.R. Page, *Illustrated Historical Atlas of the Counties of Lincoln & Welland, Ont.* (Toronto: H.R. Page, 1876), 17.

94   *Canadian Almanac, 1866* (Toronto: Scobie & Balfour, 1866), 103.

95   *St. Catharines Constitutional,* 28 February 1867.

96   Page, *Illustrated Historical Atlas of the Counties of Lincoln & Welland, Ont.*, 17.

97   Richard A. Jarrell, "Justin De Courtenay and the Birth of the Ontario Wine Industry," *Ontario History,* 103, no. 1 (2011): 81–104.

98   J.M. De Courtenay, *The Culture of the Vine and Emigration* (Quebec: Joseph Darveau, 1863); De Courtenay, *The Canada Vine Grower.*

99   Jarrell, "Justin De Courtenay and the Birth of the Ontario Wine Industry," 93.

100 De Courtenay, *The Canada Vine Grower*, 10; Reproduced in the *St. Catharines Constitutional*, 21 February 1867.

101 De Courtenay, *The Canada Vine Grower*, 24; Reproduced in the *St. Catharines Constitutional*, 31 January 1867.

102 "An act for the incorporation of the Canada Vine Growers' Association," *Statutes of the Province of Canada*, 29 & 30 Vic., c. 121 (1866).

103 Jarrell, "Justin De Courtenay and the Birth of the Ontario Wine Industry," 95.

104 Jarrell, "Justin De Courtenay and the Birth of the Ontario Wine Industry," 81.

105 *St. Catharines Constitutional*, 7 February 1867.

106 *St. Catharines Constitutional*, 28 February 1867.

107 *St. Catharines Constitutional*, 16 April 1867.

108 *The Canada Farmer* 4, no. 21 (1860): 333.

109 Company circular, 1868, Grimsby Museum.

110 *St. Catharines Constitutional*, 17 November 1870.

111 *St. Catharines Constitutional*, 2 August 1866.

112 *The Standard Trade Edition*, 29 March 1894.

113 *St. Catharines Journal*, 17 July 1869.

114 I would like to thank various people for help with this chapter, including Richard Jarrell for sending me an advance copy of his article about Justin De Courtenay, Steve Mouk for providing information about William Kitchen, and staff at the Region of Peel Archives and the Canadiana Reading Room at the Mississauga Library for information about Johann Schiller; also Loris Gasparotto for redrawing McDonell's map, Mariane Ferencevic for help with French translation, the Interlibrary Loans office at Brock University for supplying me with crucial microfilms, and personnel at St. Catharines Public Library, the Brock University Map Library, and Brock University Special Collections and Archives for general assistance. But one person stands out above all. This is John Burtniak, former Head of the Archives at Brock, who has been of immense help to me. Not only did he read numerous drafts of the chapter, but he edited the drafts and made many suggestions for improvement. In addition he played a key role in correcting and formatting the endnotes.

Two

Niagara Wine and the Influence of Government Regulation, 1850s to 1944

Dan Malleck

Introduction

In the century between 1850 and 1950, the wine industry experienced tremendous growth and change, not only in production but also in the way the industry was both perceived and regulated. The period between Confederation and the end of the Second World War saw the rise, triumph, and decline of the temperance and Prohibition movements, along with considerable expansion of the science and technology of the wine industry. The fortunes of Niagara wine must be considered in the light of the regulatory regimes of both the Dominion (federal) and provincial governments. This chapter will therefore place the development of the wine industry in Niagara in the broader social and political context of Canada's and Ontario's growing regulatory concern over the manufacture and consumption of wine. Along with charting the shifting locations of wine production in the province—which became increasingly focused in the "Golden Horseshoe" (an area around the western end of Lake Ontario, from Lake Erie in the south to Georgian Bay in the north)—this chapter argues that, while government regulation and oversight may appear to have been onerous and burdensome, the provincial government's intersecting but seemingly contradictory interest in controlling the sale of alcohol and encouraging the expansion of the industry as a revenue source helped vintners to improve both their products and their productivity.

The Emerging Industry

As Alun Hughes demonstrated in the previous chapter, there are many stories of the early years of winemaking in Ontario—some apocryphal, others rooted in more credible evidence—and it is important to be aware of the limitations of the oft-repeated stories regarding the origins of the industry. What we do know about the last half of the nineteenth century is that winemaking was in a slow transition from something farmers did for their own household consumption into a more formal and growing industry. Yet there is no hard-and-fast delineation between artisanal activity and industrial production. If a farmer presses his or her own grapes and allows them to ferment into wine, and then sells or barters it with his or her neighbours, he or she is embarking upon a sort of proto-industrial practice. At some point the actual industry emerged, but it is difficult to set a date and point precisely at one event that indicated the transition.

There are, however, some tantalizing indications of the nascent commercialization and specialization of winemaking in the province. In 1868, officials at the Provincial Exhibition in Toronto expanded their competition categories in the wine class, adding categories for professional and amateur winemakers. This change, according to Percy Rowe, was to deal with the fact that, increasingly, the competitions were being won by winemakers whose main occupation was to make wine, as opposed to farmers who tried their hand at the craft.[1] Rowe calls this "a radical change," although there is no indication that it was all that radical (yet we can appreciate the pun, since the etymology of the word "radical" is "root"). It might be more apt to characterize the development of winemaking in the region as a gradual transition from artisanal production to industrialization.

By 1880, winemaking had become important enough to merit special consideration by a provincial agricultural commission. Assembled by the provincial government in April 1880 "to inquire into the Agricultural resources of the Provinces … the progress and condition of agriculture therein, and matters connected therewith,"[2] the commission produced a detailed, five-volume report to the province, in which it enumerated the agricultural production and potential production of counties across the province. The eighteen members of the commission singled out grape growing and winemaking in a separate chapter, isolating it from other tender-fruit agriculture. No other agricultural crop merited such singular attention. The commissioners noted the complexity of the relationship between grapes and wine, in that some good "table" or "dessert" grapes did not make good wine, some grapes matured too late or too early, some did not yield enough sugar, and others were susceptible to disease. What becomes clear in this report is that, while the grape-growing and winemaking industries were still in their infancy (out of 155 people interviewed, ten discussed grape growing; of the ten, seven stated that grapes in their area had been pressed into wine), the intricacies of viticulture were topics of significant debate among farmers and vintners.

While there may have been growing interest in winemaking in the second half of the century, it was not subject to federal scrutiny until the last decade. In the Census of 1891, records appear of a wine industry that was significant enough to be enumerated. Census data were collected by electoral district, so Table 1 lists the data based upon the delineation of those electoral districts that fell within the current-day Niagara Region: Lincoln and Niagara; Welland; and

Table 2.1. Grape production, 1891

Electoral district	Acres in Vines	Grapes/Vineyards	
		Pounds of Grapes	Grapes as % of provincial total
Welland	548	1,449,367	12.36%
Lincoln and Niagara	968	2,610,752	22.27%
Monck	86	339,672	2.90%
Essex North	690	1,092,037	9.31%
Essex South	372	679,630	5.80%
Wentworth	849	2,472,055	21.08%
Provincial total	**4956**	**11,725,281**	

Source: *Census of Canada*, 1891, Vol. III, Table XI – "Fruit trees and fruit on lots."

Monck, a short-lived electoral district that straddled parts of Lincoln, Welland, and Haldimand Counties.[3] It compares them to the wine industry in Essex County (divided into two electoral districts), showing that while the Essex industries employed more people, the output from the Niagara wineries was proportionately much larger, making up almost half of the entire wine output of the province.[4] It should be noted that the inclusion of this category in 1891 is not an indication that wineries suddenly appeared in the decade between the 1881 census and that of 1891. Censuses are as much a reflection of administrative and government priorities as they are of the actual situation in the country. Nevertheless, the Census of 1881 did include "wineries" in its industrial classification. It just did not enumerate any such businesses across the country. Apparently, none of the nascent wineries bore the mark of what the enumerators would consider to be industrial wine production. The subjectivity of this administrative assessment becomes clear when, in 1933, a provincial commission listed five wineries (Fred Marsh Winery; J S Hamilton Co; Ontario Grape Growing and Wine Mfg; T G Bright & Co; and Cooksville Wine Vaults) as having been established or licensed before 1881.[5] Three of these were located in Niagara.

Niagara's proportionately large wine industry likely emerged from the large grape industry, but as Table 2.2 demonstrates, the distribution of vineyards in Niagara did not mirror the distribution of wineries, with Lincoln County producing marginally fewer grapes, but Welland County producing less wine.[6] Since we do not know which grapes were being used for wine, nor can we tell from these numbers from whence came the grapes used in the wineries, all we can know is that electoral district boundaries did not stop the movement of grapes across these imaginary lines. Moreover, the grape production in Ontario did not mirror the development of the wine industry. Finally, Essex County's vineyards appear to have been more actively dedicated to the production of native wines,

Table 2.2. Wine production, 1891

Electoral district	Number of establishments	Number of employees	Value of products	Proportion of provincial wine production
Welland	2	7	$52,000	32.07%
Lincoln & Niagara	1	25	$25,000	15.42%
Monck	1	6	$4,000	2.47%
Essex North	12	33	$37,141	22.91%
Essex South	11	21	$12,160	7.50%
Hamilton (city)	2	14	$17,200	10.61%

Source: *Census of Canada*, Vol. III, Table I – "Industrial Establishments," p. 371.

since it appears that proportionately more grapes went into the production of wine than the grapes of Niagara.

These numbers provide us with some context at the beginning of the industry. In the 1891 census, the wineries are listed as actual "industrial establishments," but given the small number of employees per actual business, these were very much small cottage industries. This general statement must, however, include the proviso that in "Lincoln and Niagara," a relatively large winery (probably the Ontario Grape Growing and Wine Manufacturing Co., later known as Barnes Wines) already existed, employing twenty-five people and producing 15 percent of total wine, by value, in the province. Another notable observation provided by this data, but that we will not examine here to any satisfactory degree, is the proportion of women working in the wine industry. In both Essex and Niagara, women made up a significant portion of the employees of wineries, further suggesting that wineries were the cottage industries that employed all members of a family, sex notwithstanding. By comparison, the ratio of men to women in all employment was significantly disproportionate to the provincial average at this time, as noted in Table 2.3 and Table 2.4.

These data, taken from one census, provide a static picture of the wine industry. Unfortunately for us, the census reports do not give such robust information after 1891. In the next census report, 1901, these data were provided only for regional industries of a significant proportion, based upon some now obscure calculation. In this and subsequent reports, the numbers were not provided for wineries in either Essex or Niagara.

Impact of Anti-Alcohol Sentiment on Wine

While the wine industry emerged, it faced significant social challenges from those who saw the consumption of alcohol as a problem. Indeed, it may be that the uneven treatment of the wine industry in the census report reflects the

Table 2.3. Employees by sex, 1891

	Over 16 Male	Over 16 Female	Under 16 Male	Under 16 Female	Total
Welland	7	0	0	0	7
Lincoln and Niagara	12	13	0	0	25
Monck	6	0	0	0	6
Essex North	21	10	2	0	33
Essex South	13	8	0	0	21
Provincial winery total	**84**	**37**	**4**	**0**	**125**

Source: *Census of Canada*, Vol. III, Table I – "Industrial Establishments," p. 371.

Table 2.4. Proportions of employees by sex

	All males		All females	
Ratios	Wineries	All industries	Wineries	All industries
Welland	100.00%	75.08%	0.00%	24.92%
Lincoln and Niagara	48.00%	74.76%	52.00%	25.24%
Monck	100.00%	92.09%	0.00%	7.91%
Essex North	69.70%	86.13%	30.30%	13.87%
Essex South	61.90%	85.73%	38.10%	14.27%
Province	70.40%	78.77%	29.60%	21.23%

Source: *Census of Canada*, Vol. III, Table I – "Industrial Establishments," p. 371, and Table II, "Industrial Establishments by District," p. 383.

heavily political and socially constructed view of the validity and advisability of nurturing a wine industry in the province. In the nineteenth century, growing sentiment against the consumption of alcoholic beverages became a mass movement; it is useful to consider how the temperance movement affected the fortunes of the wine industry. Civic-minded individuals, concerned about the apparent overuse of alcohol, especially among the working class, and the impact of drinking on the family and the social fabric, united in various organizations throughout the century. While early "temperance" movements were interested in exactly that, *tempering* the consumption of alcohol, by the end of the nineteenth century, "temperance" usually meant complete abstinence. Moreover, the temperance movement became increasingly political, with many of the anti-alcohol organizations lobbying municipal, provincial, and federal governments to enact various forms of restrictive legislation. While from a twenty-first-century perspective this movement may seem odd or even silly, the temperance movement represented a huge group of people, consisting of many individuals who considered themselves to be progressive, socially conscious, and liberal. Total abstinence and Prohibitionism were mainstream ideas.

Consequently, from before Confederation in 1867, governments were mindful of anti-alcohol sentiment. The Government of Canada enacted in 1864 legislation commonly called the Dunkin Act, after the Member of Parliament who presented it in the House. This was the first of a series of acts that permitted municipalities to vote to stop the sale of alcohol within their borders. Versions of these so-called Local Option Acts are in fact still on the books. In 1878, the Dunkin Act, which had applied only to the provinces of Ontario and Quebec, was expanded to the Canada Temperance Act, called the Scott Act. In 1892, the Dominion government created a Royal Commission to examine the liquor traffic. The Commission travelled across the country, taking testimony from residents about the impact of drinking in their communities; the impact that local option legislation might have had; the size and economic viability of the beer, wine, and spirits industries; the potential economic and social effect national Prohibition might have on the communities; and various intricacies of implementing such legislation.[7]

When taking evidence in Ontario, the members of the commission focused some of its attention on the dilemmas provided by the "native wine" industry. During its hearings in Essex County, the commission learned about local wine production. This consisted mostly of farmers who made fermented wine, apparently through spontaneous fermentation, rather than an active inoculation of cultured yeast. There was some evidence, as we have seen with the Census records of 1891, of a nascent wine industry in the area. The evidence of Essex county led the commissioners to begin asking witnesses, both from temperance ("Dry") and "Wet" perspectives, if farmers who grew grapes or apples and made wine or cider for their own use should be forbidden from doing this under Prohibition, and if such a Prohibition would "interfere with individual liberties." Moreover, the often combative commissioners would ask, if farmers are allowed to make wine, why should other people, living in cities with no access to their own wineries, be allowed to buy this native wine? Opinions of many witnesses were decidedly mixed, with some Drys (pluralized this way) admitting that they had not thought of this complication, while others argued that all intoxicants should be prohibited. But on the issue of the morality of drinking, the commission found considerable confusion. When Reverend Andrew Cunningham, a Methodist minister from Guelph and a strong temperance supporter, argued that the very licensing of the sale of wine was "an iniquity," the commission probed this assertion. The testimony led into a series of logically confusing arguments about whether drinking wine at any time was a sin, or whether it just led to sin, and therefore if it was immoral to legalize a substance that wasn't itself sinful but that could lead to sinfulness. Cunningham was clearly caught off guard, and constantly returned to the temperance assertion that drinking alcohol always leads to "immoderation."

> Commissioner Gigault: You do not know of any man who makes a moderate use of liquors?
>
> Cunningham: I do not know of any man that stops at moderation, but he continues until he uses it immoderately. I have heard of such men, read of such men, but I do not know of any. He always becomes immoderate.
>
> Gigault: Then why does not the Bible say that it is immoral to take any quantity of liquor? You said a few minutes ago that the Bible itself allows the moderate use of liquors?
>
> Cunningham: I beg your pardon, I did not say so. I said that it allowed us to use a certain class of wine. But another class it forbade, and said it was an evil.
>
> Gigault: If it allows the use of a certain class, then it allows the certain use of wine?
>
> Cunningham: Certainly, but that was not an intoxicating and fermented wine, as I understand the Bible.
>
> Gigault: Does the Bible speak of fermented and unfermented wine?
>
> Cunningham: It speaks of two different kinds of wine.
>
> Gigault: But it does not speak of fermented or unfermented wine?
>
> Cunningham: It speaks only of wine.[8]

This exchange demonstrates the complex role of wine in temperance rhetoric and in the commission's evidence. Proposed prohibitory legislation would make an exception for alcohol used in medicinal, mechanical, and sacramental purposes. Adherents to both the Catholic Church (of which Commissioner Gigault, a Quebecer, was likely a member) and many from the Church of England argued that unfermented wine was not wine. Protestant non-conformists such as Cunningham were less clear on the issue.[9] Interestingly, later that same day, Charles Raymond, also a temperance advocate from Guelph, was asked whether he would allow farmers to make their own wine because "no harm would result from it." Raymond replied, "You will not bother me as you did Mr. Cunningham. You need not try it." And when pressed as to whether he would "draw a distinction," Raymond argued that "I am not going into hair splitting."[10] The place of wine in temperance rhetoric remained complex.

Along with the philosophical consideration of wine, the evidence of the commission demonstrated that the nascent wine industry of the province had some potential for growth. Since one of the key elements of the commission's mandate was to investigate the impact of Prohibition on industry and the economy of the country, hearing of a nascent wine industry led them to ask bigger questions about its impact. While the Commission discussed the grape farmers in Essex, they did not have hearings in Niagara, so had little opportunity to speak directly to those producers. Yet, when in London, Berlin (now Kitchener), Guelph, and Toronto, they heard repeatedly about the wine that came from Niagara. Michael Roos, a liquor merchant from Berlin, noted that the Scott Act, when enacted in his municipality, had resulted in an increase in the sales of spirits, but that after it was repealed, he saw an increase in the sales of wines. Mostly, he noted, these wines were Canadian, which he would buy "by the [railway] car load" from "P G [sic] Bright & Co," whose wine was "made up in the neighbourhood of Niagara Falls."[11]

This was not the only indication of the growth of the wine industry in Niagara. David R. Wilkie, the President of Toronto's Board of Trade and the cashier of the Imperial Bank of Canada (who had lived in St. Catharines and married a woman from there), argued that the nascent wine industry needed protection from foreign imports.[12] Noting that his bank had four branches in Niagara, Wilkie agreed to a question about the increased business of winemaking, noting that not only was the wine industry increasing, but that it was raising the value of the land. He argued that the native wine industry in Ontario "tended to promote temperance" because it would encourage the consumption of lower alcohol beverages, rather than distilled spirits.[13]

Ironically, given Wilkie's perspective, one of the key issues that arose—not unsurprisingly, given the liquor focus of the Commission's mandate—was the fortification of wine with other spirits or with sugar (to increase alcohol content). Some observers suggested that this process was normal, and problematic, while others—for example, liquor merchant Michael Roos and journalist John Motz, both of Berlin—contradicted the statements that wine was fortified. Motz, speaking from personal experience, explained that wine that was fermented exclusively from the sugars naturally present in grape juice would have enough alcohol to act as a preservative: "it would not keep without fermentation, but there is very little alcohol in it."[14]

So wine occupied three cultural spaces: on the one hand, it was the drink of the elites, while simultaneously a prophylactic against drunkenness; on the other hand, it remained a sort of artisanal beverage, the manufacture and consumption of which was almost a farmer's right. Indeed, the way that wine was made on the farm, simply by not tampering with it, was something that evidence in the Royal Commission suggested set wine apart from beer and spirits, both of which required more conscious effort to create. Yet, certainly the wine consumed by the elites was something more than simply rotten grape juice. So while we cannot really unpack the cultural baggage attached to wine, it is useful to consider how Prohibitionists themselves were caught up in this elitist discourse around the less dangerous fruit of the vine.

Within the Prohibition rhetoric, wine continued to occupy an uncertain place in the hierarchy of intoxicants.[15] As early as the late eighteenth century, temperance sentiment saw wine to be at best a moderately troublesome substance, and more often a beneficial, healthful drink. The confused testimony of Reverend Cunningham of Guelph indicates how easy it was to shoot holes into the scriptural arguments against wine; similarly, physicians continued to argue that wine had healthful benefits. In any case, for most of the nineteenth century, the temperance movement was focused upon the consumption of large amounts of strong spirits like whisky and rum, rather than beer and wine. Most historians studying this period note that, after the middle of the century, the temperance movement was usually a middle-class effort at moral and social reform.[16] The working class drink of choice was spirits (usually whisky) in heavy quantities, and beer, which was probably often much stronger than the generally 5% alcohol by volume (ABV) mass-produced drink churned out by major multinational breweries today. Wine, meanwhile, was less prevalent and rarely associated with working-class consumption (farmers were not really working class in the sense of the industrialized city). So, until the later part of the century, beer and wine were often considered, if not healthful, at least not damaging. In the evidence taken in Berlin, Ontario, especially, witnesses made repeated references to the lack of drunkenness and general healthiness of residents of Germany, Italy, France, and Spain—countries, so the witnesses argued, in which strong spirits were simply not consumed in any significant way. There, "light lagers" and wine were the beverages of choice.

In the end, the Royal Commission was split in its conclusions. A majority report argued that Prohibition would be far too damaging to the economy of the country and of individual municipalities to merit its implementation. The minority report, written by Reverend Dr. McLeod, made a strong moral economy argument in favour of Prohibition.

In both reports, wine received limited attention. For the most part, the only mention of the impact of Prohibition on wine drinking in the majority report was in a section discussing the importance of the liquor industry in general to the agricultural sector, notably grains, corn, and hops production. In a sort of aside, the report mentioned that many grape farmers made wine, that the wine industry was in its early stages, and that "its stoppage would deprive many persons of their present employment, and would lessen the demand for grapes, thus probably compelling the destruction of vines in order to fit the land for other purposes."[17] Dr. McLeod had a section entitled "Beer and Light wines" in which he contradicted assertions that these could be considered temperance beverages. Most of his discussion, however, focused upon "malt beverages" and mentioned wine usually only in the phrase "wine and beer."

The commission's findings were not the end of the story for the temperance movement. A few years later, Prime Minister Wilfrid Laurier called for a national plebiscite to determine whether Prohibition was really wanted. This plebiscite was a promise he made to the temperance forces, many of whom were adherents to the liberal viewpoint. It is somewhat ironic today that anti-alcohol sentiment is considered more of a conservative view, since the temperance supporters were generally liberal in affiliation. The results of the plebiscite were fairly strongly in favour of Prohibition in English Canada, but overwhelmingly opposed in Quebec. Laurier, whose riding was in Quebec and who knew his party could not survive without the support of voters in that province, concluded that the turnout for the plebiscite was too low to be conclusive and, much to the dismay of his temperate followers, did not push prohibitory legislation.

The Prohibitionists were undeterred, and continued to advocate at the local and provincial level for some kind of restrictive legislation. In 1902, the Government of Ontario held its own plebiscite on Prohibition. The Liquor Act, 1902, had two main parts. The first was plebiscite legislation, laying out in detail the process of holding a plebiscite to enact prohibitory legislation. The second was the legislation itself. Had the voters accepted it, this would have limited to druggists the ability to sell non-industrial and non-sacramental alcoholic beverages, including wine. The right would have been granted to druggists because alcohol remained part of the *materia medica* of doctors, and would be controlled by prescription and the druggists' oversight. The plebiscite did not pass, but provincial liquor legislation became increasingly strict, until, in 1916, under the pressures of wartime resource scarcity, the provincial government passed the Ontario Temperance Act (OTA).

Unlike the 1902 legislation, however, this act did separate wine from other alcoholic beverages. Distillers and brewers were able to continue to manufacture their products, but only if they either exported it or sold it for medicinal purposes through a licensed vendor (a druggist).[18] In contrast, native wineries were allowed to continue to sell their products, providing that they did so in large, wholesale quantities, such as in kegs or dozens of quart bottles. Wine could also be sold in pharmacies, ostensibly for medical use. Given that the OTA also prohibited selling any alcoholic beverages for consumption on the premises of hotels, restaurants, and other establishments where previously alcohol could be consumed, this exception for wineries appears to have allowed wealthy customers who could afford to purchase that amount of wine to continue their genteel consumption. Indeed, for the duration of Prohibition, the exception for wine was a major sticking point for the temperance forces, which lobbied the provincial government to close this loophole.

Despite this agitation, the wine exemption was not eliminated, and its continuation suggests the persistence of several ideas about wine. First, that it was not intoxicating, or at least not as problematic as other alcoholic beverages. Second, that wine continued to be exempt from the moral associations made between spirits and problem drunkenness. These may both have something to do with the class-based perception that problematic drinking was the purview of the working class, and the working class simply did not drink wine. Rural farmers, no matter how hard they worked, were not the same as urban wage labourers. Indeed, the idea of protecting the farmer may be the reason that native wineries were allowed to continue to sell within the province. The issue of protecting local economies had been a major theme in the Royal Commission hearings a generation before Prohibition became law.

The exemption for wine suggests the grape industry's importance to the province. While grains for distilling and brewing could be redirected into other forms of production, or something else could simply be sown in the fields the next year, grapes took several years to establish themselves, and longer to yield suitable products. Grape vines were a long-term investment. So arguments from the grape growers would have been heard and appreciated by the legislators. The importance of the grape industry had seen real brick-and-mortar institutional-ization, since, in 1906, the provincial government established an experimental horticultural farm in Vineland, which included research into grapes on its vine-yard. Initially, the grape research was on "market" (i.e., non-wine) grapes; the focus on wine grapes did not begin until 1934–35. Again we see how the con-straints posed by economic arguments were important, if not actually central, to the direction of prohibitory legislation.

While Prohibition may have been a victory for the temperance movement, it also benefited in some ways the native wine industry, carving it out as an industry distinct from the demon alcohol. Subsequently, the number of wineries and con-sumption of wine increased. The fortunate upshot of the expansion of the liquor regulatory bureaucracy during and after Prohibition is that detailed records of the numbers and locations of wineries began to be kept. This started with the Board of License Commissioners (BLC), created in 1915 to manage liquor licens-ing across the province, and that enforced the rules of the Ontario Temperance Act when it became law. The Liquor Control Board of Ontario (LCBO), which was created by the Liquor Control Act (1927) to manage all forms of alcohol sales at the end of Prohibition, continued this bureaucratic tradition. The records were not exactly the same; for example, under the BLC, the maximum potential output was recorded in gallons, but this was dropped by the LCBO, which reported to the legislature the name and location of wineries. (See Figure 2.1 and Figure 2.2.)

Yet the number of wineries does not necessarily mean a growing quality wine industry. Percy Rowe argues that Prohibition created an underground, substan-dard wine manufacturing, the "bathtub gin" of viticulture.[19] Some enterprising individuals made wine in poor conditions using poor or inadequate ingredients,

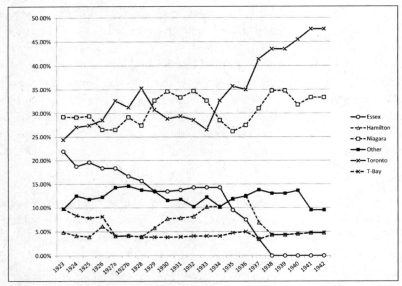

Figure 2.1 Wineries in Ontario, by region, 1923–42
Source: These numbers are calculated from the list of wineries reported by the Board of License Commissioners and the LCBO. *Ontario Sessional Papers*, 1924–1943, "Report of the Board of License Commissioners" (1924–27); "Report of the Liquor Control Board of Ontario" (1928–43). Sessional paper number varied.

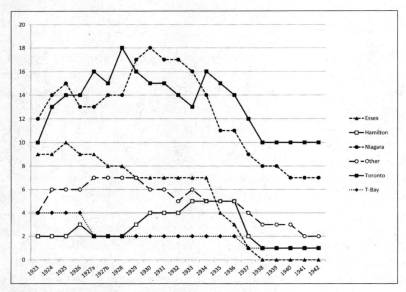

Figure 2.2 Regional wineries as percentage of all Ontario wineries, 1923–42
Source: These numbers are calculated from the list of wineries reported by the Board of License Commissioners and the lcbo. Ontario Sessional Papers, 1924–1943, "Report of the Board of License Commissioners" (1924–27); "Report of the Liquor Control Board of Ontario" (1928–43). Sessional paper number varied.

and the Board of License Commissioners did not undertake a great deal of quality assurance inspection, if any. As a result, several of the wineries listed in the BLC reports were likely the small businesses that thrived in the underground liquor economy, when people who wanted alcohol drank what they could. Rowe blames Prohibition for the consequent poor reputation of Ontario wines in the immediate post-Prohibition decades.[20] This claim would require us to be certain that pre-Prohibition wine, based upon the same grapes and grape-growing environment, with less advanced viticulture techniques, was of a higher standard than that manufactured during Prohibition, an assurance we cannot assume.

So while today most people laugh at or vilify the Prohibition impulse as a misguided attempt at social control, and view it as seriously damaging to the reputation of Ontario wine, a positive outcome emerged from this era. Its weaknesses created a situation that strengthened and expanded viable wine businesses, emphasizing quality over quantity. When the LCBO began to license wineries, it allowed them to have only one store, attached to the winery itself, from which it could sell its own wine (in addition to being sold in LCBO stores). In 1931, in its efforts to present a moderationist approach to the expansion of drinking, the LCBO decided that no more licences would be granted, thereby capping the number of licenses at 51, the number in effect that year.[21] In 1935, in an attempt to strengthen the more commercially viable wineries, the board began to allow winery owners to sell to other wineries the rights implicit in their licences. With such a transfer, the purchasing winery thus purchased the right to have another retail location. Moreover, the LCBO rescinded the rule requiring the retail store to be physically adjacent, or attached, to the winery. This was a conscious effort to permit smaller wineries that were finding it difficult to remain in business to have some resale value. Larger wineries could purchase the licence, and then move the retail part of the business to other locations in the province, thereby expanding their own retail viability.[22] T.G. Bright & Co., for example, had one address (in Niagara Falls) in 1927. After the new rules of 1935, it absorbed the Robinet Freres winery in Roseland (Essex County), and used that licence to open a store on Sparks Street in Ottawa. It also moved its sales office to Yonge Street in Toronto. By 1944, now by far the largest retailer in the province, the renamed Brights Wines had nine outlets in six cities (four in Toronto and one each in London, Ottawa, Sudbury, Hamilton, and Windsor). Its head office remained on Dorchester Road in Niagara Falls.[23] Table 2.5 lists the movement of winery licences.

The LCBO mandate was different from that of the Board of License Commissioners (BLC); while the Prohibition-era BLC was primarily interested in restriction and confinement of the liquor industry, the LCBO's mandate was to regulate but permit alcohol sales.[24] This mission included a conscious effort to promote products originating in Ontario, including wine originating from

Table 2.5 Winery transfers, 1935–47

Purchasing winery	Purchased winery	Year
Barnes Wines	Fort William Wine Co	1936–37
	Sunnybank Winery	1938–39
Brights	Robinet Freres	1934–35
	French-Italian Winery	1936–37
	Ascot Wine Co	1936–37
	Robinson Wine Co.	1936–37
	Windsor Co Ltd	1937–38
	S. Badalato, London	1938–39
	Old Fort Wine Co	1938–39
	Hillrust Wine	1940–41
Canadian Wineries	Stamford Park Wine Co	1935–36
	Lincoln Wines	1936–37
	Dominion Wine Growers	1936–37
Concord Wine Co	Parkdale Wines	1935–36
	Meconi Wines	1935–36
Danforth Wine Co	Regal Wine Co	1934–35
	Lakeshore Wine Co	1934–35
	Beaverdam Cataract Winery	1936–37
Fred Marsh	Jules Robinet (Sandwich)	1934–35
Jordan Wine Co	Canada Wine Products	1938–39
London Winery	Grantham Winery	1934–35
	Victor Robinet	1934–35
	Adelaide Winery	1936–37
	C Rossoni, Windsor	1938–39
	Royal City Winery Guelph	1941–42
	Cooksville Vaults	1946–47
Parkdale Wines	Hamilton Winery	1938–39
St. Catharines Wineries	Toronto Wineries	1936–37
Stamford Park Wine Co	Thorold Winery Co	1935–36
Turner Wine Co	Hamilton-Niagara Wines	1937–38

Source: Annual reports of the LCBO in *Ontario Sessional Papers*, 1935–1948. Sessional paper number and page numbers vary.

Ontario grapes. Indeed, in its fourth report, the LCBO reminded the legislature that "behind the wine industry and always recognized is the widespread interest of the grape growers. The board would like to see that interest conserved and proper profits secured by the growers. The board's merchandising of native wines all over the province has very greatly enhanced the interest of the grape growers and greatly advantaged grape growing."[25]

This attention to the marketing of the wine was presented as both encouraging domestic industry and framed with the idea that "light" wines were less intoxicating than "heavy" ones. In the same report it touted its efforts to increase the sale of light domestic wines, while reducing "the sale of the heavier imported wines."[26] It likened this to its dual control and marketing agenda, proudly displaying increases also in domestic beer, which was also less intoxicating and therefore less of a social issue.

It was the activities of several government agencies that enabled Ontario wineries to expand beyond the poor standards of Prohibition-era wineries. In its first report, the LCBO noted that it has "insisted upon greater sanitation and more suitable quarters than were the case in some of the smaller wineries."[27] In the next few years, the LCBO worked in conjunction with the Board of Health's Laboratory Division of analytical chemists to investigate the condition of wines produced in the province. The Department of Health itself was mandated by the 1927 Public Health Act to have control over health aspects of all winemaking facilities. In 1930, it published detailed regulations regarding the manufacture of wine, which included provisions that might be considered beyond the purview of actual health regulations. For example, the regulations set quality standards such as

a) It shall be of satisfactory colour and the odour, flavour and bouquet shall be natural and pleasing. It must be free from an excessive amount of sediment or turbidity as well as the presence of foreign matter or artificial flavour of any kind.

b) It shall conform with the Regulations under the *Food and Drugs Act* (Dominion).

c) The content of volatile acids, in terms of acetic acid, shall not exceed 0.20 per cent.

d) If designated as "dry" wine, the sugar content shall not exceed 1.0 per cent.

e) If carbon dioxide has been added, it shall have printed on the label in easily legible type the word "Carbonated".

f) If designated as Port or Port Type, Sherry or Sherry Type, it shall contain not less than 14.0 per cent, of alcohol by volume and not more than 10.0 per cent, of sugar.

g) If designated as Claret or Claret Type, the content of alcohol shall not exceed 13.0 per cent, by volume and the total amount of sugar present shall not be in excess of 1.0 per cent.[28]

Yet the Public Health Act did not permit inspection of winemaking facilities until 1932. A year later, the regulations set out the maximum amount of wine that could be extracted from a ton of grapes.[29] These agencies also co-sponsored classes for winemakers on various components of the winemaking process, and the Laboratories Division of the Department of Health assisted wineries by distributing yeast samples, apparently more suitable and likely more sanitary than what the public health officials believed were currently being used. It also annually analyzed hundreds of wine samples from the wineries, though the findings of those reports were not published. So both the LCBO and the Department of Health inspected wineries, and the LCBO claimed that these efforts were improving the manufacturing standards and quality of winery output. Yet, while they had ostensibly different motives, there was doubtless some overlap, and potentially some contradiction in the way the two agencies interpreted and enforced their rules. Winery officials may have been relieved, then, when the regulations regarding manufacturing standards, listed above, were removed from the ambit of the Department of Health, and consolidated into the work of the LCBO.[30] Thereafter, the Department of Health's rules were almost exclusively regarding sanitary and safety aspects of winery operations.

All this effort by government agencies suggests that the provincial government saw value in the native wine industry and paid attention to the matter of its viability. In 1933, the province struck a committee to investigate the state of grape growing and winemaking in Ontario. Named the Ontario Wine Standards Committee (the LCBO called it "The Ontario Grape and Wine Committee"), this panel investigated the development and current state of grape growing and winemaking in the province, arguing that taxes should be eased on grape growers, and recommended other modifications to the wine and grape industry that would improve the viability of this type of business. At the same time, the LCBO noted that it continued to "take a direct personal interest in the affairs of the grape growers," noting that it had facilitated negotiations between winemakers and grape growers to establish a fair price of forty dollars a ton for grapes. In 1935, the minister of agriculture reported that the grape research being undertaken at the Vineland Horticultural Experimental Station had undergone a "radical change." Thereafter, the research would be on grape varieties that were "better suited for wine purposes ... the Station's breeding program has had to be completely revised." He did warn, however, that "no spectacular results have been obtained as yet."[31]

The Second World War brought few changes to the industry, although the consolidation of wineries eased by 1940 (see Table 2.6). Unlike the First World War, the Second did not usher in a new era of Prohibition. While business suffered due to restrictions on trade (wineries were considered low priorities in transit quotas) and redirection of industry toward wartime production, winery output did not decline by much, and in 1944 exports shot up.

By the end of the Second World War, Ontario wineries, now located mostly around the Golden Horseshoe, had become a small but increasingly significant part of the provincial economy. Government intervention, which may appear at times to have been intrusive and problematic, managed to prune back wild growth and, however contentious and constrained, nurture new, viable expansion of an industry that, at the end of the war, still had many roots to be established.

Table 2.6. Wartime output of Ontario wineries

Year	Sold in Ontario	Exported Canada	Exported Elsewhere	Total
1936	1195166	1159010	2483	2356659
1937	1413224	1014810	4520	2432554
1938	1552481	1111769	3581	2667831
1939	1647144	1045997	1935	2695076
1940	1762426	1455432	2136	3219994
1941	1997994	1850008	5892	3853894
1942	1889400	1537579	4766	3431745
1943	1897690	1806083	5315	3709088
1944	1478735	1542432	61602	3082769
1945	1532254	1554265	49769	3136288
1946	2012879	1850650	58581	3922110
1947	2030920	2240527	28190	4299637

Source:"Report of the Liquor Control Board of Ontario," *Ontario Sessional Papers* 1937–1948. Sessional paper numbers varied.

Notes

1 Percy Rowe, *The Wines of Canada* (Toronto: McGraw-Hill, 1970), 30.
2 *Ontario Agricultural Commission*, "Report of the Commissioners," Vol. 1 (Toronto: C Blackett Robinson & Jordan Street, 1881), 3.
3 Monck was established in the British North America Act of 1867, and folded into Haldimand and Monck in 1892. Parliament of Canada, "History of Federal Ridings since 1867: Monck, Ontario (1867–1892), http://www.parl.gc.ca/About/Parliament/FederalRidingsHistory/hfer.asp?Language=E&Search=Det&Include=Y&rid=452.
4 *Census of Canada*, 1891, Vol. 3, Table 1, "Industrial Establishments," 371.
5 "Report of the Ontario Grape and Wine Committee, July, 1934," Brock University Special Collections and Archives, TP 559.C2.058 1934, 23.
6 *Census of Canada, 1891*, Vol. 3, Table XI – "Fruit trees and fruit on lots."
7 See Craig Heron, *Booze: A Distilled History* (Toronto: Between the Lines, 2004).
8 Royal Commission on the Liquor Traffic, *Minutes of Evidence*, Vol. IV, Part 1, Province of Ontario, 503–4.
9 It was the desire for non-alcoholic wine for Protestants that founded the Welch's grape juice empire.
10 Royal Commission on the Liquor Traffic, *Minutes of Evidence*, Vol. IV, Part 1, 509.
11 Royal Commission on the Liquor Traffic, *Minutes of Evidence*, Vol. IV, Part 1, 457.
12 A bank cashier at this time was a senior executive position. On Wilkie's background, see the entry in George Maclean Rose, *A Cyclopaedia of Canadian Biography, Being Chiefly Men of the Time. A Collection of Persons Distinguished in Professional and Political Life; Leaders in the Commerce and Industry of Canada, and Successful Pioneers* (Toronto: Rose Pub Co, 1886), 796. This entry calls him Daniel Robert, but the details in the commission evidence suggest it is the same man.
13 Royal Commission on the Liquor Traffic, *Minutes of Evidence*, Vol. IV, Part 1, 697–98.
14 Royal Commission on the Liquor Traffic, *Minutes of Evidence*, Vol. IV, Part 1, 459.
15 This ambiguity continued into the post-Prohibition period, when the LCBO, concerned about intoxication, was also enthusiastic about the development of a brandy industry.

Brandy is distilled wine and is stronger than regular wine (and close to spirits in alcohol level), but its association with refined, upper-class consumption relieved it of any stigma associated with working-class drunkenness. "Seventh Report of the Liquor Control Board of Ontario," *Sessional Papers of Ontario*, No. 20 (1933), 12.

16 See Heron, *Booze*; Sharon Ann Cook, *"Through Sunshine and Shadow": The Women's Christian Temperance Union, Evangelicalism and Reform in Ontario 1874–1930* (Montreal and Kingston: McGill-Queen's University Press, 1995); Dan Malleck, "Priorities of Development in Four Local Woman's Christian Temperance Unions in Ontario, 1877–1895," in *The Changing Face of Drink: Substance, Imagery and Behaviour*, ed. Jack S. Blocker Jr. and Cheryl Krasnick Warsh (Ottawa: Les Publications Histoire Sociale/Social History, 1997), 189–208; Christopher J. Anstead, "Hegemony and Failure: Orange Lodges, Temperance Lodges, and Respectability in Victorian Ontario," in *The Changing Face of Drink*, 169–88.

17 Report of the Royal Commission on the Liquor Traffic, Parliament of Canada Sessional papers No. 21, 1895, 493.

18 Ontario Temperance Act, 1916.6 Geo V, Chap 50.

19 Percy Rowe, *The Wines of Canada*, 42.

20 Percy Rowe, *The Wines of Canada*, 42.

21 "Sixth Report of the Liquor Control Board of Ontario," *Ontario Sessional Papers*, No. 20 (1932), 10.

22 "Ninth Report of the Liquor Control Board of Ontario," *Ontario Sessional Papers*, No. 20 (1935), 9.

23 "Eighteenth Report of the Liquor Control Board of Ontario," *Ontario Sessional Papers*, No. 20 (1945), 22.

24 I discuss this issue in Dan Malleck, *Try to Control Yourself: Regulating Public Drinking in Post-Prohibition Ontario, 1927–1944* (Vancouver: UBC Press, 2012); see also Sharon Jaeger, *From Control to Customer Service* (Ph.D. dissertation, University of Waterloo, 1999).

25 "Fourth report of the Liquor Control Board of Ontario," *Sessional Papers of Ontario*, No. 20 (1931), 11–12.

26 "Fourth report of the Liquor Control Board of Ontario", 7. Such self-congratulation continued the following year when, noting an increase in the gallons of imported wine sold (3213 gallons), but a decrease in the value (by $100,745.50), the board noted that this odd situation was the result of a decrease in champagnes and heavier wines, and an increase in light wines. "Fifth Report of the Liquor Control Board of Ontario," *Ontario Sessional Papers*, No. 20 (1932), 7.

27 "First Report of the Liquor Control Board of Ontario," *Ontario Sessional Papers*, No. 20 (1928), 6.

28 "Report of the Department of Health," *Ontario Sessional Papers*, No. 14 (1933), 22.

29 "Report of the Department of Health," *Ontario Sessional Papers*, No. 14 (1933), 22.

30 "Thirteenth Report of the Liquor Control Board of Ontario," *Ontario Sessional Papers*, No. 20 (1940), 9.

31 "Report of the Minister of Agriculture," *Ontario Sessional Papers*, No. 21 (1935), 15.

Three

The Growing Place of Wine in the Economic Development of the Niagara Region

Christopher Fullerton

Introduction

D riving down just about any rural route or highway in Niagara's northern tier, it's hard not to notice just how important wine has become to the region's economy. Travellers entering this part of Niagara quickly find themselves immersed in a blend of natural and built environments that wine industry scholars refer to as a "winescape," an aesthetically pleasing landscape composed of rural roads and country lanes, vineyards, farms, wineries, and small towns and villages.[1] While agricultural activities and picturesque country byways have undoubtedly long been part of Niagara's scenic repertoire, it is the emergence of a growing and increasingly reputable wine industry as the *focal point* of this setting that is especially notable. Some thirty years ago, as Hugh Gayler has often noted, the words "Niagara" and "wine" were rarely used in the same sentence, at least not in any sort of flattering way.[2] To be sure, the region did have something that could be described as a "wine industry"; nevertheless, the quality of the wines produced and the reputation they earned was nothing to be particularly proud of, let alone something that could serve as a catalyst for broader local and regional economic development. Indeed, much has changed over the past few decades.

Throughout much of the past hundred years, Niagara's economy rested comfortably atop three main pillars: manufacturing, agriculture, and tourism. Toward the end of the twentieth century and into the twenty-first, however, each of these pillars began to weaken and, in some cases, show signs of potential collapse.[3] As the manufacturing, agriculture, and tourism sectors have each suffered in their own ways from the negative consequences of economic globalization and myriad other processes of change, the relatively recent maturation of Niagara's wine sector suggests to many that the growth of this industry has the potential to compensate for many of these problems. Furthermore, regional boosters are also hoping such growth will prove instrumental in the achievement of other local and regional development ambitions, such as the revitalization of downtowns, the expansion of the homebuilding industry, and even the enhancement of local post-secondary institutions' reputations. For this to occur in a sustainable fashion, however, the region's development actors will have to adopt a number of strategies aimed at protecting the integrity of the Niagara winescape, supporting the local agricultural sector, and growing the wine industry itself.

Niagara's Changing Economy

The Manufacturing Sector

Over the past century, a diverse range of commodities has been manufactured in Niagara, including steel products, automotive components, food and beverages, and paper goods. Many companies were originally drawn to the area by assets such as a convenient proximity to Canadian and American consumer markets and the accessibility of hydroelectric power from the Niagara River. Over time, however, a large proportion of these firms have left the region, due to either their outright closure or their relocation to other points on the increasingly global economic landscape, such as the United States, Mexico, or even overseas. One source estimates that there were thirty-five plant closures in Niagara between 2000 and 2007 alone, translating into a loss of almost six thousand manufacturing jobs.[4] At the same time, downsizing has resulted in many other operations becoming skeletons of their former selves. Indeed, even a cursory review of firms that have left Niagara or scaled down considerably reads like a veritable laundry list: Atlas Steel, CanGro Foods, Cadbury Schweppes, ConAgra Foods, Cyanamid, Dana Canada, DMI Industries, Electropac, Ferranti-Packard, Ford Glass, General Motors, Gerber, John Deere, Lubrizol, Neptunus, PolyOne, Robin Hood, World Kitchen…. The list could go on, were enough space available.

The Tourism Sector

Tourism, also, has long played a central role in Niagara's economy. The most significant attractions in the region have undoubtedly been the world-famous Horseshoe Falls and the adjoining American Falls, which have been visited by well over a billion people over the past century. The establishment of other tourist attractions in the area during the same period, such as amusement parks, wax museums, and tacky souvenir shops, contributed to the creation of what could be best described as a carnival atmosphere, something to which any summer visitor to the nearby Clifton Hill and Lundy's Lane tourist districts can clearly attest. By the early 1990s, however, this tourist "honky tonk" had become somewhat tired, and visitor numbers began to stagnate.[5] The opening of Casino Niagara in 1996, followed by the much more grandiose Niagara Fallsview Casino Resort

Figure 3.1 Niagara Falls skyline, summer 2010
Photo: C. Fullerton.

in 2004, represented the most significant effort to diversify and re-energize the local tourism economy.[6] This strategy certainly worked, as it has created thousands of new jobs and stimulated a great deal of ancillary development, such as the construction of several new hotels and the opening of many restaurants (Figure 3.1); in fact, in recent years, some employers have even complained on occasion of not being able to find *enough* workers.[7] Despite this revitalization and diversification of Niagara Falls' tourism economy, events like the September 11, 2001, terrorist attacks in the United States, the SARS health crisis of 2003, and the "Great Recession" of 2007–9 showed that Niagara's tourism economy would consistently be vulnerable to negative outside forces.[8] As such, officials have increasingly realized that further tourism diversification—aimed at a broader range of tourist types—would be wise from an economic development standpoint. Furthermore, it has been widely realized that the entire Niagara region, far beyond Niagara Falls, has potential for tourism growth, but that this has largely been untapped until recently.

The Agricultural Sector

A third long-standing component of the Niagara economy has been agriculture. The region is home to the vast majority of Canada's prime tender-fruit-growing lands and, as such, much of Niagara's rural landscape is dotted with small and large farms on which grapes, cherries, plums, peaches, and apricots are produced for local, regional, national, and—to a limited extent—international markets (Figure 3.2). As in Niagara's manufacturing sector, however, economic globalization and restructuring have brought numerous changes that have placed the agricultural sector in a precarious position. Most notably, key markets for the region's grapes and tender fruits have disappeared. Major supermarket chains have largely turned to international suppliers of such products rather than those in Niagara, due to the availability of fruit from these alternate sources at a lower cost and in greater quantities on a year-round basis. The region's agricultural economy has also been adversely affected by the closure of the Cadbury Schweppes and CanGro Foods processing facilities. The former, whose operations in St. Catharines were shuttered in 2007, was a major buyer of Niagara-grown juice grapes, which were then used to make well-known products such as Welch's grape juice. After this closure, 134 juice grape growers were left without the traditional market for their produce.[9] Meanwhile, CanGro Foods' facility in St. Davids employed some 150 people in the production of Del Monte fruit products, among others.[10] It was the only fruit cannery in Canada located east of the Rocky Mountains when it was closed by its owners, Florida-based Sun Capital Partners, in 2008. Processing activities were then relocated to China

Figure 3.2 Tender-fruit orchard, Town of Lincoln, summer 2010
Photo: C. Fullerton.

and 150 Niagara tender-fruit farmers, many of whom had only recently spent thousands of dollars upgrading their operations to meet CanGro's standards, lost the market for their produce.[11]

Growth of the Wine Industry

Considering the stagnation and decline experienced throughout the Niagara economy over the past several years, it is not surprising that civic leaders and local development actors have pinned many of their hopes on the continued growth and success of the region's wine industry. As noted in several other chapters of this book, Niagara wines certainly remain in their infancy relative to those that have been produced in Old World regions for centuries; however, there is little doubt that enormous strides have been made over the past two decades in building a Niagara wine industry worthy of worldwide respect. For example, the introduction of *Vitis vinifera* grapes, the creation of the Vintners Quality Alliance (VQA) designation, and the identification of sub-appellations that assist in further delineating the best areas for growing particular types of grapes have all played roles in the Niagara wine industry's maturation. There are now approximately ninety wineries in the Niagara Region (Figure 3.3), including some 85 percent of Ontario's VQA wineries.[12] Wine grapes are grown on approximately 16,000 acres of land in Niagara, and the region's wine industry is estimated to attract about 700,000 visitors per year.

Figure 3.3 Niagara wineries: (a) The Organized Crime Winery; (b) Inniskillin; and (c) Colio
Photo: C. Fullerton.

The Importance of Wine in Niagara beyond the Wine Industry

It is not only the growing volume, variety, and quality of Niagara wines that have provided a glimmer of hope in an otherwise seemingly bleak economic landscape. Local development actors have also identified several associated industries that could well benefit greatly from Niagara's growing reputation as a world wine region. This is particularly true among industries that share a similar market to the wine industry's, one that tends for the most part to be urban based, well educated, and highly cultured. As a result, a healthy and prosperous wine industry is recognized as the lynchpin of several public and private sector development strategies aimed at everything from promoting wine, culinary, and agri-tourism and revitalizing downtown areas to stimulating new housing construction and building world-class educational institutions.

The lofty ambitions many hold for the growth of the wine industry are perhaps best exemplified by *Energizing Niagara's Wine Country Communities*, an award-winning plan released in 2008 by the Niagara Economic Development Corporation. The purpose of the study was to identify ways in which Niagara's emergence as a major wine region could be more effectively harnessed for economic development purposes than had been done up to the time the plan was created. The plan's goals included further developing Niagara into a world-class wine destination, enhancing tourism and economic development throughout Niagara's Wine Country, improving transportation links into and within the region, and encouraging the adoption of sustainable practices that would assure the long-term success of the regional wine economy. Within this framework, the plan envisions Niagara Wine Country as an "upscale destination that invites visitors to enjoy a variety of high quality cultural and heritage experiences."[13]

Wine, Culinary, and Agri-Tourism

The growing success of Niagara wineries has stimulated the diversification of the region's tourism economy by providing visitors with alternative attractions beyond Niagara's famous waterfalls and casinos. In recognizing the potential economic spinoffs that a vibrant wine economy can generate, many localities have adopted planning policies that enable the creation of wine tourism, culinary tourism, and agri-tourism operations, activities that in some cases had until only recently been forbidden due to strict municipal land use regulations. For example, it was once illegal to operate a winery in the City of St. Catharines. This is no longer the case, however, and there were five such operations within the city limits as of December 2012.

A plethora of large-scale annual events have also been developed that take advantage of Niagara wines' growing reputation and, with this, the wine tourism industry's natural linkages with culinary tourism. These include the

Niagara Wine Festival in St. Catharines; the Niagara Food Festival in Welland; Port Colborne's Flavours of Niagara International Food, Wine and Jazz Festival; and Graze the Bench, an event involving seven wineries, seven restaurants, and seven music bands along the Beamsville Bench. Other annual events include the Niagara Icewine Festival, the Niagara New Vintage Festival, and the Twenty Valley Niagara Wine Country Run.

Downtown Revitalization

The growth of the region's wine industry has also played, and continues to play, an influential role in the revitalization of downtowns throughout Niagara. The Queen Street shopping district in Niagara-on-the-Lake's Old Town, for example, attracts hundreds of thousands of tourists and shoppers throughout the year. While much of the district's popularity can be attributed to the presence of other attractions that draw people to the area, such as the Shaw Festival, the Niagara River Parkway, and the Old Town's wide array of heritage buildings, there is no doubt that many of the visitors to the Queen Street district are also wine tourists who have come to experience the region's wineries and winescape. The impressive and ongoing revitalization of Niagara Falls' downtown area, centred on that city's Queen Street (Figure 3.4), can surely also be linked to the

Figure 3.4 Queen Street, Niagara Falls
Photo: C. Fullerton.

growth of Niagara's wine industry. The rejuvenation of this commercial district has involved the adoption of an arts and culture theme, which has led to the opening of several art galleries and studios, trendy cafés and bistros, and other businesses catering to the same audience as that primarily targeted by the wine industry.[14] Thus, as more visitors come to Niagara to enjoy its wine-related attractions, more are sure to also patronize businesses within this previously neglected downtown landscape.

Moving westward across northern Niagara, the City of St. Catharines' *Downtown Creative Cluster Master Plan* also represents an optimistic appraisal of the wine industry's potential contribution to downtown revitalization.[15] In this case, St. Catharines' downtown is expected to become a cultural hub for the Niagara region. As part of this, the City has been working diligently over the years to have the well-known Wine Route rerouted to travel along St. Paul Street, St. Catharines' primary downtown thoroughfare, rather than through the suburban landscape of the city's south end.[16] The Wine Council of Ontario, which owns the proprietary rights to the Wine Route, voted in 2010 to honour the City's request and will redirect the Wine Route through downtown St. Catharines upon completion of a major bridge construction. Other wine-related ambitions stipulated in the master plan include the creation of a "Wine Country Embassy" in downtown St. Catharines that could ultimately serve as the "Headquarters" of Wine Country and, either as part of or separate from this initiative, the opening

Figure 3.5 Jordan Village
Photo: C. Fullerton.

of a landmark Liquor Control Board of Ontario (LCBO) store that would place special emphasis on Niagara-made VQA wines.[17]

Similar plans that rely on the continued growth and prosperity of the wine industry have been created for Jordan (Figure 3.5), Jordan Station, Vineland, and Grimsby. The Twenty Valley Tourism Association, an organization promoting tourism development in the Town of Lincoln (which includes the first two of these communities), has focused the bulk of its marketing energy on wine and culinary tourism. At the same time, the Town of Lincoln's Secondary Plan for Jordan, Jordan Station, and Vineland seeks to maximize the benefits of the Wine Route passing through these communities by encouraging the provision of adequate parking and pedestrian infrastructure, as well as the implementation of beautification and streetscaping strategies. Meanwhile, Grimsby's newly adopted *Official Plan* aims "to promote Grimsby as a primary gateway into the Niagara Wine Country" and, particularly, "to encourage the development of the Downtown as a signature tourism destination of the Niagara Wine Country."[18]

Housing Development

Even the local homebuilding industry has harnessed Niagara's wine industry and winescape for its own purposes. Researchers throughout North America and Europe have observed the ways in which developers market rural places

Figure 3.6 The "Hillebrand" Model, The Village
Photo: C. Fullerton.

Figure 3.7 Cherrywood Estates advertisement
Source: *Niagara This Week*, 2010, RE13.

as offering a higher quality of life than can be found in urban settings, with much of the focus being on the amenity value of the rural landscape.[19] Niagara's attractive winescape provides a perfect such setting and, accordingly, some sub-divisions throughout the region carry wine-oriented names like "The Vineyards," while others are marketed with an emphasis on their setting amidst wineries and grape-growing lands. One development in Grimsby, called Vineyard Valley, includes Cabernet Street, Vinifera Drive, Riesling Street, and Chardonnay Place. In The Village, a New Urbanist–inspired development in Niagara-on-the-Lake, several new home models are named after wines, such as the Riesling, the Merlot, and the Cabernet, while others are named after well-known local wineries, such as the Hillebrand (Figure 3.6), the Inniskillin and the Pillitteri. Other home-builders have also adopted a wine theme, one such example being the Vintage and Estate collections offered by Landmart Homes in its Cherrywood Estates subdivision (Figure 3.7). These themes help to build a particular sense of place about these residential communities that goes a long way in selling homes, espe-cially to the growing market of empty nesters who are migrating to amenity-rich areas like Niagara from urban and suburban settings.

Figure 3.8 Wine Visitor and Education Centre, Niagara College
Photo: C. Fullerton.

Education

Niagara's two major post-secondary institutions, Brock University and Niagara College, have also become active participants in the continuing development of the region's wine industry. At Brock, this has been highlighted by the creation of the Cool Climate Oenology and Viticulture Institute,[20] while Niagara College is contributing to the growth of the industry through the Niagara College Teaching Winery and the School of Hospitality and Tourism, as well as the Wine Visitor and Education Centre at the College's Glendale Campus in Niagara-on-the-Lake (Figure 3.8).[21] Developing such institutions and programs has helped each school carve a niche for itself in an era where there is increasing competition for students. At the same time, however, much of the education and training offered by these schools will, in the eyes of many potential students, have value only if there is a healthy wine industry in which they will be able to secure employment upon graduation.

Making It All Work: Implications for Future Planning and Development

Clearly, the wine industry has come to play a leading role in building Niagara's economic future and, at the same time, there are many hopes resting on this growth continuing over the long term. However, three necessary and closely interrelated conditions will have to be met for this to happen. First, the winescape upon which grape growing, winemaking, and many other associated activities take place must be preserved and protected from the ravages of urban development. Second, the region's farmers must be able to make a viable living from their activities. Third, Niagara's wine industry must be protected and promoted amid a highly competitive global marketplace.

Protect the Winescape

As noted earlier in this chapter, the word "winescape" is often used to describe the natural and human landscape within which wine country is found. This includes everything from farm fields, vineyards, and barns to wineries and quiet country roads. Protecting the winescape is seen as being critical for the growth of any region's wine industry, as it plays a vital role in building a strong and positive sense of place among residents, visitors, and wine consumers that, in turn, helps to ensure that wine takes a prominent position within the regional identity. Niagara's winescape, however, faces a considerable challenge in this sense, as it is located within one of North America's most heavily populated areas, the Greater Golden Horseshoe. For decades, this has resulted in strong development pressures that, time and again, have prompted the conversion of prime agricultural lands to urban uses, such as industrial sites and housing subdivisions. It has long been argued that the future of Niagara's wine industry is therefore heavily contingent on the preservation of Niagara's winescape from urbanization and the associated ravages of highway traffic.[22]

Significant progress has been made in this regard over the past few years, with the Ontario government leading the way. Lands within the Niagara wine region are now heavily protected from development pressures by numerous forms of provincial planning legislation, such as the Niagara Escarpment Planning and Development Act, the Greenbelt Act, and the Places to Grow Act. The Regional Municipality of Niagara's Policy Plan, as discussed by Hugh Gayler in Chapter 17, has further established limits to urban development through its delineation of urban growth boundaries for Niagara's lower-tier municipalities. However, the problem in this case may be that the development controls now in place across the Niagara Wine Country may be too rigid, as discussed in the following section.

Protect the Farmer

Agricultural lands and the various farming activities that take place on them figure prominently throughout Niagara's winescape. The various forms of planning legislation identified above certainly provide an important first step in avoiding the urbanization of this landscape, but they alone will not ensure its long-term contribution to the regional economy. It is also imperative that farmers are able to make a viable living on these lands. If there is not a willing buyer in place—and, even then, if sustainable prices cannot be charged—it will make little sense for the region's farmers to continue growing their produce. This then leads to the risk of land being taken out of production and, with few other options in terms of how it may be used, left to grow over as fallow—a scenario that could seriously harm an integral component of the region's winescape by reducing its aesthetic appeal.

Despite this reality, however, many have argued that the planning regulations currently in place go too far in their efforts to protect the landscape, in essence treating it more as an "environmental preserve" than as an "agricultural preserve."[23] That is, while these planning laws and regulations prevent sprawl, they do little to help the farmers who own much of this land. For example, the 2007 closure of the Cadbury Schweppes facility in St. Catharines more or less dried up the Canadian market for juice grapes, while the 2008 closure of the CanGro plant eliminated a huge market for tender fruits.[24] There are few large-volume buyers of tender fruits beyond CanGro; for example, the major supermarket chains have come to rely much more heavily on more exotic locales for their produce due to the year-round availability, lower prices, and ease of access (which includes free-trade policies). To be sure, the growth of the wine industry has been beneficial for many wine grape growers, but there have also been sizeable wine grape surpluses in recent years, which many growers have attributed to the larger wineries' mass importing of foreign grapes. Wineries are entitled to include imported grapes so long as they make up no more than 40 percent of the grape content in their wines, which are then marketed as "International–Canadian Blends." Niagara's five largest wineries buy most of the grapes grown in Niagara, but they also blend much of these with imported grapes from such faraway countries such as Chile and Australia. The best hopes for helping Niagara farmers to continue working the land may lie within the wine industry itself. The growth of the

Figure 3.9 Agricultural land for sale (L to R): Jordan, Beamsville, Niagara-on-the-Lake
Photos: C. Fullerton.

industry—and the continued growth of VQA wines, in particular—could lead to even further demand for Niagara-grown grapes. This could not only eliminate the annual surplus of grapes that has come with overproduction in recent years, but could also increase demand to a point where farmers who currently do not grow grapes, but who are looking for new crops to grow due to the loss of buyers of other products, may find it appealing to enter the wine-grape-growing sector. Without such a demand, entering the world of wine grape growing would prove to most a losing investment.[25] For those unable or unwilling to wait, however, the only option may be to sell their land. Interestingly, the key to selling such properties often lies in selling them as potential sites for a vineyard or winery or both (Figure 3.9).

Protect the Wine Industry

A third form of protection that will be required in order for local development actors' ambitions to be realized will be to support the wine industry itself. Although the VQA designation and more general quality improvements have certainly contributed to the strengthening of the industry, this has not yet been enough for the region's wines to exceed the market share held by foreign wines nor by those marketed as International–Canadian Blends. For example, Ontario-produced wines accounted for only about 49 percent of all wine sales in the province in 2007, a significantly lower share than that held by domestic wines in other countries (typically around 80 percent or higher). At the same time, the Wine Council of Ontario has noted that VQA wines made up about 20 percent of total Ontario-made wine sales, while International–Canadian Blends wines comprised the other 80 percent.[26] This is explained in part by the fact that blended wines are typically sold at lower prices than VQA wines and are primarily intended to compete with inexpensive (i.e., less than $10 a bottle) wines imported to Ontario from places like the United States and Chile. It is this need to stay competitive with foreign competition that has made the large corporations very reluctant to change their strategies.[27]

Conversely, many of Niagara's smaller wineries (such as Angel's Gate Winery in Figure 3.10) bottle only VQA wines. These often sell out completely, but they also don't get to sell much through the LCBO. Instead, smaller wineries must rely far more heavily on on-site sales and the occasional restaurant contract to bring in the bulk of their revenues. Accordingly, broader exposure to markets is critical for the future growth of most Niagara wineries. Acquiring greater shelf space in LCBO stores is particularly critical. Furthermore, this points to the need for a combination of planning policies that protect the winescape while at the same time supporting the region's agricultural and wine industries.

Figure 3.10 Angel's Gate Winery, Town of Lincoln
Photo: C. Fullerton.

Conclusion

Niagara's wine industry can surely not be looked upon as the *only* saviour of the regional economy. However, there is no doubt its recent growth and maturation has served to mitigate the impacts of the range of changes witnessed in recent decades within Niagara's agricultural, manufacturing, and tourism economies. There is great potential for the wine industry to further strengthen Niagara's economy, but what many people likely do not realize is the extent to which the growth of the wine industry also has the potential to stimulate growth in other sectors. From tourism and housing development through to downtown revitalization and the advancement of local post-secondary institutions, the spinoffs that could emanate from the success of Niagara wines are numerous. This further strengthens the importance of ensuring that appropriate policies are in place that will help to protect Niagara's winescape, farmers, and the wine industry itself.

Notes

1 Barbara Carmichael, "Understanding the Wine Tourism Experience for Winery Visitors in the Niagara Region, Ontario, Canada," *Tourism Geographies* 7, no. 2 (2005): 185–204; Gary Peters, *American Winescapes: The Cultural Landscape of America's Wine Country* (Boulder, CO: Westview Press, 1997).

2 Hugh Gayler, "The Niagara Fruit Belt: Planning Conflicts in the Preservation of a National Resource," in *Big Places, Big Plans*, ed. Mark B. Lapping and Owen J. Furuseth (Hampshire, UK: Ashgate, 2004), 55–82; Hugh Gayler, "Niagara's Emerging Wine Culture: From a Countryside of Production to Consumption," in *Covering Niagara: Studies in Local Popular Culture*, ed. Joan Nicks and Nick Baxter-Moore (Waterloo, ON: Wilfrid Laurier University Press, 2010), 195–212.

3 Don Fraser, "Jobless Rate Climbs to 10.8%," *Welland Tribune*, 9 January 2010, A1.

4 Canadian Auto Workers Local 199, "Manufacturing Matters," Presentation to St. Catharines City Council, May 7, 2007, www.caw199.com/ppt/niagara-presentation.ppt.

5 Karen Dubinsky, *The Second Greatest Disappointment: Honeymooning and Tourism at Niagara Falls* (Toronto: Between the Lines, 1999); Gayler, "The Niagara Fruit Belt."

6 Jeffrey Stewart, Linda Bramble, and Donald Ziraldo, "Key Challenges in Wine and Culinary Tourism with Practical Recommendations," *International Journal of Contemporary Hospitality Management* 20, no. 3 (2008): 302–12.

7 Angus Scott, "Plenty of jobs ... but no employees," *Welland Tribune*, 4 March 2004, A7.

8 Don Fraser, "Thirsting for a Rebound: Tourism Sector Confident 2004 Will See Start of Recovery," St. Catharines *Standard*, 26 April 2004, A6; Monique Beech, "Tourism Season Highly Erratic," St. Catharines *Standard*, 8 July 2008, A1.

9 Tiffany Mayer, "Niagara Farmers Taking Federal Pullout Cash," St. Catharines *Standard*, 22 November 2008, A7.

10 "Niagara Has Lost Something Special," St. Catharines *Standard*, 23 April 2008, A14.

11 Christina Blizzard, "Bitter Harvest for Niagara's Fruit Farms: Closure of CanGro Is a Sad End to What Was Once a Mighty Industry," St. Catharines *Standard*, 3 May 2008, A14.

12 Wine Council of Ontario, *Backgrounder* (St. Catharines, ON: Wine Council of Ontario, 2010).

13 Peter J. Smith and Associates, Ltd, *Energizing Niagara's Wine Country Communities*, Report prepared for the Niagara Economic Development Corporation, 2007, 20.

14 John Robbins, "Weathering Winter," *Niagara Falls Review*, 23 February 2010, A1.

15 Samantha Craggs, "Businesses Welcome Creative Cluster Plan," St. Catharines *Standard*, 30 April 2008, A3.

16 Erik White, "Downtown Link for Wine Route? Two-Way St. Paul Street Touted as Wine Council Leader Proposes Luring Visitors into Taking Longer Look at Sights of St. Catharines," St. Catharines *Standard*, 20 March 2004, A3.

17 Marlene Bergsma, "Study Recommends Creating Wine Embassy Downtown," St. Catharines *Standard*, 12 March 2007, A2; Joseph Bogdan Associates, *St. Catharines Downtown Creative Cluster Master Plan*, Prepared for St. Catharines Downtown Creative Cluster Master Plan Steering Committee, 2008, 50.

18 Town of Grimsby, *Draft Official Plan*, 2009, 24.

19 Gerald Walker, "Class as Social Formation in the Exurban Fringe," in *The Rural–Urban Fringe in Canada: Conflict and Controversy* (Brandon, MB: Rural Development Institute, 2010), 2–9; Paul Cloke, "Conceptualizing Rurality," in *Handbook of Rural Studies*, ed. P. Cloke, T. Marsden, and P.H. Mooney (London: Sage, 2005), 18–28.

20 Colleen Turner, "New Wine Institute Attracts First Students," St. Catharines *Standard*, 21 July 1997, B3.

21 Peter Downs, "A Leader in Tourism Education: Niagara College Has Found a Niche That Works with Its Community," St. Catharines *Standard*, 29 April 2002, 16.

22 Lori Littleton, "Pressure Mounting to Preserve Niagara Farmland," *Niagara Falls Review*, 2 September 2000, A5; Marlene Bergsma, "Council Urges Ban on Farmland Development: Napa Valley Cited as Example of How Restrictions Can Work," St. Catharines *Standard*, 27 July 2000, A3.

23 Matthew van Dongen, "Greenbelt Not Saving Farms, Summit Hears," St. Catharines *Standard*, 1 April 2010, A1.

24 Blizzard, "Bitter Harvest for Niagara's Fruit Farms," A14.

25 Don Fraser, "Grape Glut Feared if Plantings Continue: Do Not Plant without a Firm Winery Contract, Growers Are Advised," St. Catharines *Standard*, 26 February 2003, A5.

26 KPMG, *Study of the Ontario Economic Impact Content of Ontario Wines vs. Foreign Wines and Ontario VQA Wines vs. Ontario Cellared-in-Canada (CIC) Wines*, Report prepared for the Wine Council of Ontario, 4 September 2008.

27 Peter Downs, "Dying on the Vines: Cheaper Imported Wines Are Putting Pressure on Ontario Growers to Accept Lower Crop Prices," St. Catharines *Standard*, 28 July 2001, A1.

Business and Bottles

Four

The History of the VQA

Linda Bramble

Introduction

T he Vintners Quality Alliance of Ontario (VQA) is an appellation system, the marque of which signifies that 100 percent of the grapes used to make the wine were grown in Ontario. It also means that the finished wine, made from only authorized grape varieties, has passed provincially regulated minimum standards of production and the wine is fault free. The VQA's early establishment came about during one of the most turbulent periods in the history of the Ontario wine industry, but not because of it. With impeccable timing, the rules and regulations that were framed by what were considered, at the time, a group of rogue young winemakers helped to save the very existence of what many assumed was a moribund industry. Its evolution is a story of vision, collaboration, and enormous determination.

The Early Years: 1982–88

Although some early students of viticulture would claim otherwise, the idea and initiative for the establishment of a controlled appellation system[1] in Ontario began with Donald Ziraldo (Figure 4.1), the co-owner of Inniskillin winery, the first Ontario winery since Prohibition to be granted a licence in 1975 to manufacture and sell wine. He and his partner, Karl Kaiser, had been making wine for seven years when, in 1982, Ziraldo attempted to sell his Maréchal Foch to the French. They liked the wine but refused to buy it on the grounds that it had no provenance. How did they know that the grapes in the wine actually came from Ontario, let alone Ziraldo's vineyard, as he had claimed? What proof could he offer that he hadn't blended in an imported wine to boost his Ontario-grown wine's appeal? Ziraldo's hands were tied because, in fact, he had no proof. The wine industry at the time was governed by the Wine Content Act, which allowed

Figure 4.1 Donald Ziraldo
Photo: Victoria Gilbert.

wineries to blend a percentage of local product with imported wine yet still call it a Product of Canada.

Ziraldo had been attending, as an observer, the meetings of the International Organization of Vine and Wine (*Organisation International de la Vigne et du Vin*), or OIV, to learn about international markets, hear about the latest research, and network with the most significant producers in the world. Part of the organization's mandate was the harmonization and adaptation of regulations by all their members, which included France, Italy, Germany, and Austria, plus many others. Canada was not a member because it did not have an appellation system that guaranteed origins, authenticity, or quality of any sort. Ziraldo could understand why the French were reluctant to do business with him. He knew that every serious wine region in the world had a system that identified the places where better quality grapes could fully ripen; each also had a list of which varieties would grow well and how they could best be grown. They also had a set of standards that would assure a baseline for quality in winemaking.[2]

Most of Ontario's wines up until this time were made from various crosses and hybrids of native species of grapes from the *Vitis labrusca* family of grapes, which the Europeans disallowed in their wine appellation laws. Grapes such as Dutchess, Concord, Niagara, Seibel, and Isabella, according to the

Europeans—and, increasingly, many Canadians—were not capable of making flavourful table wines. In order to mollify the grapes' high acidities, the wines had to be watered down, sweetened, fortified, or blended with imported wines to make them palatable. As fortified wines, there had once been a market, but the consuming public had shifted its preference to drier table wines with lower alcohol levels and more acceptable flavours.

After being rebuffed by the French, Ziraldo returned from Europe with a plan to initiate talks among his colleagues in the industry to create a system that would guarantee origins and standards of production—an appellation system for Niagara.[3] To rally other like-minded wine producers, most of whom had newly established cottage wineries like his own, he pressed the importance of differentiating their "new" wines, made from the better quality *Vitis vinifera* family of grapes, from the older, more stodgy portfolio of wines made from *Vitis labrusca*–based grapes and imported blends.

At first, Ziraldo's volunteer committee would gather after meetings of their trade association, the Wine Council of Ontario (WCO), to discuss Ziraldo's proposals. The larger wineries saw his ideas as no threat to their businesses. In the beginning, they didn't even send a representative to join in the discussions. Those who did attend agreed to form an association of vintners who would produce only top-end wines. Their first formal meetings began in 1983 with discussions revolving around whether they should be an association of wineries or of wines. Only a few really understood how an appellation system worked. Paul Bosc, Sr. (Figure 2), from Château des Charmes and born in Algeria, was a fourth-generation winemaker and had studied winemaking in France, so he understood how the system worked. Peter Gamble, who was making wine at Hillebrand Estates, as well as teaching Wine Appreciation at Brock University, also grasped the intricacies involved. Early conversations revolved around the question of what the group would be called. The model the committee members best understood was that of the WCO, where membership was granted to a winery; this was the model many assumed their new appellation system would also follow. They wanted to reflect the fact that they generated the project. It was not a policy driven by the government. The term "vintners," loosely referring to wine producers in the title, had great appeal, but it also had its shortcomings.

"I kept saying," recalled Gamble, "if we try and do it as an association of wineries as the Wine Council was structured, then all of a member's wines would have to fall into the quality category. How many wineries were we going to get? Appellations are based on wines, not wineries. It wasn't a winemakers' club."[4] Gamble lost the battle on the name of the organization, but won the war. All eventually agreed that it was a membership of eligible wines, not wineries. The wines from a small winery that produced *vinifera*-based Ontario wines had

Figure 4.2 Paul Bosc, Sr., on the right.
Photo: Château des Charmes.

as much potential for acceptance as a large winery that also produced *vinifera*-based Ontario wines as part of its portfolio. Nevertheless, it would be called the *Vintners* Quality Alliance.

Although much of the work on the standards was done by Dr. Vicki Grey at the Horticultural Products Laboratory at the Horticultural Research Station at Vineland, the group had to fully understand and agree to what she had suggested. Among their first tasks was to define terminology so they could share common ground on what they meant by such terms as estate bottled, vineyard designation, vintage, viticultural areas, and the percentage of the varietal listed on the label that must be from the designated variety. Their guidelines followed the European model, with the goal of eventually harmonizing with the OIV.

The push for quality through better grapes was becoming a common theme in Ontario in the 1980s. But the implications of such a transformative move would be catastrophic for the majority of growers in Ontario, who grew only *labrusca*. The livelihood of nearly eight hundred growers was at stake. Ziraldo was sympathetic but unswayed. Labruscas could be grown for juice and jam, but they would not be part of the appellation system he had in mind for Ontario.

With the help of the late Dr. Peter Peach, Earth Science professor at Brock University, Ziraldo and his group defined the geographic boundaries where historically grapes, albeit *labruscas*, could be grown, then identified the natural geographic boundaries that would delimit the area. They also needed to have a legal structure. Ken Douglas, a St. Catharines lawyer and grape grower, stepped in to take the place of lawyer Tony Doyle, a pioneer grape grower and early winery owner (*Willowbank*) when Doyle could no longer spend the volunteer time it took to give the new organization a corporate footing. Douglas helped the Vintners Quality Alliance set up Letters of Patent, with no shares or profit.[5] By 1987, the VQA of Ontario had its own board, with Ziraldo as founding chair; every member winery had one vote. When it began, it was completely member driven, and was to remain so until it was legislated in 1999. The board's role was not to initiate but to oversee.

In 1987, when Len Pennachetti (Figure 4.3), president of the newly established Cave Spring Cellars, joined the group as a bona fide member, he was soon to take on a central role in the codifying of the VQA rules as they were being developed. "Len was my sidekick," said Ziraldo. "Thank God he was there because I didn't have the patience for the details." Pennachetti did. With a year of law school in his background and an affinity for fundamentals and legal form, Pennachetti, along with another newcomer, Paul Speck, Jr. (Figure 4.4) from Henry of Pelham, joined forces first with Bob Downey from the LCBO, then, by 1988, with Leonard Franssen, also from the LCBO, to establish the specific standards that members of the new organization were fleshing out. Franssen would end up working with them for the next twenty years.

Figure 4.3 Len Pennachetti
Photo: Dwayne Coon.

Figure 4.4 Paul Speck, Jr.
Photo: Henry of Pelham.

The process they used was creative, democratic, and open. At meetings, the attending winemakers would brainstorm such questions as "Should we have rules about how many hectolitres per hectare we can harvest?" No, they decided. They didn't want a highly regulated industry like Bordeaux. They reasoned it just led to people cheating anyway. "Should they have ripeness guidelines determined by the amount of Brix (sugar) at harvest?" Yes, the committee members determined. They needed guidelines for Icewine and late harvest wines as winemakers did in Germany. On questions that would test international acceptability, they would defer to Gamble. He was the professor in the group. It almost went without saying that only *vinifera* and *vinifera* hybrids would be allowed, but which ones? They had jettisoned all *labrusca*-based varieties. But there were still tonnes of French *vinifera* hybrids in the ground. Bosc wanted to disallow all hybrids entirely, but the reality was that until more *vinifera* could be planted, many of the new and smaller producers needed hybrids to stay afloat. But not all hybrids were created equal. De Chaunac and Michurinetz, for instance, did not have the same potential as Baco or Maréchal Foch. How would the winemakers decide which ones to include or disallow? Two strategies emerged. The first strategy was based on Bosc's influence. He became the conscience of the group and helped to define where they should draw the line. He had the benefit of having worked both sides of the industry—as winemaker for Chateau-Gai for over a decade, then, since 1978, as winemaker at his own estate, Château des Charmes. If they set the rules at a point that could be contentious, they would ask Bosc how the big wineries would respond. "Usually," recalled Gamble, "Paul would let us know where we stood with an expletive every fourth word in highly elevated volumes, which was just great."

When the large wineries started to push for a much broader VQA, Bosc would let his views be known, and his influence was considerable. He was, however, only one voice, and a subjective one at that. The issue of ruling on which wines could reflect a quality standard that consumers would support required a more objective strategy. They resolved that challenge by creating an independent panel of experts who could rule on the acceptability of a variety from a consumer and retail sales point of view. With the support of the LCBO, the panel would be set up using their highly trained product consultants to assess the wines. The panel would remain as a significant component of the system. The LCBO provided services in sensory evaluation, lab analysis, and packaging reviews, which included a label examination. It also reviewed packaging compliances according to VQA regulations, such as location of terms on the label, whether the appellation was correctly used, and whether the labels complied with federal requirements. The VQA itself was responsible for determining whether a wine would be certified to carry the VQA medallion if the producer was a member in good standing and passed all the necessary requirements.

In the end, there were thirteen acceptable *vinifera* hybrids included in the list of permissible grape varieties that could produce a varietal wine, with Baco Noir, Maréchal Foch, and Vidal, the Icewine grape, among them.[6] Most *viniferas* were accepted. If a variety was not on the list, and a grower or winery could prove that the variety they wished to try to grow was *vinifera*, and was willing to experiment with its potential for growth, it was automatically approved. Syrah, Nebbiolo, even Zinfandel were eventually added to the list by adventuresome producers. Some varieties proved their merit; others simply couldn't survive the experiment. The varieties chosen were also part of an authorized list from the OIV. The hybrids that would be allowed could achieve only the broader "Ontario" designation, as opposed to the smaller designated viticultural area designation as Niagara Peninsula, Lake Erie North Shore, or Pelee Island.

Another issue with which the vintners had to reach consensus was the permissibility of adding water to ostensibly ameliorate the acidities of some varieties, which resulted in cost-saving stretched volumes. The larger wineries argued that adding water didn't necessarily make a wine faulty, just diluted. Unmoved, both Gamble and Pennachetti were adamant on this issue. It was important that VQA wines be made of 100 percent pure grape juice—no water or sugar,[7] for that matter, added. Since the independent taste panel could not rule on a watery wine, they had to devise another objective way to monitor amelioration. Here was where the early drafters of the VQA turned a perceived vice into a virtue. The LCBO already had regular winery audits in place. To reconcile the amount of grapes purchased with the amount of wine bottled to prevent bootlegging, the LCBO had created an audit procedure. All the auditors had to do was add the VQA audit to its assessments. They would record what quantity of potential VQA wine the winery had in its tanks and compare that to what the winery was bottling. It was an additional task, but the auditors accepted it. It was in the LCBO's interests to help elevate the industry's wines. Money could be made at the retail level if the wines were authentic and typical of their type.[8] In addition to the regular audit, the winery had to keep separate records for VQA wines.

By 1988, the committee had drafted its completed rules and regulations. Pennachetti, Gamble, Franssen, and Speck had done yeoman's work, with Allan Schmidt, Bosc, and others assisting when they could. The group of four would listen to discussions, then codify them, bring their notes back to the members for approval, and highlight issues to be further discussed. The process was completed just as the Ontario wine industry entered its darkest moment.

As a result of free trade, the markup on imported wines that had protected the domestic industry for several years had been traded away. Within ten years (front-end-loaded by 24 percent the first two years and then 10 percent every year after that), the 66 percent the LCBO marked up on imported wine would be reduced to 60 percent and Ontario wines would go from 1 percent to 60 percent. Ontario wines would thus reach parity with imports, with both at 60 percent. The larger wineries knew they couldn't compete. As Jeff Ward, President of Barnes Wines, expressed to a CBC interviewer who was reporting on the effects of free trade on the wine industry, "We may well become bottlers of Californian wine."[9] Not enough *vinifera* grapes were in the ground. The new rules also stipulated that no table wine could be sold through the LCBO if made from *labrusca* grapes. The Quality Control lab of the LCBO had easy tests to analyze the presence of the defining compound that was found only in *labrusca* wines—methyl anthranilate. The wineries couldn't even fudge their way through.

As counterintuitive as it may have seemed at first, the smaller estate wineries, especially those of the VQA, were not as threatened by the prospect of going head to head with California or France, since the General Agreement on Tariffs and Trade (GATT) was also amended to reach parity at the same time. While the larger wineries were talking of mergers and acquisitions as the competitive answer to salvage their businesses, the smaller ones were growing more and more confident they could survive. They had already put their stake in *vinifera* vines that they were growing themselves, and in doing so had become the darlings of the wine media.

Peter Ward of the *Ottawa Citizen* wrote on May 19, 1989:

> It's pretty impressive stuff when a team of young winemakers can run off with the prize for the best Canadian wine in a major show, just three years after they've gone into business. That's what the team of Jordan's Cave Spring Cellars has accomplished. Their 1987 oak-aged Chardonnay was named best Canadian wine and won double gold as best value in wine from anywhere in the $10–$20 bracket at the recent Toronto wine and cheese show.[10]

A month later, David Lawrason of the *Globe and Mail* wrote:

> Ontario wineries rolled out barrel samples of 1988 chardonnay Monday at Nekah Restaurant in Toronto for the first public tasting of what may be the finest vintage Ontario has seen in many years.... As in all new wine regions, chardonnay, the world's most popular white wine, is the flagship of the Ontario industry (even though many feel Riesling is our strongest suit). To make good chardonnay is to arrive, and with so much negative free-trade fallout, Ontario has arrived in the nick of time.[11]

After so many years of lassitude and debate, the free trade talks were accelerating such profound changes it left many in a state of fitful uncertainty. From one day to the next, affected individuals weren't sure whether their business would survive or flourish. During the same years as the estate producers were meeting to raise the bar on quality in order to differentiate themselves from the commercial products of the larger wineries, the Provincial Secretary of the Cabinet, Robert Carman, had alerted Valerie Gibbons, the Deputy Minister for the Ministry of Consumer and Commercial Relations, that the Wine Content Act was about to be terminated and needed to be renegotiated. In 1987, he had asked Gibbons to assemble a multiparty group of wine industry stakeholders to participate in a series of strategic planning meetings to see if they could come together to isolate the difficulties each saw with the other and "find ways to forge solutions under a common vision."[12]

The intrepid Gibbons and her committee met over the course of a year from 1987 to 1988, but produced no final report. Free trade was to take over the process. The Liberal premier of Ontario, David Peterson, was interested in making sure that whatever deal was struck by the federal Conservative government, the wine industry could live with it and not be outfoxed by the Americans and lose market share, or worse, be annihilated entirely.

Negotiators had to be nimble and creative enough to be able to place reasonable options on the table, but not give them away too soon. By the time the wine industry was under negotiation, it seemed the tenor of the talks had changed quickly, which set the provincial government scrambling. Gibbons was now under the gun. Recalled Gibbons:

> [Premier Peterson] asked me if I could make a deal with the wine and grape industry to bring them together in a more helpful accord. However, he asked me to do this over the course of a weekend! Because the committee had much of the work already under way, we were able to come to an agreement. The most painful concession was the acceptance that the *labrusca* grapes were not going to be the foundation for a thriving industry. They would have to be pulled out and new grapes planted and better varieties. This was a huge sacrifice on behalf of the grape industry and one they didn't undertake lightly.... I told them, "We're not going to be perfect but we're going to have two industries that are going to survive."[13]

Peterson also asked Gibbons and her committee to quantify how much money it would take to pull out, replant, recapitalize, and promote the renewed wine industry. With their figure in hand, Peterson called the provincial treasurer to see if the money could be found. The federal negotiators, who were once willing to trade the industry away for more protection in another sector, eventually agreed to contribute funds as well.[14]

To persuade the federal negotiators that the industry was, in fact, worth saving, the province needed a demonstration that a new industry was emerging, dedicated to change and quality. Ziraldo got a call one night from a provincial official asking him whether the vQA rules and regulations that his committee had been working on were completed to a sufficient stage that they could be put on the table as part of the negotiations. Although the provincial officials could not be at the table, they could sit outside the room. The negotiators could justify the inclusion of the vQA appellation system by demonstrating that if Ontario was going to export to the United States, it had to meet international expectations. The vQA standards could accomplish that task.

The endorsement of the as yet untested vQA standards as credible enough to serve as a bargaining chip for the industry boosted the morale of their framers and supporters. It acknowledged that their idea, motivated by market differentiation but driven by the belief that it was the right thing to do, had paid off. They were vindicated. Pennachetti and Franssen co-authored the first published set of Rules and Regulations, which the industry nicknamed "The Little Black Book," not only because it contained the list of the standards that would eventually lend legitimacy to their efforts, but because of the shiny black cover and handy small format. For the next decade, the vQA operated as a volunteer organization, with no legitimacy (except for a trademark on VQA) with which to enforce their own rules other than the force of their mutual commitment to comply and to make it work. When one member did not comply and sued the organization in the early nineties, the other members—both small and, by now, large—pulled together and contributed money in legal fees respective to their size. Schmidt, winemaker from the small estate winery Vineland Estates, remembers making out cheques for $10,000 a year.[15] Speck, president of Henry of Pelham, also then a small cottage operation, recalls paying more to the VQA lawyer than to his own lawyer. The case unified the group and built social capital during the traumatic years of adjustment that were to follow until final parity was reached in 1997 with imports and new *vinifera* plantings coming into fruition.

The Voluntary Years: 1990–2000

What was ironic about the framing of the VQA was the fact that the members were trying to regulate themselves. With Ziraldo at the helm and Speck from Henry of Pelham, along with Pennachetti from Cave Spring Cellars, Franssen from the LCBO, and Gamble, they all put thousands of hours into the process of refining the standards. However, they were also acutely aware of the paradox of their efforts. "We were a bunch of entrepreneurs sitting around the table who essentially didn't want anybody telling us what to do," recalled Speck, "and here we were putting regulations in place to make it more difficult to make wine in this province. We were all just trying to figure it out. I always thought that was a struggle. How much should we regulate ourselves?"[16] Curiously, that same sentiment was felt at Queen's Park. This was the time of Premier Mike Harris's "Common Sense Revolution," which included deregulating business and freeing the hands of entrepreneurs by minimizing red tape, so when the VQA board went to the government to ask to be regulated, the response from politicians and bureaucrats alike was "Are you nuts? You want us to regulate you?"[17]

During the voluntary decade, which could justifiably be called the experimental decade, the members amended their rules as they went along. "We'd spend a lot of time on the phone," recalled Paul Speck. "'This isn't working. We're thinking of changing it to this. What do you think?' By the time the board met, each member would have been canvassed and had reflected on the issue to be able to make a decision on the change."[18]

Standards Development Committee – the Technical Committee

During the non-regulatory period, the group had put in place a change process that would enable a member to modify a rule. The member had to submit the proposed change to the Standards Development Committee (later called the Technical Committee). Any change relative to a technical aspect of the rules would go through this committee. It would make a recommendation to the board, which would then vote on it. Even if the board was considering a change of its own, it would refer this to the Technical Committee for review and consideration. This same democratic process remains in place today.

From the very beginning, members knew they had to get the legal force of government behind their Rules and Regulations. The process couldn't continue forever as a "gentlemen's agreement." The earlier lawsuit was evidence enough. That meant getting the legislation passed into provincial law. Gamble had been appointed the original executive director of the VQA. When he left in the mid-nineties, Linda Franklin, then president of the Wine Council, took on both roles. "It was unbelievable how much work Linda could get done, willingly," recalled Pennachetti.[19]

With wit and an enormous reservoir of quick intelligence, Franklin escorted the inexperienced vintners through the machinations of provincial politics. She had worked at Queen's Park (1981–87), holding such significant positions as chief of staff for Treasurer Ernie Eves. She knew premiers Mike Harris and Bill Davis well, but more importantly, she knew how government relations worked. She understood what kind of policy requests should go to the civil service or go directly to Cabinet. Franklin taught the vintners how to get a hearing and how to present themselves with clarity and a unified focus. "We were cowboys," recalled Pennachetti, "loose individualists with a scatter-shot approach. Training us was Linda's greatest gift to us all."[20]

David Tsubouchi had been newly appointed as Minister of Commercial and Consumer Relations, the ministry responsible for beverage alcohol in the province, when representatives from the wine industry (Pennachetti, Speck, Bruce Walker, and Franklin) met him in his Queen's Park office. Rather than going to him and asking for something, Franklin simply laid out their key points, giving Tsubouchi every reason to say yes. She even provided a road map on how the government could accomplish the legislation. As he listened, Tsubouchi realized the significance of what they were presenting. "VQA, to me, was a no-brainer," said Tsubouchi. "It helped us to be more competitive and would give a compelling reason worldwide for people to buy Ontario wines. Plus, the cost of it was minimal.... It was one of those things—only good news. A win-win for the wine industry, the grape growers, the government, and the people of Ontario. It was a very pragmatic issue, not political. No political party would oppose it on any measure."[21] The legislation was drafted in 1988; VQA Ontario (VQAO) was incorporated in November 1988.

The VQA Act was passed in the legislature on May 4, 1999. Minister Tsubouchi signed off on the VQA legislation in June 1999. Once the members got the green light that it was going to happen, they seconded Catherine Burns from the LCBO policy section and Joe Hoffman from the Ministry of Commercial and

Figure 4.5 VQA logo
Photo: Vintners Quality Alliance of Ontario.

Consumer Affairs to help them organize how their Rules and Regulations could be converted into a legal statute. Speck and Pennachetti, with Franssen from the LCBO quality control section, were the architects who met weekly for the next year, putting the proposed statute into legislative form. It is interesting to note that once the first draft of the regulations was produced in the spring of 2000, the document was deemed confidential and could not be circulated to industry. VQA Ontario's lawyer and executive director were the only ones outside government to see the final wording of the regulations before they were proclaimed. On January 15, 2000, Laurie Macdonald, the first executive director of VQA Ontario, was hired. The regulations were finally proclaimed on June 28, 2000, and became operational (see Figure 4.5).[22]

Today, the organization is fully regulated by the Province of Ontario. No longer enforced by just the goodwill of its members, it is a delegated administrated legal authority with the power to invoke sanctions, impose penalties, and even lay criminal charges, as set out in the act and regulations. As stated in its 2010 annual report, "As an administrative authority, VQA Ontario is an independent corporation which exercises delegated authority from the Province to administer and enforce the VQA Act. VQA Ontario is managed by the wine industry but is ultimately accountable to the Province and the public."[23] Its mandate is to establish, monitor, and enforce the appellation of origin system, to control the use of specified terms, to participate in any discussions that relate to quality wine standards, and to educate and promote the value and benefits of VQA-approved wines. VQA Ontario is considered Ontario's wine authority, "responsible for maintaining the integrity of local wine appellations and enforcing wine-making and labeling standards," and is now accountable to the Minister of Small Business and Consumer Services.[24] As the industry's watchdog on behalf of

Figure 4.6 Laurie Macdonald, VQA Ontario
Photo: Steve Elphick.

the consumer, it is led by an executive director (still Laurie Macdonald) and governed by a thirteen-member board of directors who manage it as a not-for-profit corporation under a stringent agreement with the provincial government. Nine directors are elected from winery members and one each from the Ontario government, Grape Growers of Ontario, and the Ontario Restaurant Hotel and Motel Association.[25] Revenue is generated mainly through member bottle and approval fees.

A violation of the VQA Act or regulations is a provincial offence; as a result, the VQAO can lay charges under the Provincial Offences Act (section 10). Charges are heard in court and the judge can levy a maximum fine of $100,000 per charge. "We rarely lay charges," explained Macdonald, "as we have a progressive warning policy that gives wineries an opportunity to comply before a charge is laid."[26] But charges *are* laid. In 2009, a non-participating member was charged for using VQA-regulated terms on his label without approval. The compliance orders eventually led to two convictions and a fine of $10,000.[27] The VQAO can refuse to approve a wine, or can revoke its original approval if a wine has been misrepresented. According to Macdonald, compliance levels are very high because of the frequent audits conducted by her office as well as by the LCBO. "To date, we have had relatively few cases where we have had to take severe enforcement actions.

We issue about four or five compliance orders per year and have laid charges in only three incidents, with four convictions of fines ranging from $1,000 to $5,000, which goes to the provincial government."[28]

The VQA was founded on three fundamental values: authenticity, origin, and quality. Although they are equally important, the emphasis has shifted slightly over the years. The guarantee of authenticity meant that everything on the label was what it said it was. Origins reflected the assurance that the grapes were grown in the designated area claimed on the label. Quality has meant that the wine met the minimum growing and winemaking standards required for approval. In the early years, the value of quality took priority over the other values. It was necessary to demonstrate that the quality of the wine in the bottle was substantially different by a magnitude of ten from the Ontario wines of the past. In retrospect, however, it was never about high quality, but minimal quality. The framers had set the bar that a winemaker couldn't go below. Some growers and producers were concerned about achieving the bar; others felt it was set at a level that the industry would take higher. Over twenty years, with continuous improvements, there has been an immense jump in quality and the bar now looks more like a minimum. The growth of VQA wines has steadily increased in market share (10.8 percent), in volume produced (from 1,473,000 litres in 1992 to 16,617,005 in 2012), and in dollar value of sales (in 1992 it was $16,691,000 and by 2012 it had grown to $309,019,118, which corresponds to 13.8 percent market share).[29]

Today, the focus has shifted from minimal quality to the importance of the origins of where the grapes are grown. The largest appellation, the *Niagara Peninsula*, has been divided into ten sub-appellations defined by their respective soils and climate variations. The vision is that someday the consumer will start to get a sense of place and will buy his or her wine according to where it grows best. "If you really like Cabernet Sauvignon," envisions VQA chair Douglas, "try the ones from Four Mile Creek; they're turning out to be the best, grown in the hottest part of Niagara. If you like citrusy Rieslings, maybe try the ones from the Bench."[30]

In November 2009, the Government of Ontario announced to the members of the Wine Council of Ontario that it would support only VQA wines, and would no longer provide grants to support imported blended wines. In an expected but nonetheless divisive move, the government placed a 10-percent surtax on Cellared in Canada wines, the imported blends sold from private retail stores; the proceeds would go to supporting VQA initiatives. The wineries on which this hefty levy fell (Vincor, Colio, Andrew Peller Ltd., Diamond Estates, Magnotta, Kittling Ridge, Pelee Island, and Vinoteca) withdrew from the Wine Council (except Pelee Island, which belongs to both organizations) and formed their own organization, the Winery and Grower Alliance of Ontario (WGAO), so they could focus on their mutual interests. Their portfolio of imported blends now constitutes 79 percent of all the wine produced in Ontario; however, between them they also produce 55 percent of all vQA wines. Their interest in vQA remains unabated. Their objective is to increase the market share of both vQA and imported blends as critical to growing the entire wine industry of Ontario.[31]

The imported–Canadian blends wine segment has been one of the greatest sources of confusion and often derision in the Ontario wine marketplace. The practice allows a small number of wineries (all the members of wGAO, except Diamond Estates) to produce a product that consists of wines made from locally grown grapes that are blended with offshore wine. This has been an allowable practice since the early 1970s that has benefited both growers and those wineries that engage in it; however, it has miffed the media and confused the consumer mainly because of the misleading way in which the wines have been labelled and shelved in government retail stores.

The Wine Content and Labelling Act and Federal Labelling Laws

Until 1972, all Ontario wine was produced from 100 percent Ontario-grown grapes. In 1972, the grape industry suffered a devastating vintage, harvesting a small percentage of grapes compared to its usual expectation of supply. Wineries, in order to sustain their mainly *labrusca*-based businesses yet compete with a surge of European imports, negotiated with the provincial government to change the Wine Content regulations to allow them to supplement their wine with imported wine, "the equivalent of 10 percent of their domestic purchases and blend up to 25 percent imported product in any one bottle of wine."[32] In 1980, the wineries persuaded the government to change the regulations again to allow them to "blend up to 30 percent imported product in any one bottle of wine."[33]

By this time, the market was demanding premium French hybrids and *vinifera*-based wines, yet there was not a sufficient supply available in Ontario to satisfy the demand at prices that were competitive with imports. This time the amount wineries could blend was not linked to the percentage of domestic grapes purchased. The grapes they needed weren't in the ground or, if they were, supply was limited or sold at premium prices. The Wine Content and Labelling Act has been a time-limited act subject to termination and renegotiation, given the nature of changing circumstances of supply and demand. When the drastic measures were about to come into effect as a result of the North American Free Trade Agreement, in order to compensate and enable wineries to stay afloat until the replacement of *labruscas* with *viniferas* came to fruition, the government changed the act to allow a minimum of 30 percent Ontario-grown product to be part of the blend, down from the previous 70 percent domestic grown requirement of the previous stipulation. The immense dislocations of this turbulent period were exacerbated by a severe crop failure in 1993, which triggered yet another change to the act, seven years earlier than its original sunset. This time the wineries were allowed for one year to put a minimum of only 10 percent Ontario-grown product, which meant their international blends could be composed of up to 90 percent of imported product. By 1994, the Wine Content and Labelling Act was changed again to increase the minimum of Ontario-grown product from 10 percent to 25 percent; however, this was still lower than it had been prior to 1993. By 2001, the percentage was increased to 30 percent of Ontario-grown product and a maximum of 70 percent imported blend.

Many jurisdictions around the world engage in the practice of importing wine or juice from outside their regions to bolster their portfolio, particularly at the lower price-point categories. (A "Product of USA," for instance, can be a blend of 75 percent US wine.) The controversy in Ontario has centred on the way in which these wines were labelled and shelved at the LCBO. The fact that only those wineries that existed prior to 1993 could engage in this rather lucrative practice and sell the wines at their private retail stores (another edge that wineries established after 1993 were not permitted to enjoy) added distance to the widening gap between the VQA-only producers, which mainly constituted the small to medium-sized wineries, and the large wineries that engage in the practice.

When the Wine Content and Labelling Act was first legislated, it was legal to call the imported–domestic blended wine a "Product of Canada." As long as it was bottled in Canada, it could be made with a minimum of 75 percent Ontario-grown grapes and 25 percent imported wine. In 1996, to appease the critics of the labelling practice, who argued that consumers thought they were buying an all-Canadian product, the wineries agreed to label their imported blends with the term "Cellared in Canada," followed by the phrase (albeit in small letters) "This bottle contains a mixture of imported and local wine." The name change gave little solace to the wine media, which asserted that the terminology still implied that the grapes were grown in Ontario. In a 2007 article that reviewed Canadian wines, influential UK wine critic Jancis Robinson wrote, "I was rather horrified ... to see in the LCBO's flagship store in Toronto how many of these blends were displayed on shelves all mixed up with VQA wines under the large banner 'Ontario.' No Canadian I showed these bottles realised that they contained anything other than their own wine."[34]

Since the international blends constituted 80 percent of the domestic wine market, and since the VQA was in the business of making a profit for the provincial treasury, it shelved the blends prominently in its six hundred stores. Once again, pressures from the wine media, the wineries that produced only VQA wine, and a better educated wine consumer have resulted in the gradual shift of this category to a separate position in most stores, replaced by prominent displays of VQA wines.

Much of the controversy may be laid to rest when the Wine Content and Labelling Act is repealed, it is estimated by 2015. Concurrent to that proposed legislation will be a suggested increase in the percentage of local product in the blend, to be raised to 40 percent from the current 25 percent. The labelling term "Cellared in Canada" has been replaced with the phrase "International–Canadian Blends."

The wine industry has matured in the past twenty years and is now able to demonstrate that it makes fine wines, not just in one year or by one winery, but in several years and by many wineries; not with one grape, but with several. Since the standards are minimum standards, members now debate whether they should raise the bar yet again with even stricter standards, or whether they should allow the wineries to develop their own brands to enhance the standards and let the marketplace sort it out. Another debate is about packaging. The rules are very restrictive. No bag-in-the-box, no plastic corks for sparkling wine, no Tetra Paks—ostensibly for quality reasons, but also to protect the image of VQA wines as a premium product. Should the VQA get out of the packaging business or continue to be the arbiter of what constitutes a good package?

Since the Rules and Regulations are not as easy to change as they once were, a key issue will be the extent to which the VQA remains relevant. It must stay relevant to producers and consumers and it must continue to be the most leading-edge appellation system in the world. As it stands, the framers of the VQA are proud of the way the industry has grown and is maturing. "We were all so young," states Speck from Henry of Pelham. "It's crazy, when you think about it. We weren't writing legacies at the time. We were just doing whatever made sense and it's still that way today."

Notes

1 Controlled appellation system is a "method of labeling wine and designating quality that is modeled on France's Appellation Contrôlée system. It embraces geographical delimitation and is the principle on which quality wine schemes such as the DOC of Italy and Portugal, the DO of Spain and the AVA system of the United States are based." *The Oxford Companion to Wine*, 2nd ed., ed. Jancis Robinson (Oxford: Oxford University Press, 1999), 198.

2 Interview with Donald Ziraldo, co-founder, Inniskillin Wines, 1974–2005, 14 August 2007.

3 At first, their discussions included only Niagara. Within a few years, this broadened to include all of the existing grape-growing regions in Ontario, which at the time included Pelee Island and Lake Erie North Shore near Windsor. Prince Edward County was added later.

4 Interview with Peter Gamble, Executive Director, VQA, 1991, 15 December 2007.

5 The idea at the beginning was to garner national credibility by getting British Columbia included as well. Most of the newer estate wineries supported the idea, but for several reasons, each province was to structure its own course.

6 A longer list of fifteen hybrids identifies those varieties that can be used up to 15 percent in a blend. It was originally twelve hybrids, but some moved from the allowed list to the blending list over the years and some were deleted.

7 Adding sugar after fermentation was illegal. Adding sugar before fermentation was part of the process known as *chaptalization* and was permissible.

8   Interview with Peter Gamble, Executive Director, VQA, 1991, 15 December 2007.
9   "Free Trade: Crisis or Opportunity?" CBC Television Archives, 21 October 1987. Barnes was to last only a year after this interview, purchased by Chateau-Gai in 1988, which itself was purchased by a management buyout the following year, starting a chain of acquisitions and mergers.
10 Peter Ward, wine review, *Ottawa Citizen*, 19 May 1989, B5.
11 David Lawrason, wine review, *Globe and Mail*, 3 June 1989, C11.
12 Interview with Valerie Gibbons, Deputy Minister of Consumer and Commercial Relations (1987), 20 December 2007.
13 Interview with Valerie Gibbons, 20 December 2007.
14 The grape growers received $100 million through the Grape and Wine Assistance Program, and the wineries received $45 million in forgivable performance loans. "8,578 acres of vineyards vanished in the span of six years. More than five million vines were affected; 94 percent were varieties grown for wines. Qualifying varieties were those the wineries judged as having no future demand." Grape Growers Marketing Board. "Enterprise: 50 Years of the Ontario Grape Growers' Marketing Board, 1947–1997" (Vineland Station, ON: 1997), 34.
15 Interview with Allan Schmidt, General Manager/President, Vineland Estates, 4 June 2008.
16 Interview with Paul Speck, President, Henry of Pelham Family Estate Winery, 24 September 2010.
17 Interview with Paul Speck, 24 September 2010.
18 Interview with Paul Speck, 24 September 2010.
19 Interview with Leonard Pennachetti, President, Cave Spring Cellars, 4 December 2007.
20 Interview with Leonard Pennachetti, 4 December 2007.
21 Interview with David Tsubouchi, former Minister of Commercial and Consumer Relations, June 2008.
22 Interview with Laurie Macdonald, Executive Director of VQAO (via email), 5 October 2010.
23 VQAO Annual Report, 2010, 5.
24 www.vqaontario.com/Home
25 www.vqaontario.com/AboutVQA
26 Email communication from Laurie Macdonald, 7 October 7 2008, re: VQA penalties for infractions.
27 VQAO Annual Report, 2010, 17, www.vqaontario.com/Resources/Library.
28 VQAO Annual Report, 17.
29 Based on Wine Council of Ontario "Sales of Ontario wine in Ontario (VQA only) from 1992 to present"; Annual Sales per LCBO (Vintages not included) from 2002 to 2009; and VQAO Annual Report, 2012.
30 Based on Wine Council of Ontario "Sales of Ontario wine in Ontario (VQA only) from 1992 to present"; and Annual Sales per LCBO (Vintages not included) from 2002 to 2009.
31 Telephone conversation with Patrick Gedge, President and CEO of the Winery and Grower Alliance of Ontario, 25 October 2010.
32 www.grapegrowersofontario.com/the cellar/whatsinabottle.html
33 www.grapegrowersofontario.com/the cellar/whatsinabottle.html
34 Jancis Robinson. "Canadian Wine? Accept with Curiosity," *Financial Times* (London, UK), 13 January 2007, 4.

Five

Ontario Wines in the Marketplace

Astrid Brummer

Introduction

This is an exciting time to be part of the Ontario wine industry. The current climate is positive for retailers enjoying sales increases, marketers engaging new consumers with their brands, and producers crafting mouth-watering treasures. In these positive times, it is possible to forget that we comprise a tiny industry. In 2008, the entire country of Canada produced 540,000 hectolitres of wine, or 0.19 percent of world production.[1] Compare this to top-ranking Italy's production of 51,500,000 hectolitres, or China's 14,500,000 hectolitres, or Peru's 480,000 hectolitres. Ontario wine production may be relatively small, but our wines are seeing acclaim at home and abroad and are enjoying solid sales growth. Success feeds new endeavours and in turn leads to more success. The current sales growth and positive consumer response are leading the industry into greater development and further innovation.

This chapter describes the recent sales performance and trends of Ontario wines sold at the Liquor Control Board of Ontario (LCBO). After looking at LCBO sales of Ontario wines, I will highlight notable demographic trends, share two examples of ground-breaking products, and conclude with a summary of the successful 2010 LCBO goLOCAL promotion.

A good place to begin this discussion is to place LCBO sales into context. As shown in Figure 5.1, the LCBO is responsible for the sale of 84.3 percent of the volume of wine sold in Ontario.[2] It is the only retailer of imported wines (61.5 percent of the volume of wine sales) and shares the sale of domestic wines with winery retail stores (13.4 percent) and winery direct deliveries to licensees, primarily restaurants (2.3 percent of wine sales).

All figures and statements that follow relate only to the LCBO Wines business unit and do not include Vintages (the fine wine and premium spirits business unit at the LCBO), except where noted, or any Ontario winery retail sales.

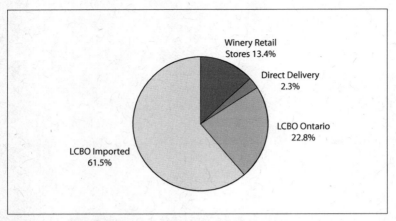

Figure 5.1 Wine – total market volume sales – market share percentage by channel

2009–10 Sales and Trends

The LCBO Wines business unit is made up of three buying groups: Ontario Wines and the two imported wine buyers, New World Wines and European Wines. Each is further subdivided into smaller sets or subsets by country of origin, varietal, or style.

The Ontario Wines buying group includes Vintners Quality Alliance (VQA) wines, International–Canadian Blends (ICB) and a group designated in Figure 5.1 and Figure 5.2 as "Other" that includes fortified, fruit, specialty, and sparkling wines. In the 2009–10 fiscal year, this portfolio included 456 active product listings or stock keeping units (SKUs). In the same year, New World Wines offered 448 SKUs and European Wines offered 527 SKUs.

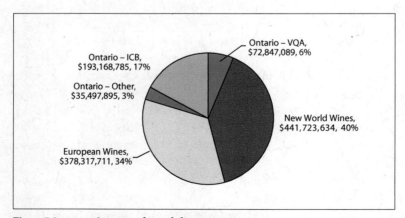

Figure 5.2 LCBO wines net sales and share percentage

Ontario wines sold at the LCBO have a significant portion of the provincial wine market, at 26 percent, and VQA wines in particular are experiencing strong growth.[3] Figure 5.2 shows the relative size in net sales dollars of these sets compared to the New World Wines (at 40 percent) and European Wines (at 34 percent). Table 3 details their respective sales and volume trends.

The data in Figure 5.2 and Table 5.1 show that Ontario Wines is the smallest of the three buyers in the Wines business unit. While its total share is 26 percent, it is enjoying the most sales growth, at 4.5 percent. VQA wines are the driving force behind this growth. In fiscal 2009–10, VQA wines experienced a 19.6 percent net sales increase over the previous fiscal year. It is important to note that this robust sales growth occurred during a period of global economic instability that, in Ontario, resulted in a general shift in consumers purchasing less expensive beverage alcohol products. Ontario VQA wines bucked this trend.

There are several possible reasons for this increase: The "go local" promotional messaging that is being used in the retail sector may be responsible for driving some interest in purchasing local wines. Customers have also shown enthusiasm for discovering new wines, especially when these offerings are unique and dramatically differ from previously held perceptions of Ontario wine. We know that customers often look to New World countries, including Ontario, for simplified labelling by grape variety. The popularity of local tourism creates an emotional attachment to the Ontario wine country; this is also likely a factor in Ontario's success. The change is quite palpable: ten years ago, it was not always easy to interest LCBO customers in Ontario wine recommendations. However, in the past five years, things have changed—few customers are adamantly opposed to trying Ontario wines and many have a strong affinity for local gems or are willing to try.

All of the LCBO Wine subsets in Table 5.1 experienced net sales growth during fiscal 2009–10, with the exception of Ontario sparkling wine, which decreased by 1.8 percent, and European wines, which decreased by 4.3 percent.

Table 5.1. LCBO buyer/set net sales and litre sales trends

Buyer/Set	Net $ Sales	Litre Sales
ON – VQA	19.60%	19.30%
ON – ICB	0.20%	−0.30%
ON – Fruit Wines	7.40%	9.50%
ON – Fortified	2.50%	−0.90%
ON – Specialty	5.00%	2.70%
ON – Sparkling	−1.80%	−5.00%
New World Wines	2.40%	8.60%
European Wines	−4.30%	−4.50%

In Table 5.1, the only LCBO Wines buyers/sets that did not experience growth in volume were ICB (down 0.3 percent), Ontario fortified (down 0.9 percent) and European Wines (down 4.5 percent).

In Figure 5.2, ICB wine sales were $193,168,785 in 2009–10, or almost twice the size of VQA, but sales growth in Table 5.1 was minimal, at 0.2 percent above the previous fiscal year. There are a few factors contributing to this trend. The traditional demographic for ICB wines is not increasing in size or consumption (see Demographic Trends). The traditional ICB brands have largely remained just that: traditional. Producers have not chosen to invest in these brands to make them attractive to new consumers, and those new brands that have been developed have yet to achieve high-volume success.

The size of the Vintages contribution to Ontario wine sales and volume is not shown in these figures because it is relatively small. Vintages, as the fine wine and premium spirits business unit of the LCBO, releases a constantly changing assortment of wines every two weeks. Releases typically include seven Ontario VQA wines. Vintages does not sell ICB wines. During the 2009–10 fiscal year, Vintages Ontario Icewine had net sales of $5,722,071 and table wines had sales of $11,472,469. At $17 million in annual net sales, Vintages Ontario's portfolio is equivalent in value to roughly one-quarter of the LCBO Wines Ontario portfolio and contributes an even smaller portion to overall volume sales.

Figure 5.3 shows the relative size of the litre sales of these same sets. The three LCBO Wines buyers are almost the same in size when measured in litre sales, with an average volume of 34 million litres. Ontario Wines grew by 2.3 percent in volume but increased 4.5 percent in net sales, indicating that customers have been purchasing at higher price points during this year. It is interesting to note that while New World Wines grew by 2.4 percent in net sales, volume sales grew at the much higher rate of 8.6 percent. This indicates that customers increased their purchases of lower-priced wines.

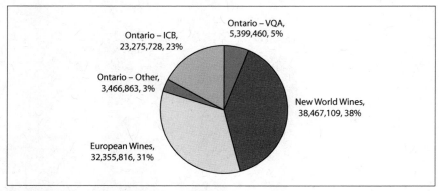

Figure 5.3 LCBO wines litre sales and share percentage

Price Band Performance

As stated earlier, the value of purchases within the Ontario wine portfolio has risen due to the increased sales of higher-priced VQA wines. The two figures below compare the price band size and growth for the Ontario portfolio versus an amalgamated Imported (New World Wines and European Wines) portfolio.[4]

Figure 5.4 shows that both Ontario Wines and the Imported buyers have a large volume of sales in the inexpensive under $8 price band. The large size of this price band in Ontario Wines reflects the large volume of ICB sales. Figure 5.5 allows us to see that this price band is decreasing in size for both domestic and imported wines: Ontario Wines is decreasing by 0.9 percent and Imported by 8.5 percent.

In the $8 to $12 band, Ontario Wines is much smaller compared to the largest single price band for LCBO Wines—Imported had an impressive 28.2 million litres sold in the $8 to $12 range, while Ontario Wines sold just under 10 million litres. Ontario Wines is experiencing higher growth, at 9.6 percent, versus the Imported buyers' 5.5 percent growth. In the $12 to $15 band, Ontario Wines still has a proportionally smaller presence, but again is experiencing higher growth, at 11.4 percent, versus the Imported buyers' 8.0 percent.

Both Ontario Wines and Imported have a relatively small presence in the over $15 price band compared to the lower priced bands: Ontario is roughly a quarter of a million litres, while Imported is 3.7 million litres. However, Ontario Wines is experiencing its most dramatic growth on this small base: 32.3 percent versus the Imported buyers' small 5 percent growth. Ontario Wines enjoyed this increase due to the growth of VQA wines. As noted in the Sales and Trends section, it is impressive that during a time when many customers were choosing

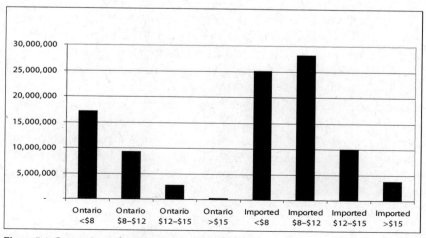

Figure 5.4 Comparison of retail price band size – litres

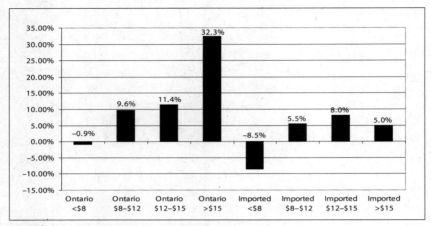

Figure 5.5 Comparison of retail price growth – litres

to purchase less expensive wines and European Wines experienced sales decline, it was Ontario Wines—driven by VQA wine sales growth—that both increased sales in higher price bands and saw overall net sales and volume growth.

Demographic Trends

LCBO sales results show that Ontario purchases (this includes both VQA and ICB) are higher in the LCBO's Central Region.[5] The highest indexing areas for VQA purchases are Toronto's north and core, Ottawa, and Thorold, while the highest indexing areas for ICB purchases are Sault Ste. Marie and Timmins.

Customers who purchase Ontario Wines exclusively are primarily ICB shoppers. This customer group is responsible for a minor portion of LCBO revenue and their dominant age range is 45 to 64 years. The shoppers who purchase the most VQA wine are non-exclusive Ontario customers. This group purchases wine from all countries, including ICB, in addition to the large percentage of VQA wines they add to their shopping carts. The primary age range is 45 to 54 years.

Notably absent from the non-exclusive Ontario customer group are legal drinking age Millennials, suggesting that an opportunity for the future could include bringing younger adult wine consumers into this purchasing group.[6] Sources generally agree that this cohort includes people born from the mid-1970s to early 2000s. They are also called Generation Y, Generation Next, and Echo Boomers. This demographic is very large and its members are interested in wine; therefore, they can be very influential in the Ontario wine industry's future.

Brand Concept and Packaging Success

LCBO wine shoppers have demonstrated a profound interest in new brands and have been very responsive to innovative brand concepts. Notably, based on net sales performance measured at the end of December 2010, three of the top ten VQA wines had been on shelves for only the previous two years. In contrast, the other seven most successful VQA wines entered the market between 1988 and 1996.

In the past five years, Ontario's VQA producers have recognized the importance of concept and package design and had tremendous success by bringing to market strong brands. The best of these brands leverage emotion, quality, fun, and great label design and resonate strongly with customers. The most noteworthy examples of this type of innovation are two that led the movement: Colio Estate Wines' Girls' Night Out—which grabbed customer attention and became a catalyst for industry change, and Henry of Pelham Family Estate Winery's Sibling Rivalry, which found its own loyal customer base and became an industry guiding light.

Girls' Night Out

This very successful brand launched in the summer of 2008 with a chardonnay, merlot, and chardonnay–merlot rosé wine. It was significant because it was the first premium-priced VQA wine ($12.95) that used a striking package and boldly targeted a female customer (Figures 5.6a and 5.6b).

Figures 5.6a and 5.6b Girls' Night Out Chardonnay and detail of Girls' Night Out Merlot label

The brand now includes five VQA wines and two fruit-flavoured wines. In fiscal 2009–10, four of these rank in the top five wines in their respective subsets. Girls' Night Out Strawberry Samba and Girls' Night Out Merlot Chardonnay Rosé VQA each lead their respective subsets. Girls' Night Out Chardonnay VQA not only leads the chardonnay subset, but is one of the top ten performing VQA wines measured both by volume and net sales dollars.

In 2009, the superior efforts of Colio Estate Wines were recognized by the LCBO. This innovative winery was given the VQA Award of Excellence at the annual LCBO Elsie Awards, which celebrate the contributions of LCBO trade partners. The brand has also received recognition from the packaging and design industry for excellence. More information about this recognition and about how Colio is using events and charitable works to bring customers together around the brand is available at www.girlsnightoutwines.ca.

Sibling Rivalry

This brand launched in the spring of 2009 with a white (Chardonnay-Riesling-Gewürztraminer) blend and a red (Merlot, Cabernet Franc, Cabernet Sauvignon) blend. It was significant because it was the first brand to popularize the VQA varietal blend concept. It did this with premium pricing ($13.95), a colourfully modern package (Figure 5.7), and a concept that was both true to the founding brand and unique. Within a year of release, the net sales of both wines ranked in the top fifteen of all VQA wines. The success of this brand has been noted by every Ontario winery doing business with the LCBO. In 2010, Henry of Pelham Family Estate Winery was presented with the VQA Award of Excellence at the LCBO Elsie Awards in large part because of the success of Sibling Rivalry.

The winery has received acclaim for this brand from many international sources. Sibling Rivalry was featured in fall 2010 at the San Francisco Museum of Modern Art Evolution of Wine exhibit. This homegrown star was part of a display that examined the changes in wine labels and stories that has occurred over the past three decades. More about the ways this brand is connecting with its consumers can be found at www.siblingrivalrywine.ca.

Figure 5.7 Sibling Rivalry red and white

These are just two examples of brands that have worked very hard to engage customers in new and interesting ways. They have led a movement that is characterized by stylishness and innovation—many are using these as guides in developing their own unique brand stories and marketing plans.

Customer Response to Promotions: International–Canadian Blends

In 2009, the LCBO collaborated with several stakeholders to change the name and signage in the Cellared in Canada section to International–Canadian Blends (ICB). This change was initiated by the huge media interest in this category and a desire to find a new term that would be clear, understandable, and accurate. Customers have responded positively to the change. A large portion of customers who were unfamiliar with the term or unsure of its meaning had an improved opinion when they learned that these products contain both domestic and international wine and that a significant proportion of Ontario-produced grapes are used to create these blends. Few customers had a negative opinion upon learning the definition.

For many years, ICB wines have been the go-to products for customers looking for exceptional value. This has been the primary consideration for the loyal ICB shopper and continues to be an important purchase consideration for younger and newer segments. New brands that have just entered the market may challenge this. We will see if there is a sizeable customer base that wants both great value (i.e., wine that delivers an experience that is worth more than its retail price) and a compelling story about the wine's origins.

Customer Response to Promotions: 2010 goLOCAL Campaign

Each fall, the LCBO focuses its promotional efforts on Ontario wines. For the past two years, the message has centred very successfully on an invitation to "go local." The 2010 goLOCAL campaign had an ambitious list of goals that included increasing VQA and ICB sales, promoting the quality and value of locally made wines, providing education about the cool-climate style, introducing the concept of key Ontario varietals or styles (Chardonnay, Riesling, Baco Noir, varietal blends, and rosé), promoting new wines, and striving to entice more 35- to 40-year-old shoppers to the section.

The 2010 campaign increased the net sales of Ontario wines by 8.2 percent.[7] The portfolio of ICB products had an increase of 3.0 percent and VQA products a whopping 18.7 percent. This result is even more exciting because it was on top of the year-over-year increase of 17.5 percent that occurred in 2009. ICB occupied 63 percent of the market share compared to VQA's 37 percent (an increase of 4 percent over 2009).

Litre sales of VQA increased by 13.8 percent compared to the previous year's promotion. This shows that while volume had increased year over year and customers purchased higher-priced products compared to last year, they were purchasing lower-priced VQA wines during this year's promotion. A large number of new VQA wines were launched before and during the promotion and many of the most successful of these wines were priced in the $10 to $11.99 range.[8]

Perhaps customers have been responsive to the "go local" message because they are hearing it from many other sources, too, and it appeals to them emotionally right now. Perhaps with its implicit message of supporting local economies, it is especially resonant during and following a global economic downturn.

Conclusions

Compared to global production, wine production in Ontario is extremely small. Yet, at the LCBO, volume sales in the Ontario Wines category account for almost one-third of LCBO Wines' revenues. It is fair to say this is the direct result of the integrated marketing and promotional efforts that take place on the home turf, quality products with memorable names and eye-catching packaging being brought to market by the wineries, and consumers choosing to "go local." Ontario VQA wines have been experiencing double digit growth. Exciting products, innovative concepts, and packages that are all designed with cultural relevance in mind are meeting up with consumers eager for tastes of home. Ontario is clearly poised for even greater success.

Notes

1 Wine Institute, "World Wine Production by Country," www.wineinstitute.org/resources/worldstatistics/article87.
2 Rolling 13 net sales measured as of financial reporting period 7 of 2010–11.
3 All net sales, litre sales, and trend information are measured at the end of fiscal year 2009–10.
4 Rolling 13 net sales measured at end of financial reporting period 10 2010–11.
5 From sales data measured at 2009–10 fiscal year end.
6 Evelyne Resnick, *Wine Brands: Success Strategies for New Markets, New Consumers and New Trends* (Hampshire, UK: Palgrave Macmillan, 2008), 37.
7 Measured at the end of financial reporting period 7 of 2010–11.
8 The top five new VQA wines were Flourish Riesling Vidal $10.95 and Flourish Merlot Pinot Noir $10.95 from Vincor Canada, Union Red $13.95 from Generations Wine Co., Pelee Island Lighthouse Riesling $11.95 from Pelee Island Winery, and Fresh Cabernet Gamay $11.95 from Diamond Estate Wine & Spirits.

Six

The Ontario Wine Industry: Moving Forward

Maxim Voronov, Dirk De Clercq, and Narongsak Thongpapanl

Introduction

O ver the past 20 years, the Ontario wine industry has progressed in leaps and bounds. The number of wineries is growing, many wineries have gained critical acclaim and won awards both domestically and internationally, and Ontario consumers are becoming increasingly enthusiastic about Ontario wine. However, many challenges remain. In this chapter we reflect on some of the key business challenges that Ontario wineries face and how they might overcome them. Specifically, we note the positioning of Ontario wine in the marketplace as a function of an optimal balance among three goals: complying with artistic and commercial rationalities; pursuing the traditions and styles of winemaking that are most compatible with the climate and geography of Ontario; and initiating conversations about the unique and distinctive identity of Ontario wine. These reflections are based on an ongoing research study, initiated four years ago.

Methodology

Because Ontario wineries and grape growers face unique climatic, geographical, and regulatory conditions, we have adopted a methodology that allows us to capture these unique conditions. Rather than merely applying results of previous research in other regions to Ontario, where the context is likely too different to make those findings useful, we seek a rich, detailed understanding of the wine business in Ontario, with a focus on pursuing "depth" rather than "breadth."

We have attempted to grasp multiple facets of the Ontario wine business by interviewing a variety of winery employees, observing multiple meetings, and noting other aspects of the daily work at these organizations. We also have had ongoing discussions with a number of wine writers, restaurateurs, Liquor Control Board of Ontario (LCBO) executives, and other industry insiders. In addition, we have collected quantitative data through a large-scale survey of

Ontario restaurants, assessing their perceptions of the Ontario wine industry and their relationships with wineries. In presenting our findings, we avoid singling out individual wineries for either praise or criticism; though we use occasional quotes to illustrate important points, we do not identify individuals or organizations by name.

Art and Commerce Go Hand in Hand

It is commonly believed that there is an inherent conflict between art and commerce. But our research indicates that what we call artistic and commercial rationalities are very closely linked in the wine business and should not be seen as irreconcilable, or even contradictory.

When we refer to *rationality*, we mean the underlying motivation, rules, and norms that govern behaviour (see Table 6.1 for a detailed explanation of the differences). Thus, *artistic rationality* indicates attempts to produce wines that are worthy of critical acclaim and national and international awards and that are sophisticated and complex. In contrast, *commercial rationality* refers to the motivation to run a profitable business.

First, commercial rationality is an inevitable reality of any business, and wineries must focus on running profitable businesses. Yet, because of the Ontario wine industry's focus on producing premium and ultra-premium wines, Ontario wineries also must meet the demands of artistic rationality. Any winery that aims to command a high price for its offerings must cultivate images of artisanal,

Table 6.1. Artistic and commercial rationalities in the wine industry

	Artistic	Commercial
Dominant value	Aesthetics	Wealth creation
Supporting values	Handcrafted, small-scale, personal	Efficient, economies of scale, professional
Sources of identity	Winemaking as an art	Winemaking as a business
Sources of legitimacy	Artistic acclaim	Business expertise as expressed in sales volume, revenues and profits
Arbiters of legitimacy	Critics, high-end restaurateurs, connoisseur customers, peer wineries	Liquor Control Board of Ontario, peer wineries
Production focus	Express sense of place (terroir), slower is better	Speed to market, cost reduction
Marketing focus	Explain terroir; cultivate exclusivity, specialness, romanticism, sophistication, scarcity	Make wines easy to obtain, simple to appreciate, and affordable
Basis of mission	Build prestige of winery, express terroir	Be economically successful
Basis of attention	Develop and maintain artistic reputation	Develop and maintain market share
Basis of strategy	Win awards, obtain good reviews	Secure broader distribution, establish superior market position

personal winemaking and distance itself from the perception that its methods or products are "industrial" or for mass consumption.

Second, the two rationalities, artistic and commercial, are meaningful only in relation or in opposition to each other. That is, artistic rationality is meaningful only to consumers and industry insiders because it is defined in opposition to commercial rationality, and vice versa. Thus, a winery seeking to cultivate an artisanal image can do so only by shielding itself from customers' perceptions that it engages in overtly commercial activities.

The implication is that wineries that seek to compete as producers of premium or ultra-premium wine must have both a commercial and an artistic strategy, and the former depends partly on their ability to undertake the latter strategy successfully. Figure 1 summarizes the relationship between consumers' perceptions of artistic value and price. The more expensive the wine (or the more expensive a particular winery's products, on average), the greater the perceived artistic value should be. Most wineries in the Niagara region aim to develop a *differentiation* or *focused differentiation* approach, with some wineries having different brands or product lines aimed at multiple segments.

Because the two rationalities (artistic versus commercial) place conflicting demands on wineries, it is impossible for any winery to meet their demands consistently. Therefore, wineries tend to choose the optimal balance between complying with the demands of each rationality. For example, a winery that

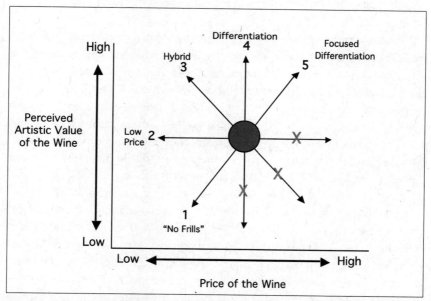

Figure 6.1 Recommended relationship between artistic acclaim and price sought for wines
Source: Based on Johnson et al. (2007).

wants to focus on high-end wines might safeguard its artistic identity at all costs, even if that means compromising its short-term profitability. But a winery that hopes to position itself more broadly, offering wines across a variety of price ranges, may be less preoccupied with projecting artistic images, even though it undoubtedly has to take care to avoid appearing too commercial.

In addition, the wine business is unlike most "conventional" businesses. Of course, wineries pay attention to the volume of sales or the shelf space at the LCBO, but, equally importantly, some of the most successful wineries keep track of their artistic reputation in the industry, as measured by the number of awards, reviews by noted wine writers, inclusion on prestigious restaurants' wine lists, and word of mouth. To accomplish this strategy, wineries position themselves, or certain of their brands, in contrast with the rationality that contradicts that winery's (or brand's) positioning. For example, if the winery wants a reputation as an artisan, it may actively educate consumers about what it is doing to meet this goal and describe how those actions distinguish the winery's offerings from other, more commercial products, such as those from Chile or low-end European producers. In other words, because those artistic and commercial rationalities are meaningful only in opposition to each other, a winery cannot simply position itself as artisan; it also has to differentiate itself from more commercially oriented activities by others, whether domestically, abroad, or both.

To demonstrate that their wines are sufficiently artistic to justify a price premium, wineries tend to maintain close connections with the guardians of artistic rationality—wine writers, bloggers, prestigious restaurants, and other opinion leaders.

Wine writers and bloggers. Wine writers are most interested in the artistic aspects of the wine, such as where the grapes are grown, what winemaking techniques are used, and, most important, the extent to which the wine reflects the local terroir. Thus, wine writers and award ceremonies play important roles in verifying that a winery complies with artistic rationality; arguably, they have more power than anyone else to do so. Given the plethora of wine choices (and the relative lack of confidence among many consumers in choosing the "right" wine), consumers are fairly reliant on wine writers to help them make choices. As one writer observed, "I think the role of the writer is really to explain and to interpret the wines for people." A number of wineries report that positive reviews—especially from high-profile wine writers—can result in substantial increases in sales. As one sales manager observed, "You only have to influence one wine writer with your story and if he writes about it ... *The Globe and Mail* will affect, you know, a hundred thousand readers a day, or maybe two hundred thousand." However, wine writers also can generate negative publicity. One winemaker, for instance, observed that a particular high-profile critic generated a great deal of visibility for the industry and its offering, but added, "but then on the other hand, when we produced ... an oxidized or a ladybug contaminated chardonnay ... he was just brutal on the ladybug by claiming that those wines should never have hit the shelf."

It should also be noted that famous wine writers are not necessarily more important than the various bloggers who write about wine. Whereas the writers for major newspapers have a great deal of visibility and a devoted following, their readership is increasingly fragmented. In the same way that people obtain their news from a variety of sources, people obtain information about wine from disparate sources, too. Thus, sommeliers, educators, and grassroots enthusiasts with blogs can be very influential in acclaiming a winery as a producer of high artistic value. Furthermore, whereas most famous wine writers attempt to cover both foreign and domestic wine, an increasing number of bloggers write purely about Vintners Quality Alliance (VQA) wines or local wine and food. Thus, in some cases, they may be especially connected to the readership that is more receptive to VQA wines.

Restaurateurs. Prestigious restaurateurs, especially those in Toronto, play important roles in verifying that a winery complies with artistic rationality. Not only do restaurants represent an important sales channel (especially for smaller wineries with limited access to the LCBO), but when prestigious restaurants select a winery's products for inclusion on their wine list, it signals that the wine is of superior quality. The artistic focus of high-profile restaurants is reflected in their attempts to place a personal touch on the restaurant and the wine list, seeking personal satisfaction from crafting an excellent wine list, and searching for critical and peer acclaim. We find that the stronger this artistic focus is, the more likely such a restaurant is to support VQA wines.

In summary, because the ability to demonstrate compliance with artistic rationality plays such an important role in wineries' ability to successfully compete in the premium and ultra-premium segments, wineries develop strategies to reach out effectively to these individuals.

Wineries often send wine to select wine writers or bloggers. However, many wineries fail to grasp that the most famous writers are not necessarily the most important ones. These renowned writers receive wines from a great many wineries and may not be able to write about all the wine they taste. Emerging bloggers are more appreciative of samples. In addition, wine writers outside of the Greater Toronto Area (e.g., in Ottawa and elsewhere) are often less connected to the industry, mainly due to their distance. They tend to be particularly grateful for samples and thus are more inclined to write about those wines. In addition, different writers and bloggers have different tastes and preferences. Therefore, it can be beneficial to send samples to those writers or bloggers whose tastes appear to match the styles of the wines being presented (e.g., send bone-dry Rieslings to those writers who tend to like and write about such wines, rather than to those who do not write about such wines often). Another strategy involves inviting the writers and bloggers to the winery. Again, wineries often fail to recognize that such opportunities are especially important for lesser-known writers and bloggers, even though their reviews may be as beneficial as those of the more prestigious writers. The more writers and bloggers know about a winery's approach to winemaking, the more able they are to explain it to their readers. It can also be beneficial to seek out writers and bloggers who favour the winery's winemaking approach (e.g., if producing biodynamic wines, seek out writers/bloggers who seem to care a great deal about such winemaking practices or about environmental sustainability).

Not all wineries understand the importance of educating restaurateurs about their approaches to winemaking. Restaurants are more likely to support local wine when wineries develop strong relationships with them. Such relationships can expose restaurants to broader knowledge about the opportunities underlying the sale of VQA wines—including the role of terroir and "heroic" family stories of how a winery has been launched—and in doing so provide them with the tools to promote such wines to customers more effectively. Thus, wineries tend to benefit from hosting restaurateurs at their premises, showcasing their wines, explaining their winemaking philosophy, and describing how the wines would match with a particular chef's food. It is also crucial to have sales representatives in Toronto, though many smaller wineries have difficulties affording quality representation in Toronto. Yet many restaurateurs report that though they get approached by representatives of larger wineries, they do not receive much contact from smaller wineries.

Premium and ultra-premium wines often evoke romantic images for consumers (e.g., small-scale production, hands-in-the-dirt winemaking) that are perpetuated by wine writers, bloggers, and restaurateurs. Thus, in managing relationships with these opinion leaders, it is important to avoid conveying images that are inconsistent with such perceptions. For example, unless a particular writer or restaurateur is very knowledgeable about the winemaking process, a winery may try to avoid providing too many technical details about the winemaking techniques for a certain wine, because those details might unintentionally convey the image of "tampering" with grapes. Instead, it is more valuable to show off vineyards and emphasize the terroir of the wine, as well as the link between the specific locale and the wine in the bottle. For this reason, wineries sometimes try to showcase as much of their artistic side as possible, while keeping commercial and technical activities out of sight. Although more and more Ontario wineries are diversifying and launching multiple brands, they often aim to keep the identities of those brands as distinct from one another as possible. Multiple brands that are obviously linked to the same winery can convey the sense that the winery or a certain brand is "too large" or "too commercial." Thus, a winery that owns multiple brands tends to carefully cultivate and safeguard each brand's distinct identity.

An Old World Wine Region Located in the New World

Fine winemaking is codified within two competing traditions, each of them emphasizing and prescribing different features of winemaking and defining high-quality winemaking differently. Although much of modern winemaking philosophy, conventions, and know-how can be linked to France, that country has not always enjoyed international acclaim as a superior fine wine producer.[1] In addition, the so-called Old World wine regions, such as Burgundy and Bordeaux, are increasingly challenged by the New World wine regions, such as California, both with respect to export growth and market share in a variety of markets[2] and as standard setters and perceived epicentres of superior quality and innovation.[3] But "Old World" and "New World" are not simply regions. They are traditions of winemaking. The adoption of one of these two traditions, or a combination of them, is important for a region seeking to establish itself as a legitimate winemaking region. As a part of this research, we have examined how the two traditions of Old World and New World are used by wineries, wine critics, and restaurants in Ontario in their attempts to establish a legitimate local winemaking tradition and promote the region's offerings.

In essence, the difference between the two traditions can be summarized as follows (see Table 2): "In the Old World, with its centuries of winemaking tradition, Nature is generally regarded as the determining guiding force. In much of the New World, however, it may be regarded with suspicion, as an enemy to be subdued, controlled, and mastered in all its detail, thanks to the insights provided by science."[4]

Thus, the Old World tradition defines wine as a deeply tradition-bound object, connected to the particular geographical and climatic conditions of the particular wine-producing region. As such, it is not easily accessible in terms of either production or consumption. The winemaker's role is somewhat mythical and passive, with the soil, climate, and particular locality—the terroir—as the key determinants. Such wines are to be cherished. They are not easily drinkable, due to their high acidity, lower fruitiness, and greater complexity. As such, they reward aging and being paired with the right food. With respect to the image of the wine, it is romantic, quaint, and somewhat sacred.

In contrast, the New World tradition defines wine as a fun and easygoing product that places a great deal more emphasis on the winemaker's skills and scientific advances and less on nature or terroir. It is unbound by tradition. Such wines are meant to be more easily accessible and do not imply a personal sacred bond between the consumer and the wine. They also are easy to drink, even

Table 6.2. Old World versus New World traditions of fine winemaking

Dimensions		Old World tradition	New World tradition
Geography and Climate	• Countries typically associated with the template	France, Germany, Italy	US (California), Australia, Chile
	• Climate associated with the template	Cool	Warm or hot
	• Importance of (variation in) climate	Higher	Lower
Viticulture and Winemaking	• The relative role of tradition versus fashion	Tradition	Fashion
	• Number of grape varietals	Limited	Varied
	• Importance of terroir	Very important	Important (?)
	• Relative focus on appellation versus varietal	Appellation	Varietal
	• Relative importance of nature versus science	Nature and oneness with nature more important	Latest scientific advancements more important
	• Greater importance of winemaker or vineyard	Vineyard more important	Winemaker more important
	• The use of machinery	Lower	Higher
	• Planting density	Higher	Lower
Wine Characteristics	• Taste and aroma	More acid-driven Less sweet, less fruity More built on complexity Lower alcohol	Less acid-driven Fruit-driven More built on concentration Higher alcohol
	• Demanding particular food pairings	Higher	Lower
	• Approachability versus age-worthiness	Age-worthiness	Approachability

without food. They are more fruit forward, comfortable, bold, and powerful, and romance is replaced with glamour and luxury.

In understanding geography/climate, winemaking tradition, and the resulting wine profiles of Ontario wine, wineries selectively rely on the two traditions to develop their particular styles, and they (as well as the aforementioned guardians of artistic rationality) rely on these traditions to explain the suitability of the local conditions for fine winemaking and endorse the resulting wines. The geography and climate of Ontario, the style of winemaking, and the resulting wine profiles tend to be characterized as Old World, even though by virtue of its location in the New World, Ontario is a New World wine region. This characterization is somewhat surprising. New World wine regions, such as California and

Australia, have been very successful, and there is great demand for big fruit-forward (red) wines that are most closely associated with those regions. As in many other markets, Ontario consumers increasingly prefer New World wines over European wines.[5] However, there is a general recognition that the region is not positioned to produce fashionable New World wines. Thus, the offerings are often characterized as "not New World." Geographically, Ontario is located in a cooler climate with variable vintages that is more akin to European wine regions, such as Burgundy, Germany, and parts of Bordeaux. This cooler climate dictates a more terroir-driven rather than science-driven approach to winemaking. It requires abandoning attempts to make wines with consistent profiles from year to year or to produce styles that are more fashionable; instead, the soil and the climate (terroir) dictate what kind of wine will emerge, with a relatively passive role played by the winemaker.

The resulting style of Ontario wines tends to be described as more European, more refined and complex (as opposed to powerful, as in the New World tradition), and acidic or mineral (as opposed to fruity, as per the New World tradition). As such, the wines are claimed to be more food friendly. This food friendliness might set these wines apart from and make them superior to the New World wines, which often are characterized as being too high in alcohol and too sweet to be effectively matched with food. As one wine writer summarized,

> A lot of the New World wines from Australia, Chile, and so on are very high in alcohol, they're huge in flavour, and they kill anything that you try to eat with them. The wine just dominates everything. And you don't get a very nice pairing of food and wine. Whereas in Ontario wines, like a lot of French wine and Italian wine, a lot of, you know, Spanish wines and so on, with Ontario wines you're much more likely to get this nice match in which, you know, you can enjoy the food and the wine and one of them isn't killing the other.

However, New World traditions certainly are present in Ontario, most significantly in the form of the relative lack of restriction with respect to the grape varietals that can be planted. In contrast to France and Germany, and more consistent with Australia and California, a variety of grapes can be planted, with the exception of *labrusca* grapes, which are deemed unsuitable for high-quality winemaking. Furthermore, there is recognition that Ontario can produce some wines that may appeal to customers who favour New World wines, as achieved by a select few wineries or, on a larger scale, for certain vintages (e.g., 2007). This scenario appears to further underscore that the prevalence of the Old World tradition is by default, rather than by choice. As one wine writer put it, "[Y]ou know, most of the wines in Ontario are not the big fruity style that you get from warm places like Chile, Australia, South Africa, and California. And people happen to like big, fruity, fairly high alcohol wine. And it's just not what Ontario makes."

Our key observation is that the tradition (i.e., New World) that does not fit the local conditions, at least most of the time, is not helpful in establishing the region as a legitimate producer of fine wines, no matter how popular that tradition may be with a certain segment of consumers or restaurants. Wineries should take care to follow the tradition that better fits the climatic and geographical conditions and clearly communicate the nature of Ontario wine to consumers and other stakeholders, so they do not expect New World–style wines. As one restaurant owner pointed out, "It's a hell of a lot harder to sell an Ontario wine to clientele that's expecting California."

In Search of Identity

At the industry level, we believe a conversation is necessary to determine what the distinctive identity of Ontario wine should be. Established wine regions appear to have distinctive identities (e.g., France and "sophistication," Australia and "playfulness" and "easygoingness"). At present, Ontario wineries aim to demonstrate that the quality of Ontario wine complies with established standards of excellence. But to stand out in the crowded wine marketplace, Ontario wineries must demonstrate to consumers, critics, and restaurateurs what makes Ontario wine different from more familiar and established wine regions. In other words, the most fundamental question is this: *What is the distinctive character of Ontario wine?*

Part of the answer is some kind of "character" or personality (as with examples of France and Australia). But there are other components, such as varietals and winemaking styles. Although experimenting with a range of grape varietals and winemaking styles is healthy and valuable, the Ontario wine industry should start focusing on a select number of grape varietals and styles. In all, we discourage attempts to produce New World styles of wine—no matter how fashionable they may be—because they can succeed only during rare vintages. Even the 2007 vintage, often claimed to be "the best ever in Ontario," did not allow most wineries to produce wines that rivalled those from California or Australia. Critics and restaurateurs have remained critical of many 2007s, especially the whites. Even the rare successful attempts may not necessarily be beneficial to the reputation of the region, because subsequent vintages may disappoint consumers who grow accustomed to the unusual (for Ontario) profile of the 2007s. Thus, rather than pursuing the short-sighted strategy of over-emphasizing a particular (atypical) vintage, Ontario wineries should educate consumers about the variable nature of Ontario vintages. Similarly, a greater appreciation for European wines among Ontario consumers would likely lead to an increased receptiveness to Ontario wines, due to the common "challenge" of vintage variability.

In a similar vein, since wine writers and bloggers tend to respond more positively to Ontario wines that are made in the Old World rather than the New

World tradition, the focus on the former approach will likely allow wineries to benefit more from the positive publicity that writers and bloggers can generate, thereby increasing sales. This effort is especially important for smaller wineries with limited marketing budgets.

Focusing on grape varietals that do well in Ontario should be an increasing imperative for smaller and medium wineries, too. Because the cost of production in Ontario is quite high, the only possibility for small wineries to exercise economies of scale is to plant fewer varietals and focus on those that are more likely to result in quality wines year after year. We would not necessarily discourage wineries from planting the grapes that do not typically do well in Ontario (e.g., Cabernet Sauvignon), but we suggest that small wineries should plant such varietals only on a small scale and focus instead on those that are more reliable.

Introducing more proprietary blends can also be valuable. First, such blends offer wineries more flexibility in responding to Ontario's variable vintages, allowing them to blend the varietals in a way that may showcase the strengths of a particular vintage without rendering the weaker varietals unusable. Second, it will reduce direct competition with both domestic and foreign wineries, as there is a great deal of competition for each single varietal (or even Meritage, a brand for red and white Bordeaux-style wines) product, but it is more difficult to determine direct competitors for a proprietary blend.

International critical attention is very important for the industry. However, it is more valuable to pursue European wine critics than American ones, as the European critics are more likely to favour the more Old World style of wines produced by Ontario wineries. Moreover, even though Ontario produces relatively small amounts of wine, it is important for the region to start developing a concerted exporting strategy. Because Canadian consumers tend to look more favourably on products that succeed internationally, success in export markets is likely to result in more enthusiasm for Ontario wine among Canadian consumers.

Conclusion

Our findings suggest that much work remains to be done, on the business side, to ensure that the Ontario wine industry continues to thrive. Many wineries must learn to balance artistic and commercial pressures. Ontario wineries have yet to accept that the climate is rarely suitable for successful attempts at New World–styled wines and to get comfortable with the more "European" style of Ontario wine. Finally, and most important, it is essential for the industry to define a unique identity to ensure that its offerings stand out from those of its foreign competitors.

Notes

1 R.C. Ulin, "Invention and Representation as Cultural Capital: Southwest French Winegrowing History," *American Anthropologist* 97 (1995): 519–27.
2 Euromonitor, "Old World Still Facing Struggle in UK Wine Trade," *Global Alcoholic Drinks: Wine: Maturity Constrains Growth*, October 2004, http://www.foodand drinkeurope.com/Products-Marketing/Old-World-still-facing-struggle-in -UK-wine-trade.
3 J. Robinson, *The Oxford Companion to Wine*, 3rd ed. (Oxford: Oxford University Press, 2008).
4 Robinson, *The Oxford Companion to Wine*, 476.
5 S. Sayare, "French Diss: Wine Sales Sour at Home and Abroad," Toronto *Star*, 28 May 2009, http://www.thestar.com/business/article/641672--french-diss-wine-sales-sour -at-home-and-abroad.

Seven

The Hands behind the Harvest: Migrant Workers in Niagara's Wine Industry

Janet McLaughlin

T he preparation for the production of a bottle of Niagara wine begins thousands of miles away.

Carlos, a 40-year-old father of five from a small village in Hidalgo, Mexico, is making preparations for his tenth trip to Canada to work in a Niagara vineyard. His flight is set to depart a week from Friday, but his family does not yet know this. Carlos knows that the weeks leading up to his departure are sad and stressful for his family, particularly for his young daughters, who are normally distraught when they realize that he is leaving. He tells them only two days before, when the inevitable preparations make it obvious anyway, to lessen the period of their sadness and to enjoy his final moments with them as much as possible.

When Friday finally comes, he wakes up after only a few hours of sleep to start his journey by bus to the airport in Mexico City, facing the crisp air of the early morning. His wife sees him off, offering one final lingering embrace, and weeps as he departs. Carlos must be strong for her. He fights back his own welling tears, managing to maintain his composure, but is unable to speak due to the lump in his throat.

Like every year for the past decade, it will be another eight months until they can be reunited. In the meantime, they will be bonded by occasional phone calls and biweekly remittance cheques, which Carlos will send from a convenience store in Virgil, Niagara. They accept the tough goodbyes and long absences because the money he earns allows them to live better—to build and maintain a home, educate their children, and put food on the table. Due largely to global economic restructuring, related to policies and practices far beyond his control, Carlos, and many small farmers like him in countries like Mexico and Jamaica, are no longer able to make a living working their own land. They must instead go abroad to work as farm labourers simply to survive. His children do not yet appreciate the global and structural context of his departure, or the noble motivation for his absence—they just feel, and resent, its painful sting.

When Carlos arrives at the airport, he is comforted by the sight of old friends and co-workers who are boarding the same flight, who have just endured, but do not speak of, similar goodbyes. They are all wearing their best clothes, emotions a mix of pride, anxiety, and sadness as they head off to work abroad once again. They are the providers for their families and communities, and they welcome the opportunity to work in Canada. But this opportunity comes with immense sacrifice.

As the workers fly over Toronto, they notice the familiar but dreaded sight of a climate much colder than the cactus-spiked landscape they left behind. It is February, and snow still blankets the ground. They shudder at the thought of entering the cold from the airplane—they have no winter coats on their backs. (To save space on their return journey home last fall, the jackets and boots they had purchased at a St. Catharines thrift store were left behind at their bunkhouse.)

When they arrive at Toronto's Pearson Airport, a young English-speaking customs agent asks one of men how many seasons he has spent in Canada. "Eighteen," he replies in awkward English. "My God," the woman responds, shaking her head and motioning him forward. "It's like you spend more time here than there. Welcome home," she jokes. Such is the life of Niagara's migrant workers, living between two places, but belonging fully to neither.[1]

Discussions and celebrations of the Niagara wine industry seldom include a mention of the paradox that this "local" industry depends largely on the labour of imported workers to tend the vines throughout all of their stages of growth. This chapter delves into the context of this largely invisible workforce, highlighting the lives and contributions to Niagara's wine industry of temporary foreign agricultural workers, more commonly known as migrant workers, discussing the programs through which they come, the degree to which they are socially integrated, and some of the key debates surrounding their employment. The findings are based on nearly a decade of research, involving living and travelling alongside migrant workers in Niagara, and in Jamaica and Mexico, and several studies involving surveys and focus groups with workers as well as qualitative interviews with workers, employers, government officials, and others, supported by literature reviews and archival research.[2] I argue that migrant workers are the backbone of Niagara's viticulture industry, contributing a unique labour force that is valued for its productivity, reliability, and flexibility. For all of this, migrant workers remain on the margins of society, largely invisible to wine consumers and to the public. Migrant workers' employment in Canada, lauded publicly as a "win-win" solution to growers' needs for reliable labour and migrant workers' desire for a better income, in fact generates contradictory outcomes, with significant benefits and challenges. I propose that structural changes to migrant employment schemes, aimed at empowering and recognizing workers' contributions, and providing improved social and political integration, can benefit not only the workers, but also the Niagara wine industry as a whole.

Migrant farm workers come to Canada each year through two streams of the federal government's Temporary Foreign Worker Program (TFWP). The oldest of these, the Seasonal Agricultural Worker Program (SAWP), is based on long-standing bilateral agreements between Canada and Mexico as well as several English-speaking Caribbean islands (primarily Jamaica) to provide labour for up to eight months each year. In place since 1966, the SAWP provides the majority of workers in Niagara's wine industry. The second initiative, the Pilot Project for Occupations Requiring Lower Levels of Formal Training, National Occupations Classification C & D (now called Stream for Lower-skilled Occupations, hereafter NOC C & D Pilot), has been in place since 2002. In 2011, the federal government announced an Agriculture Stream of the aforementioned Pilot Project, which includes more specific guidelines around housing, wages, and contracts. The NOC C & D Pilot allows greater flexibility, employing workers from any country for visas up to two years, with extensions possible up to four years. There is also less government oversight and regulation, particularly on the part of sending countries, which, in contrast to the SAWP, do not play a formal role in program management or mediation.

The use of migrant workers is extensive in Niagara viticulture, and Niagara is among the top regions of migrant worker employment in Canada, following only Haldimand-Norfolk and Essex counties in Ontario. Fruit in general, and viticulture in particular, employ the largest number of migrant workers in the region. Migrant workers are legally employed in Canada. As such, they are covered under most of the same protections as other domestic agricultural workers, including minimum wage laws. Workers in the SAWP also have access to consular representation to help mediate disputes. Workers in both program streams pay provincial and federal taxes, and contribute to the Canada Pension Plan (CPP) and Employment Insurance (EI). They generally do not benefit from their contributions to the social contract, although they are eligible for pension benefits upon retirement (these are normally meagre). They used to be eligible for special EI benefits, and many workers received parental benefits through this program. This right, however, was revoked in 2012 by the Harper government, despite that migrant agricultural workers contribute about $25 million a year to the EI program. Although they have temporary access to Ontario Health Insurance Plan (OHIP) coverage for medical care in Ontario, this expires at the end of each contract, and they do not have any long-term form of health insurance to cover treatment for serious or lasting health conditions, despite the substantial health risks of working within agriculture, one of Canada's most dangerous industries.[3] They also have deductions from their pay for program administration fees, supplementary health insurance, travel and visa expenses, and housing costs, though these vary by program stream and participating country. As a result of these deductions, what may have seemed like a lucrative minimum wage in Canada is in fact quite reduced, and workers are often dismayed at their take-home wage.

Employers, too, must make contributions on behalf of workers for taxes, the Workplace Safety Insurance Board (WSIB), EI, and CPP, which also affect their bottom line. In this way, the federal and provincial governments can collect contributions for these workers, but rarely have to pay them benefits. Since temporary workers are not citizens, they are ineligible for most services available to Canadians, and are even ineligible for services such as language training and integration programs, which are available to new immigrants. In spite of the number of years they have spent working in Canada, foreign migrant workers are considered permanently temporary.

Such labour arrangements have become ubiquitous wherever labour-intensive agricultural industries predominate. The demand for temporary migrant labour emerged in southwestern Ontario in the 1940s, as small family farms consolidated into larger operations at the same time as there was a decreased desire among urbanizing, and more highly educated, Canadians to work under the difficult and variable conditions in agriculture.[4] In light of similar migrant labour arrangements in competitor countries—the United States, Australia, and European nations—farmers in the Niagara Region were among the original group of lobbyists who asked the Canadian government to establish a foreign worker program, arguing that reliable domestic workers could not be secured.[5] The federal government finally conceded to a pilot project in 1966, with some of the first Jamaican workers arriving in Niagara that year. The workers were carefully selected by the Jamaican government for agricultural aptitude and willingness to work hard without complaint, a process that has continued throughout the program's history.[6] The project was immediately hailed as a success by both Canada and Jamaica, and has continued to expand since then.

As of 2011, there were nearly 38,000 temporary foreign agricultural worker positions approved throughout Canada, with workers employed across nine provinces from a wide range of countries.[7] Although increasingly diverse, the majority of workers are Mexican, male, and employed in southwestern Ontario. Most male workers are married with dependents, while the 3 percent of women participants are principally single mothers.[8] The proportion of women is higher in Niagara, where "feminized" industries, such as viticulture, nurseries, and floral production, predominate. Despite migrant workers' long-standing contributions to Canadian agriculture, and their value to the industry and general economy, they are refused the opportunity to immigrate to Canada. Thus they live their lives continually in limbo, working and residing in a country in which they will never be recognized as citizens, and in which they are separated from their families. In 2011, Niagara employed approximately 3,000 agricultural migrant workers, the vast majority in the fruit and grape industry (followed by nurseries, flowers, and greenhouses). The majority of workers in Niagara (56 percent) came from Mexico, followed by Jamaica and other Caribbean countries (38 percent), Guatemala (4 percent), the Philippines (1 percent), Honduras (1 percent), and

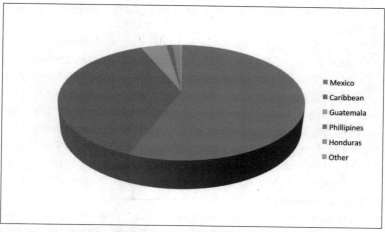

Figure 7.1 Country of origin of temporary foreign workers in agriculture in Niagara, 2011 (SAWP and NOC C & D)

small numbers from various other countries, ranging from Thailand to South Africa (Figure 7.1).[9]

The Value of Migrant Workers

Interviews with employers of migrant workers in the Niagara region paint a consistent picture—migrant workers are valued for their reliability, flexibility, and productivity. As the following quotes from interviews with farmers demonstrate, migrant workers have become a central feature of the success of agriculture in the region.

> "The farmers know that we need those guys as much as they need us—it's a co-dependence."

> "Offshore workers are probably 25 percent more productive than Canadian workers, but much more reliable.... These workers are hand-picked."

> "Without them we'd go under.... You just can't find Canadians to do this work."

> "They're the backbone, all I can say is if the program dropped there would be a number of farmers that just couldn't handle it. I would say most farmers, because most farmers are large farmers, way back even when we started farming in the early 60s, you had a lot of small farms and 15-acre farms, and you could survive. But today with these larger 200, 250 acres, I'm not kidding, they could not survive without bringing those workers over here."[10]

Contrary to popular belief, migrant workers are not necessarily "cheaper" than Canadian employees. They are supposed to be paid minimum wages, which are higher in Ontario than many other jurisdictions. (Of course, some labour advocates contend that employers should offer higher wages to make these jobs more attractive to Canadian workers, but employers argue that doing so in the face of globalized competition with lower wage standards would drive at least some of them out of business.) Employers must also provide housing that meets basic standards and incur other expenses associated with migrant worker employment. Migrant workers' distinct advantage, instead, rests in their reliability, flexibility, and productivity.[11] Each of these traits is facilitated not by the inherent "nature" of migrant workers, as is often believed, but rather by the conditions of their contracts, which are extremely restrictive and precarious.

With respect to *reliability*, farmers continually suggest that Canadian workers cannot be depended upon, and when they find the work too difficult, they simply leave. Migrant workers, by contrast, arrive on the day they are needed, and work as long and hard as employers demand. They rarely, if ever, quit a job because it is too difficult or leave before the growing season is complete. Migrant workers depend on their employment to support their families, but unlike Canadian citizens, their work visa—and their right to stay in Canada—is tied to a contract with a specific employer. Thus, even if migrant workers are unhappy with their employment conditions, they have little freedom to leave or to seek alternate employment. Their one option is to leave Canada and return home, but to do so would forfeit their family's only hope for a living wage. This restriction on the freedom of their employment conditions renders them an extremely dependable labour force. Further, if workers become sick or injured, or for any reason cannot continue working, other workers waiting in the queue easily and quickly replace them. Thus, not only individual migrant workers, but also the labour force from which they come, are constructed to be extremely reliable.

Separating migrant workers' labour and familial and social obligations provides distinct advantages for Canadian producers. Unlike many Canadian workers, temporary migrant workers tend to be willing to work no matter what hours or tasks are asked of them. This is a particularly important trait within agriculture, as working conditions fluctuate with weather, crop development, and other variable factors, many of which are beyond employers' control. Because they normally live on farm property and have minimal social obligations, migrant workers can be called upon to work unfavourable hours—evenings, weekends, and holidays—during peak periods, and then may be expected to work few or no hours during periods of relative inactivity. In Canada, without their families, migrant workers are less likely to object to working weekends or holidays; their primary objective is to work as many hours as possible to earn money to support their families for the rest of the year. These conditions set them apart from domestic employees who generally value evenings, weekends, and holidays to spend with their families and expect a steady and predictable pattern of employment.

Finally, many employers praise migrant workers' high levels of *productivity*; some growers insist that migrant workers accomplish as much as two or more Canadian workers in the same amount of time. In part, workers' exceptional productivity is a result of a strong work ethic and adaptation to hard physical labour based on the circumstances of their countries of origin, but even more so, it relates to their placement of competition with each other for coveted, but precarious, positions of employment. Workers know that if they do not perform adequately, they can and will be replaced by workers from their own country or from other countries participating in foreign worker programs.[12] This threat of competition drives them to perform at or even beyond their capacity, sometimes to the detriment of their health and well-being.[13] With neither job security nor collective bargaining rights, migrant workers' only guarantee of a future job is their diligent performance and consistent acquiescence to employer demands. For these reasons, all mediated by the restrictive nature of their employment conditions and rights, as well as by notions of gender, race, and racialization, migrant workers are considered an ideal labour force in Canadian agriculture. Indeed, many employers go so far as to invoke racialized explanations of workers' attributes and abilities, suggesting that racialized workers are better "suited" to agricultural work. Some employers even assign different ethnic and gender groups specific tasks or crops based on their perceived inherent abilities or limitations.[14] In this context, a racialized workforce comprised primarily of male black and Latino workers has quietly become institutionalized as the invisible backbone of labour-intensive agriculture, including viticulture.

A World within a World: Social Inclusion and Exclusion

Many tourists, theatregoers, and food and wine lovers visit the village of Niagara-on-the-Lake. In the summer months, the municipality's byways bustle with visitors. Vineyards surrounding the quaint town offer regular tours and wine tastings throughout the day, and restaurants, theatres, hotels, and pubs fill with revellers in the evenings. The town appears to be cohesive, vibrant, and friendly, tied together by a history and culture based in large part around pride in its locally produced food and wine. Migrant workers, though present in the thousands, are largely invisible to these visitors. Despite their residence in this tourist hotspot for months at a time, workers do not have the opportunity, disposable income, or social connections to frequent the fine restaurants, museums, or theatrical productions.

In the Niagara Region, migrant workers' contributions to the local agricultural and wine industries are generally overlooked or ignored. As a researcher and tourist who frequents the region and enjoys its offerings, I commonly and informally inquire about production processes and labour conditions during wine tours and tastings. Any mention of migrant workers' presence and contribution to the industry is normally conspicuously absent from tour guide responses. Of the many aspects of which Niagara is celebratory and proud, the presence of migrant workers does not appear to be one of them.

Migrant workers lead a different existence in Niagara. Their days are spent working in vineyards or farms, occluded from public view. They are not usually employed in public positions—as tour guides, wine tasters, greeters, or vendors at wine shops, farm stands, or markets. Instead, they live and work behind the scenes, providing the manual labour necessary to grow and cultivate the grapes. They typically reside on farms in trailers or bunkhouses, or sometimes in purchased or rented converted homes nearby. Their nights are spent preparing meals and calling their families or sending money home. Friday evenings and Sunday afternoons, often their only time off, may allow some time to do laundry, buy groceries, deposit money, and run other errands. Those who work in more isolated parts of the region must rely on their employers, or expensive taxis, for transportation to access shopping and amenities.

Motivated workers may try to learn some English or visit the public library. Others ride their bicycles around the rural roads, enjoying the fresh air. For the most part, however, they do not participate in the community's affairs, and keep interactions largely within their own networks. Vendors, often immigrants from similar ethnic backgrounds, arrive in vans and set up makeshift stores to sell the workers culturally familiar food, clothing, long-distance phone cards, and other items. On occasion, a St. Catharines bar hosts theme nights targeting the workers. At such times, one feels as if one has stepped into "little Mexico." Such scenes are characterized by blaring ranchero music, people eating tacos, and, most notably, a large group of lonely-looking Mexican men, many in cowboy hats, competing for a salsa dance with the rare—and revered—women in attendance.

Workers are keenly aware of their "present absence" within the region. As one Mexican worker observed:

> When they talk about coming to Canada, they say "have a great time in Canada." I've wanted to come to Canada for many years, and now here I am. But we're not here to have fun. How can you have fun in a country where you only get one day off a week, and where no one else speaks your language, and you're away from your family and friends, and face discrimination from other community members? We don't come here for a holiday; we come to work. That is the mentality among Mexicans. We work very hard to earn as much money as possible to support our families.[15]

Others appear more hostile about their marginalized position, citing ambivalence or even racism on the part of residents. Clendon, a Jamaican worker who has been in Canada for several decades, explained: "We are just recognized as workers, nothing else … they don't include us in anything here. No one asks for our opinion. They just see us as workers."[16]

It is hardly surprising that workers of colour who lack citizenship rights and integration into the predominantly all-white rural areas in Canada where they work report experiencing discrimination and a sense of isolation. In the Niagara Region, 2006 Statistics Canada data reveal that just 6.2 percent of residents identify as part of a "visible minority" group—including 1.2 percent who identify as "black" and 1.0 percent who identify as Latin American. There are even lower proportions of visible minorities in Niagara-on-the-Lake, where the majority of migrant workers live and work. Many workers, especially those from the Caribbean, say they feel that some of their interactions with Canadians have been tainted with racism, either on occasion or commonly. While incidents of racism may be more explicit among darker-skinned Caribbean workers, some Mexicans also feel that they are discriminated against, in part because they do not speak English. As one asserts, "We face a lot of discrimination. People see us and they think we're stupid because we don't speak English. They don't even realize how much English I understand."[17]

My interviews and interactions with both Canadians and workers reveal that Canadians' views of and interactions with migrant workers range from what could be categorized as fearful and discriminatory; indifferent and ignorant; showing pity and sympathy; grateful and respectful; to expressions of solidarity, friendship, and active support. Diverse attitudes across this spectrum are exhibited by community members, business owners, and employers alike. It is not surprising, then, that workers' own internalization of Canadians' views toward them varies, too, and is based largely on personal experience. When Canadians actually get to know workers, more positive views and meaningful relationships often emerge. Some Niagara residents and employers alike, who have formed long-term relationships with migrant workers, invite them for dinners or parties in their homes or even visit them in their countries of origin during the winter.

Over the last decade, as awareness of the workers' existence and plight has spread, various groups have organized to make concerted and targeted efforts to advocate for and support migrant workers. Niagara is home to several active support and advocacy organizations. *Enlace*, a group from Toronto, hosts an annual bicycle safety event and soccer tournament in Niagara-on-the-Lake for Mexican workers. The locally based Caribbean Workers Outreach Program (CWOP) arranges for workers to participate in church services and social events such as domino and cricket tournaments. In conjunction with the CWOP, a particularly passionate local bed and breakfast owner, Jane Andres, spearheads a large welcome concert for Caribbean workers each spring, which the mayor, among other community members, normally attends. The Toronto-based activist collective *Justicia for Migrant Workers* also does active outreach and organizing in the area, promoting workers' rights and arranging trips to Toronto. Local churches hold specialized Spanish-language services.

Most recently, the Niagara Migrant Workers Interest Group (NMWIG) has formed as a collective of groups and individuals concerned with the well-being of migrant workers in the region. Its three committees—health, legal, and social/cultural support—have generated an impressive list of accomplishments in a short time, including specialized health clinics partnered with local community health centres and bolstered by volunteers from the Niagara campus of McMaster University's medical school, an annual health fair, large theatrical productions featuring the words and work of migrant workers, and legal challenges and support. Member organizations such as AIDS Niagara and two Niagara-based legal clinics have begun targeted outreach and service efforts among migrant workers. Migrant workers have contributed ideas and input into some of the initiatives. Upon the suggestion of *Dignidad Obrera Agrícola Migrante,* a group of Mexican migrant workers in the Niagara Region, a scholarship program for migrant children with Brock University and Niagara College has been launched. The first recipient, Sayuri Gutierrez, the daughter of a St. Catharines nursery worker from Mexico, began studies at Brock University in 2012.[18] While workers are by no means well integrated into the community and its local affairs, the considerable—and largely unpaid—efforts of such groups and individuals have collectively made a major difference in providing a sense of belonging and support to the workers in the region.[19]

Finally, an Agricultural Workers Alliance support centre in Virgil offers workers information and support on accessing their rights and benefits, as well as social and educational events. Many workers rely on the centre for assistance with everything from benefit programs to medical appointments. Yet the presence of the centre, which is funded and organized by the United Food and Commercial Workers (UFCW), has created tension in the community. Most employers resent the union's involvement in workers' affairs and fear the implications of unionization. Many workers have received threats from employers and even their own government officials to stay away from the union and any of its affiliates. However, any concerns about unionization in Niagara, for the time being at least, are unwarranted. All agricultural workers, including migrant workers, are legally forbidden from collective bargaining as part of a union in Ontario. The UFCW and others have challenged this exclusion, launching numerous cases against the provincial government and farmers' organizations over the past decade. In 2011, the exclusion was upheld in the Supreme Court of Canada, much to the dismay of union, labour, and social justice movements across the country and beyond, who criticized the decision as a major denial of rights to one of the country's most vulnerable groups of workers.[20]

In the absence of these rights, workers have little recourse to fight against poor working and living conditions, low wages, and other unfavourable conditions of their contracts. When workers have protested such issues in the past, they have often been quickly, and quietly, fired, sent home, and removed from the program. This threat is ever-present, serving as an effective mechanism of control and compliancy.[21] Essentially, workers are told to take or leave the conditions they are offered. If they choose to leave, there are many eager workers in line waiting to take their place.

What's Behind the Glass?

Although few consumers know it, most glasses of Niagara wine are enjoyed at least in part thanks to the reliable, flexible, and productive labour of migrant workers. As viticulture has increasingly been seen as a more lucrative or enticing venture in the Niagara Region, it has expanded, subsuming land previously occupied by other crops, particularly tender fruits. Grape growers have become a leading employer of migrant workers in the region. Emerging industries such as viticulture are only the most recent to benefit from an already well-established labour arrangement that is more widely used throughout the region and beyond.[22]

Grape and wine producers did not create the Temporary Foreign Worker Program (TWFP), but they do benefit from its existence. Without question, these well-established programs of "just in time" labour have greatly contributed to the success and expansion of viticulture in the region. Although growers face many variables and unknowns in their production process, ranging from local weather to global competition, reliable, productive labour has been one factor about which they need not worry. Migrant workers who come from economically depressed regions normally welcome the opportunity to work under conditions of legality in Canada. There is immense competition to enter these programs and to remain in them. While farmers say they can barely get Canadian workers to stay on the job a week, there is a near endless supply of young, healthy workers from developing countries ready to accept and commit to positions as agricultural labourers. For many workers, the program has become a lifeline, providing them with an income to house, feed, and educate their families.

These substantial benefits to both employers and workers, however, have not come without challenges. Long and painful family separations, health issues, social exclusion, and limited rights are all issues that taint the reputation of the program and the way workers experience it. It is in part due to these controversies that the contributions of migrant workers are seldom highlighted, and are sometimes even actively hidden, within the wine industry. To be sure, some individual employers go to great lengths to treat their workers well, far beyond the minimum standards required of them. Southbrook winery, for example, already celebrated for its environmental innovation with biodynamic and organic vine-yards, not only arranges Internet and Skype connections for its Mexican workers, but also provides them with cooking lessons. Other wineries share in the pride of their finished product with the workers, providing them with a complimentary bottle along with each paycheque.

Yet the structure of the migrant worker programs, which places an inordinate amount of power in the hands of the employers while denying job security or mobility rights to workers, leaves those who encounter less favourable employ-ment conditions with little recourse, and threatens the image of the industry as a whole. Sadly, examples of abuse and neglect are also ripe for the picking. While the majority of workers continue to return to Canada year after year and appreciate the opportunity for employment, many have experienced difficul-ties such as unfair dismissals, abusive employers, major injuries and illness, or even death.[23] Such issues, which had previously been largely neglected, have increasingly aroused the interest of the Canadian media, social justice groups, and the public. Advocates and scholars have argued that structural changes to the TFWP—such as a pathway to permanent residency and family reunifica-tion, open work permits, an appeals process when fired, collective bargaining rights, accessible services, long-term trans-border health insurance, equal and full access to employment insurance benefits, and expanded community inte-gration—would provide workers with more empowerment and rights.[24] Others have called for labels to identify and celebrate exemplary labour practices among those employers who consistently treat workers well. In the United States, for example, wineries utilizing union-grown grapes contain labels boasting of this contribution.[25]

Increasingly, socially conscious consumers are concerned not only with the taste and quality of wine, but also the environmental, social, and labour condi-tions surrounding its production. Given the centrality of brand reputation, win-eries arguably have the most to gain from model employment conditions of all the industries employing migrant labour. Regardless of the efforts of individual employers, which, as in any industry, vary widely, it is in everyone's interest to ensure Niagara's wine producers are associated with only the highest standards of integrity, from grape to glass. Structural changes that improve workers' empower-ment and social integration are perhaps the best way to ensure uniform improve-ments. Accordingly, migrant workers' contributions to the industry, as well as their treatment within it, could be celebrated instead of hidden. This would be a true "win-win" outcome, further strengthening a dynamic and responsive grape and wine industry.

Notes

1 Adapted from field notes and Janet McLaughlin, "Trouble in Our Fields: Health and Human Rights among Mexican and Caribbean Migrant Farm Workers in Canada," doctoral thesis, Department of Anthropology, University of Toronto, 2009. All names used in quotes and vignettes are pseudonyms.

2 This research has been funded at various stages by the International Development Research Centre (IDRC), the Social Sciences and Humanities Research Council of Canada (SSHRC), the Public Health Agency of Canada (PHAC), the Canadian Institutes for Health Research (CIHR), and the WSIB Research Advisory Council. The majority of data in this chapter are based on research conducted for my doctoral research, which was supported by SSHRC, IDRC, University of Toronto graduate fellowships, and the Institute for Work and Health.

3 Canadian Agricultural Injury Surveillance Program, *Agricultural Fatalities and Hospitalizations in Ontario 1990–2004* (2007). Available online at www.caisp.ca; Eric Tucker, "Will the Vicious Circle of Precariousness Be Unbroken? The Exclusion of Ontario Farm Workers from the Occupational Health and Safety Act" in *Precarious Employment: Understanding Labour Market Insecurity in Canada*, L.F. Vosko (Montreal and Kingston: McGill-Queen's University Press, 2006), 256–76.

4 Victor Satzewich, *Racism and the Incorporation of Foreign Labour: Farm Labour Migration to Canada since 1945* (New York: Routledge, 1991).

5 McLaughlin, "Trouble in Our Fields."

6 Janet McLaughlin, "Classifying the 'Ideal Migrant Worker'": Mexican and Jamaican transnational farmworkers in Canada," *Focaal: Journal of Global and Historical Anthropology*, 57 (2010): 79–94.

7 http://www.ufcw.ca/index.php?option=com_content&view=article&id=3155%3A harper-government-deals-another-blow-to-vulnerable-migrant-workers&catid=6%3 Adirections-newsletter&Itemid=351&lang=en.

8 This applies to workers from Mexico. Selection criteria vary by country of origin. See Janet McLaughlin, "Classifying the 'Ideal Migrant Worker'" for further discussion.

9 Based on 2011 data provided by Human Resources and Skills Development Canada for the SAWP and agricultural workers in the NOC C & D Program.

10 As cited in Janet McLaughlin, "Trouble in Our Fields."

11 See similar arguments in Tanya Basok, *Tortillas and Tomatoes: Transmigrant Mexican Harvesters in Canada* (Montreal and Kingston: McGill-Queen's University Press, 2002); and expanded arguments in Janet McLaughlin, "Trouble in Our Fields."

12 Janet McLaughlin, "Classifying the 'Ideal Migrant Worker.'"

13 Jenna Hennebry and Janet McLaughlin, "The Exception that Proves the Rule: Structural Vulnerability, Health Risks and Consequences for Migrant Farmworkers in Canada" in *Legislated Inequality: Temporary Migrant Workers in Canada*, ed. Patti Lenard and Christine Hughes (Montreal and Kingston: McGill-Queen's University Press, 2012); Janet McLaughlin and Jenna Hennebry, *Backgrounder on Health and Safety for Migrant Farmworkers* (IMRC Policy Points, issue I, 1 December 2010), http://www.wlu.ca/imrc; Janet McLaughlin, "Trouble in Our Fields."

14 See related discussions in Jenna L. Hennebry, "Globalization and the Mexican-Canadian Seasonal Agricultural Worker Program: Power, Racialization & Transnationalism in Temporary Migration," doctoral thesis, Department of Sociology, University of Western Ontario, 2006; Janet McLaughlin, "Classifying the 'Ideal migrant worker'"; Kerry Preibisch and Leigh Binford, "Interrogating Racialized Global Labour Supply: An Exploration of the Racial/National Replacement of Foreign Agricultural Workers in Canada," *Canadian Review of Sociology and Anthropology* 44, no. 1 (2007);

Nandita Rani Sharma, "On Being Not Canadian: The Social Organization of 'Migrant Workers' in Canada," *The Canadian Review of Sociology and Anthropology* 38, no. 4 (November 2001): 415–40; Satzewich, *Racism and the Incorporation of Foreign Labour.*

15 As cited in Janet McLaughlin, "Trouble in Our Fields."

16 As cited in Janet McLaughlin, "Trouble in Our Fields."

17 As cited in Janet McLaughlin, "Trouble in Our Fields."

18 See http://www.studyinniagara.ca/migrantaward/the-award and http://www .theglobeandmail.com/news/national/mexican-girls-college-dream-in-danger-of -being-undermined/article2428291.

19 See Jenna Hennebry, *Permanently Temporary? Agricultural Migrant Workers and Their Integration in Canada* (Ottawa: IRPP, 26 February 2012); Kerry Preibisch, *Social Relations Practices between Seasonal Agricultural Workers, Their Employers and the Residents of Rural Ontario* (Ottawa: North-South Institute, 2003); Kerry Preibisch, "Migrant Agricultural Workers and Processes of Social Inclusion in Rural Canada: Encuentros and Desencuentros," *Canadian Journal of Latin American & Caribbean Studies,* 29 issue 57, no. 8 (2004): 203, for further analysis of social inclusion/exclusion.

20 See *Constitutional Labour Rights in Canada: Farm Workers and the Fraser Case,* ed. Fay Faraday, Judy Fudge, and Eric Tucker (Toronto: Irwin Law, 2012).

21 Basok, *Tortillas and Tomatoes*; Preibisch, "Migrant Agricultural Workers and Processes of Social Inclusion in Rural Canada." 203; Hennebry, *Permanently Temporary?*; Janet McLaughlin, "Trouble in Our Fields."

22 Much attention has been paid, for example, to Mexican workers in the Leamington greenhouse industry. See the Internet-accessible NFB documentary *El Contrato,* by Min Sook Lee (National Film Board of Canada, 2003); and Basok, *Tortillas and Tomatoes.*

23 Basok, *Tortillas and Tomatoes*; Hennebry; McLaughlin, "Trouble in Our Fields"; Veena Verma, "The Mexican and Caribbean Seasonal Agricultural Workers Program: Regulatory and Policy Framework, Farm Industry Level Employment Practices, and the Future of the Program under Unionization" (Ottawa: North-South Institute, 2003).

24 See Fay Faraday, *Made in Canada: How the Law Constructs Migrant Workers' Insecurity.* Toronto: Metcalf Foundation (2012), http://metcalffoundation.com/publications -resources/view/made-in-canada; Hennebry, *Permanently Temporary?*; Jenna Hennebry and Janet McLaughlin, *Key Issues & Recommendations for Canada's Temporary Foreign Worker Program: Reducing Vulnerabilities & Protecting Rights (IMRC Policy Points,* Issue II, 1 March 2011), http://www.wlu.ca/imrc; Janet McLaughlin and Jenna Hennebry, *Backgrounder on Health and Safety for Migrant Farmworkers.* Also, several advocacy organizations, such as Justicia for Migrant Workers (www .justicia4migrantworkers.org) and the Agricultural Workers Alliance (www.awa-ata.ca) have made similar and further recommendations.

25 See http://www.ufw.org/_page.php?inc=orga_label.html&menu=organizing.

The Vineyard to the Bottle

Eight

The Use of Geospatial Technologies for Improved Vineyard Management Decisions in the Niagara Region

Marilyne Jollineau and Victoria Fast

Introduction

T hose involved in grape growing and wine production, from viticulturists and winemakers to vineyard owners and managers, continuously strive to better understand the interaction of several factors—such as geographic location, mesoclimate, underlying geology, soils, and local topography (slope, aspect, and elevation)—known to influence the composition and quality of wine grapes.[1] In this context, they have recently shown greater interest in obtaining precise (detailed) and accurate (correct) spatial information about their vineyards, as this information has the potential to improve vineyard management decisions. For example, spatial information about well-established vineyard and fruit compositional variables that are indicative of grape quality, from soil moisture and vine vigour (a measure of the vine's vegetative growth) to grape sugar (°Brix) and pH,[2] and the spatial and temporal variability of these variables, is increasingly of interest. This information is significant since vineyard management practices can be adapted if these variables, and their underlying variability, are well understood.

Based on existing studies, geospatial technologies, such as global positioning systems (GPS), remote sensing, and geographic information systems (GIS), can be used to obtain and analyze information about vineyard variables.[3] In addition, they can be used to characterize the spatial and temporal variability of these variables.[4] This is significant since variations can influence grape quality and yield, and can ultimately impact wine quality.[5] Using geospatial technologies to complement traditional field methods of data collection to inform vineyard management decisions offers several advantages. These include a standardized data-collection procedure over a range of spatial scales; a digital library (or dataset) of valuable spatial information that can be consistently updated throughout the growing season and from season to season; a spatial dataset from which a range of vineyard maps can be displayed and subsequently analyzed, both spatially and temporally, in order to develop new information about vineyards; and a baseline dataset against which assessments of the efficiency and effectiveness of vineyard management strategies can be conducted over time.

As noted in other chapters of this book, the Niagara Region has had a relatively short history in which vineyard managers have had to develop vineyard management strategies, compared to well-established wine regions, such as Greece, where wine production was profitable by the eighth century AD.[6] In a globally competitive wine industry, current research suggests that vineyard managers in newly established wine regions, such as the Niagara Region, should focus on developing strategies that optimize grape quality and yield with the aim of producing higher-quality wines. In this context, vineyard managers and other viticulture experts in the region have recognized that vineyard management decisions must account for spatial variability in grape quality and yield in order to produce higher-quality, higher-value products. However, these decisions depend on the availability of detailed, accurate, and reliable datasets that describe the spatial variability exhibited by vines.[7] Geospatial technologies have been identified as powerful tools for this purpose.

Geospatial technologies are increasingly being used to study vineyards in the Niagara Region. Although such studies are relatively new, there is clear evidence of their value when applied to vineyard management practices.[8] For example, existing studies have shown that detailed maps can be used to identify specific vineyard zones for more precise management, including segmented harvesting and differential irrigation.[9] In the Region, several collaborative research projects, such as those funded through the Ontario Centres of Excellence (OCE), have been undertaken by researchers (e.g., from Brock University, University of Guelph, and Niagara College) as well as industry personnel with local wineries and various grape growers over the last five years. Complementary in nature, these studies have involved the use of geospatial technologies, in varying capacities, to obtain precise and accurate spatial information about vineyards to better inform management decisions. In addition to their active participation in these projects, vineyard managers have also taken part in the development of a grape-growing database (i.e., the Vitis Management System), in which vineyards in the region have been precisely mapped, according to grape variety, using a hand-held GPS unit. Maintained by the Grape Growers of Ontario, this database contains information about vineyard characteristics such as their geographic location, the number of vines, the year they were planted, total vineyard acreage, and soil conditions.[10] Collectively, these projects demonstrate significant interest in the value and use of geospatial technologies for improved vineyard management in the Niagara Region. This is important since relatively few research studies have focused on this region, despite the fact that improvements in wine quality can better establish Niagara, locally and internationally, as a producer of premium wines.

Within this context, this chapter provides a review of the value and use of geospatial technologies in viticulture for vineyard management purposes. In particular, the question arises as to what extent these technologies, and their associated datasets, can be used to characterize vineyards (and their spatial and temporal variability) to improve vineyard management decisions. The ability to characterize vineyards and their variability is especially important in the context of grape quality and yield, given their overall influence on wine quality. To illustrate the capabilities of geospatial technologies for improved vineyard management, examples are provided from an OCE-funded collaborative research project (headed by Dr. Jollineau) with Stratus Vineyards, a winery in the Niagara Region.

The Use of Geospatial Technologies in Viticulture: A Brief History

In 2001, the American Society of Enology and Viticulture organized an international symposium to explore specifically the use of geospatial technologies in the grape and wine industry. The symposium, entitled *Space Age Winegrowing*, suggested that the use of geospatial technologies for vineyard management was a space-age concept. While these technologies have been successfully used to observe, analyze, and manage other vegetated environments (e.g., agricultural fields, forests, and wetlands), it was not until the twenty-first century that they were recurrently used to study value-added crops, such as vineyards. The practice of using geospatial technologies to manage vineyards is known as precision viticulture, a term that "encompasses the use of a range of tools and technologies that allow viticulturists and winemakers to make more informed, targeted management decisions in the vineyard."[11] Precision viticulture applications increasingly make use of three predominant geospatial technologies: GPS, remote sensing, and GIS. Although each of these technologies provides different types of spatial information that can be useful for vineyard management purposes, GIS provides an environment in which disparate data sources can be incorporated into a single database for visual display, as well as advanced spatial and temporal data analyses.[12] These technologies have been used for vineyard management purposes, such as vineyard mapping and monitoring, across vineyards at a range of spatial scales, from entire wine regions to within-vineyard sub-blocks. They have also been used to study vineyards in different geographical settings, most notably in Australia, New Zealand, California, and Canada.

Understanding Vineyard Variability

The practice of precision viticulture is largely based on the understanding that vineyards are by nature spatially and temporally variable. In particular, variability exists in vineyard variables that are ultimately linked to wine quality, such as soil moisture, plant water status, and vine vigour. Geospatial technologies (such as remote sensing) permit the simultaneous acquisition of information about vineyard variables and their natural variability. Using remote-sensing technologies, for example, both vine vigour and its spatial variability can be captured with the acquisition of a single satellite image (Figure 8.1). Quantitative assessments of vine vigour over time can also be made when multiple images are acquired throughout the growing season. By understanding this variability, vineyard managers are better able to incorporate large-scale variability information (i.e., at the geographic scale of an individual vineyard sub-block) into their management decisions—decisions that can lead to precision practices, such as segmented harvesting and differential irrigation. While this information can be useful to vineyard managers, it is important to note that without the use of geospatial technologies, many wineries in Niagara and around the world are still able to recognize portions of their vineyards that consistently produce grapes with distinctive characteristics. Grapes are often selectively treated and harvested in order to produce the best vintages with limited production, leading to an increased value of the wine and a diversified wine portfolio for the winery. However, not all vineyard variability is visually apparent or easily quantified. In addition, mapping and monitoring important vineyard variables requires a substantial amount of data; traditional methods of generating these data are both expensive and time consuming.[13] Without the appropriate tools or technology to detect, record, and measure variability in a cost-effective and timely manner, uniform vineyard management has typically prevailed.[14]

At present, the use of geospatial technologies in viticulture provides an opportunity to digitally capture vineyard characteristics, and their inherent variability, with unprecedented precision and accuracy. A detailed review of current literature suggests that the rationale for using geospatial technologies in vineyard management has increasingly been for the purpose of delineating vineyard zones of characteristic performance, especially in terms of yield and grape composition, in order to conduct zonal viticulture.[15] Zonal viticulture (also known as zonal management or targeted management) provides an opportunity to separately manage and monitor selected vineyard zones at various stages of grape production.[16] These zones are frequently identified through visual inspection of a range of vineyard maps produced largely from the use of geospatial technologies.

Mapping and Monitoring Vineyards Using Field-Based Data

Vineyard managers in the Niagara Region often collect and maintain detailed field-based data about their vineyard(s) throughout the growing season and from season to season in order to better understand, and subsequently manage,

Figure 8.1 False-colour composite image showing the spatial variability of vine vigour within vineyards located in the Niagara Region of Ontario.
This image was acquired by the QuickBird satellite on September 20, 2007. Vegetation is readily detected in this image. Vegetated surfaces in this image appear in different shades of red, depending on the type and condition of the vegetation, while non-vegetated surfaces appear in grey, blue, or black, depending on their composition.
Source: Courtesy of DigitalGlobe.

the interaction of various factors that ultimately contribute to wine quality. These records normally include information related to grape variety, year planted, acreage, and canopy management. More recently, these records have also included information about other physical, chemical, and biological factors (e.g., soil type, spray schedules, and plant condition) related to overall vine performance. Some vineyard managers, especially those working on collaborative research projects, have also conducted further analyses on harvested grapes from their vineyards.[17] Some of this information was recently compiled in a computerized map database system, the Vitis Management System; this system has become a useful planning tool that can be used to determine ideal locations for growing certain types of grapes in the Niagara Region.[18] To date, however, spatial and temporal analyses of these and other types of vineyard data have not been widely conducted, even though variability is a well-known fact in viticulture.

The adoption and use of GPS technology in precision viticulture meant that traditional field-based measurements, as described above, could be more easily linked to precise geographic locations within a vineyard. Today, GPS is commonly used to obtain precise and accurate location information (mainly latitude and longitude) from vineyards, such as the geographic location of individual vines (Map 8.1). While there are a plethora of GPS units available for daily navigation today, mid-range units (those that cost between $6,000 and $12,000 CAD) are most often required to delineate vineyard boundaries and demarcate individual vines more accurately.[19] Plotted as individual points on a map, where each point (e.g., an individual vine) has a precise unit of latitude and longitude attached to it, these data are referred to as spatially georeferenced data. These data alone are useful for producing maps of vineyard vines, sub-blocks, and boundaries. However, the power of geospatial information comes from attributing visual observations or detailed measurements (or both) of important variables (such

Map 8.1 GPS data points from Stratus Vineyards, Niagara-on-the-Lake. Map showing GPS data points obtained from three vineyard sub-blocks at Stratus Vineyards, located in Niagara-on-the-Lake, Ontario. Each point on the map represents sentinel (sample) vines within each sub-block. To provide some geographical context, these GPS data points have been superimposed onto an airborne remote-sensing image acquired in spring 2006.
Source: 2006 orthoimagery courtesy of The Regional Municipality of Niagara, Area Municipalities and their Suppliers.

as vine vigour, disease, grape composition, and soil moisture) to these geo-referenced data. With GIS mapping software, such as ESRI's ArcGIS, detailed and accurate maps of these important vineyard variables, and their spatial relationships, can be produced from GPS data.

An example of a vineyard variable that is a good indicator of subsequent grape quality is soil moisture. Soil moisture measurements obtained at sample vines corresponding with GPS points can be mapped to reveal patterns that might not be obvious through visual assessment of a vineyard, as shown in Map 8.2. While visual inspection of a soil moisture map, such as the one shown in Map 8.2, allows a vineyard manager to quickly identify zones of similar soil moisture values (e.g., deficiencies), it also permits a visual assessment of the spatial variability of soil moisture within and between different vineyard blocks. Combined with other spatial data, such as a digital elevation model (DEM), this information has the potential to improve vineyard management decisions as the interaction between vineyard components (e.g., soil moisture and local topography) are

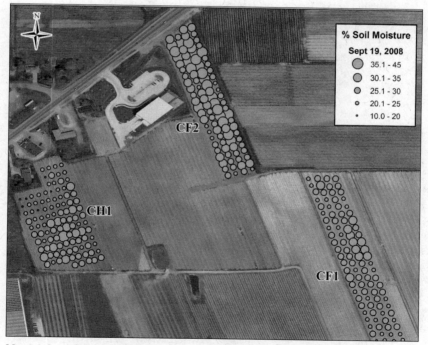

Map 8.2 Spatial variability of soil moisture at Stratus Vineyards, Niagara-on-the-Lake.
Map showing the spatial variability of soil moisture within and between three sub-blocks at Stratus Vineyards. Soil moisture data were acquired on 19 September 2008 using time-domain reflectometry (TDR), measuring volumetric water content as a percentage. Each soil moisture measurement was obtained at an existing GPS location.
Source: 2006 orthoimagery courtesy of The Regional Municipality of Niagara, Area Municipalities and their Suppliers.

better understood (Map 8.3). Further, using a GPS unit to revisit these exact locations provides an opportunity to measure and monitor changes in other vineyard variables over time. In fact, subsequent field-based data collected at these locations can be used to produce maps of a range of vineyard variables, including soil type, texture, and moisture; vine water status; grape yield; fruit composition (e.g., sugar, titratable acidity, and pH); and aroma compounds, such as monoterpenes.[20] Within a GIS environment, more sophisticated spatial data analyses may be performed on these mapped data, to develop new information about vineyards.

In the recent past, two spatial data analysis techniques have been increasingly used to develop new information about vineyards. These include spatial interpolation (i.e., the process of estimating the value of a variable at an unsampled

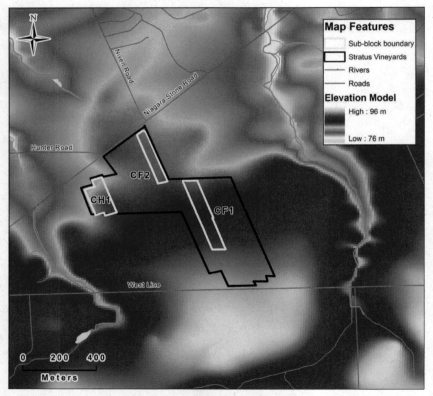

Map 8.3 Local topography surrounding Stratus Vineyards. Map showing local topography (produced from a DEM) and other features (i.e., rivers and roads) surrounding Stratus Vineyards. Using data obtained from a hand-held GPS unit, the vineyard sub-blocks and boundaries at Stratus Vineyards have been superimposed onto this map.

Source: Digital Elevation Model data courtesy of Ontario Ministry of Natural Resources under the Ontario Geospatial Data Exchange Program @ Queen's Printer of Ontario; CanMap Water and CanMap Roads courtesy of *DMTI Spatial Inc.*

location based on known values at sampled locations within the same area) and spatial clustering (i.e., the process of grouping similar values of a variable in order to divide a vineyard into zones of similar characteristics).[21] For example, using the advanced spatial data-analysis tools (e.g., spatial interpolation) commonly available in most GIS environments, field-based point-data values can be spatially interpolated to create a continuous surface (or map) across an entire vineyard sub-block (Map 8.4). These maps can then be used to visually examine the spatial variability of vineyard variables, such as soil moisture, over time. They can also be used to visually identify zones of similar soil moisture values. More importantly, automated spatial data analysis techniques (e.g., spatial clustering) can be applied to these interpolated data to produce new information, such as clusters (or zones) of similar soil moisture values, as illustrated in Map 8.5. These zones can subsequently be used to guide zonal management activities, such as differential irrigation. Given that these techniques are based on statistical calculations, as opposed to visual observations, the potential for error or misinterpretation is also reduced.

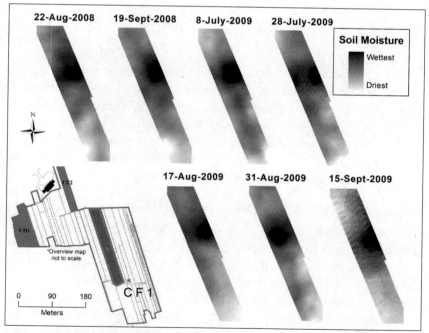

Map 8.4 Spatial and temporal variability of soil moisture at Stratus Vineyards. Map showing the spatial and temporal variability of soil moisture within a Cabernet Franc block (CF1) at Stratus Vineyards (from August 2008 to September 2009).
Source: Inset map produced by Loris Gasparotto, Cartographer, Brock University.

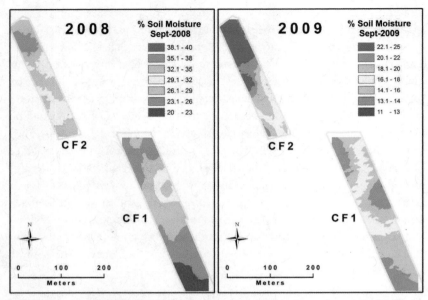

Map 8.5 Clusters of similar soil moisture values. Map showing clusters (or zones) of similar soil moisture values. This map can be used to compare the spatial and temporal variability within and between two Cabernet Franc blocks (CF1 and CF2) at Stratus Vineyards.
Source: Authors.

Using a combination of advanced spatial data analysis techniques (e.g., spatial interpolation and clustering) over different vineyard sub-blocks also provides an opportunity for vineyard managers to compare the spatial and temporal variability within and between these sub-blocks, also illustrated in Map 8.5. This is relevant since these data can provide vineyard managers with extremely valuable information that can be used to better inform vineyard management decisions. Remote-sensing technologies can also be used for this purpose.

Mapping and Monitoring Vineyards from a Distance

While the integration of GPS and field-based data within a GIS provides an excellent opportunity to map and monitor important vineyard variables, as well as analyze their variability, not all vineyard variables are visually apparent or easily quantified on the ground. An aerial view of a vineyard can provide a better perspective on vineyards and their overall condition.[22] In this context, remote-sensing technologies are particularly useful for obtaining critical spatial information about vineyard variables, especially those related to the vines themselves (e.g., vine vigour) over time. This technology can also be used to develop new information about vineyards. For example, Johnson et al.[23] segmented remotely sensed image data into uniform zones of canopy vigour in order to maximize the production of high-quality wines. Currently, geospatially based research studies aimed at improving vineyard management activities are primarily focused on the development and use of remote-sensing technologies.

Remote-sensing data are normally obtained from sensors onboard air- and space-borne platforms. In terms of their technical specifications, the spatial resolution (or pixel size) of remote sensors (a measure of the smallest object detectable on the ground) can range from 1 kilometre to 10 centimetres. In the past, satellite spatial resolutions (i.e., geographic areas of 10 metres by 10 metres on the ground) were inadequate for detailed examination of small vineyard blocks. With recent advances in remote-sensing technologies, however, much smaller pixel sizes (i.e., 10 centimetres) allow for more detailed spatial analyses of such vineyards, as shown in Figure 8.2. Onboard space-borne platforms, these sensors can usually view the same location on the Earth's surface once every three days. These sensors are often designed to measure and record sunlight reflected by objects within vineyards (such as vines, cover crops, and soil), for example, across different portions (or wavebands) of the electromagnetic (EM) spectrum. While all photosynthesizing plants, including vine canopies, strongly

Figure 8.2 Improvements in pixel size for the same vineyard block. False-colour composite images showing improvements in pixel size for the same vineyard block from (a) SPOT satellite imagery with a ten-metre spatial resolution and (b) aerial imagery with a forty-centimetre spatial resolution.
Source: (a) © [2007] CNES, Licensed by the Alberta Terrestrial Imaging Center, www.imagingcenter.ca; (b) imagery courtesy of Dr. R. Brown.

absorb incident sunlight from the visible blue and visible red wavebands of the EM spectrum, they strongly reflect sunlight from the visible green portion of the spectrum. As a result, healthy plants appear green when viewed by the human eye. However, healthy plants most strongly reflect incident sunlight (i.e., greater than 65 percent) in the near-infrared (NIR), a portion of the spectrum that is undetectable by the human eye. It is within this region of the EM spectrum that important information about plant health or condition (such as internal leaf structure, photosynthetic activity, and water content), can be obtained using these sensors.[24] In the same way that X-ray technology can be used to detect tooth decay long before that decay is noticeable to the human eye, near-infrared technology provides an opportunity to detect changes in plant condition long before those changes are detected by field scouts within a vineyard.

The use of remote-sensing technologies for mapping and monitoring vine-yards, especially the vine canopy itself, offers several advantages over traditional field-based methods of data collection, including a bird's-eye view over large geographic areas, repeat coverage, and cost-effectiveness. It can also be used as an early warning system that detects changes in plant condition as they emerge. This is significant because collecting vineyard data in the field can be time consuming, expensive, and inaccurate; therefore, such collections are usually restricted to small geographic areas.[25] Recent advances in remote-sensing technologies, combined with increased availability of such technologies and reduced costs, have resulted in a proliferation of remote-sensing studies in grape-growing environments globally.[26]

Today, the use of remote sensing to map and monitor the grapevines themselves (from plant growth and productivity to canopy development and architecture) has become particularly important to vineyard managers, given that a crucial part of vineyard management is canopy management. Vineyard managers know that plant growth and productivity are often significantly influenced by other vineyard variables, such as topography, soil type, and soil moisture.[27] Changes in these variables typically manifest themselves as changes in plant growth and productivity; in the remote-sensing community, these are normally expressed in units of leaf area (i.e., the area of leaves per unit area of ground, expressed in m^2/m^2). Similarly, changes in a plant's ability to uptake water or nutrients from the soil, perhaps due to disease or pest infestation, also manifests itself in a reduction in leaf area.[28] Remote-sensing technologies can be used to produce a number of plant growth and productivity-related indices, such as the leaf-area index (LAI), over vineyards. Quantitative measures, such as LAI, are particularly important because they allow comparisons within and between vineyards over time. Furthermore, a strong positive linear relationship has been established between field-measured LAI and remote-sensing data. Thus, digital maps of LAI can be easily produced from remote-sensing data. Once in map form, however, vineyard managers must use ground-based data (e.g., mesoclimate, topography, soil) to explain the LAI variations. To date, vineyard leaf area has been related to infestation and disease, water status, grape characteristics (such as the ripening rate), and wine quality.[29] This information can better inform vineyard management decisions, especially those related to canopy management activities such as vine pruning, shoot thinning, leaf removal, and irrigation.

Canopy management is also an important part of vineyard management since the canopy (unlike climate) is one of few variables within a vineyard manager's control. Remote-sensing data collected from the red and NIR portions of the EM spectrum are of particular interest given they provide specific information about the vine canopy, such as plant chlorophyll absorption and internal leaf structure. Typically, changes in plant condition are first detected by changes in plant reflectance from the NIR; thus, remote-sensing technology can also act as an early warning system that identifies changes in the vine canopy as they emerge. Various mathematical combinations of these two portions of the spectrum, such as vegetation indices, have proven to be extremely sensitive to vegetation cover and density. One such index, called the normalized difference vegetation index (NDVI), is related to vine canopy LAI and is known to be a very useful measure of vine vigour. These indices can subsequently be used to produce maps of vine vigour (Map 8.6). These maps are especially important since vine vigour is frequently reported to have a significant effect on fruit quality and yield.[30] Vine vigour has also been linked to wine quality. Therefore, understanding vine vigour has the potential to improve canopy management decisions. Canopy management is particularly important in vineyard zones of high vigour in order to control vine disease, grape quality, and yield, given their overall influence on wine quality.[31]

While mapping and monitoring vine vigour is an important management activity, digital maps of numerous other vineyard and fruit compositional variables are also being produced using remote-sensing technologies, given their overall influence on wine quality. Vineyard managers use these maps to identify and delineate vineyard zones of characteristic performance (e.g., yield). Maps of local canopy shape and size,[32] grape variety,[33] vine water stress,[34] and disease infestation[35] are of particular interest. The integration of maps derived from remote sensing, along with GPS and other field-based vineyard data, within a GIS environment further provides an opportunity to map and monitor important vineyard variables and their spatial and temporal variability. Visualization of these maps allows vineyard managers to respond quickly to changes in vineyard variables over relatively large geographic areas. Using this information, vineyard managers are also able to conduct zonal management.

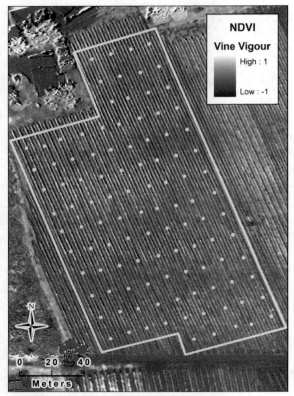

Map 8.6 Vine vigour at Stratus Vineyards. Map showing vine vigour at Stratus Vineyards. This map was produced by applying a normalized difference vegetation index (NDVI) algorithm to the red and NIR bands of QuickBird data acquired on September 20, 2007. The lightest areas on this map represent healthy vegetation, while the darkest areas represent non-vegetated surfaces, such as exposed soil or roads. Tonal variations within this vineyard block represent differences in vine health and vigour.
Source: Courtesy of DigitalGlobe.

Future Research Directions

While the rationale for using geospatial technologies to conduct zonal management within vineyards has largely remained unchanged since the Space Age Winegrowing symposium in 2001, improvements in the methods used to extract detailed and accurate spatial information from vineyards using these technologies have themselves permitted new research opportunities. New research studies are largely aimed at developing more sophisticated automated, rather than manual, approaches to detecting, delineating, and differentiating zones of characteristic performance within vineyards.[36] While the need for field-based vineyard data cannot be completely eliminated, there is increasing interest in obtaining this information from a distance. There is also a need to develop a systematic method for integrating these technologies into existing vineyard management systems for their operational use by vineyard managers themselves.

Conclusions

The purpose of this chapter was to provide a review of the value and use of geospatial technologies in viticulture for improved vineyard management. In particular, the question arose regarding to what extent these technologies, and their associated datasets, can be used to better inform vineyard management decisions. The ability to characterize vineyards and their variability was especially important in the context of grape quality and yield, given their overall influence on wine quality. To illustrate the capabilities of geospatial technologies for improved vineyard management, examples were provided from an existing collaborative research project with Stratus Vineyards.

Although geospatial studies of vineyards are relatively new, especially within the Niagara Region, there is clear evidence of their value when applied to vineyard management practices. Existing studies have shown that detailed maps can be used to identify specific vineyard zones for more precise management. This is important since these variations are known to influence grape quality and yield, and can impact wine quality. These digital technologies complement traditional field methods of data collection to better inform vineyard management decisions and they offer several advantages. In particular, they can provide a baseline dataset against which assessments of the efficiency and effectiveness of vineyard management strategies can be conducted over time. As those involved in grape growing and wine production continue to strive to better understand the interaction of several factors known to influence grape quality, the use of precise and accurate spatial information about vineyards can facilitate this understanding and potentially improve vineyard management decisions. Spatial information about vineyard variables, such as soil moisture and vine vigour, is increasingly important since vineyard management strategies can be adapted if vineyard variables, and their underlying spatial and temporal variability, are well understood. Current research suggests that the ability to identify and respond to changes in vineyards as they emerge, using geospatial technologies, could significantly influence the efficiency and effectiveness of farming operations; this could also have a positive economic impact on the wine industry. Improved vineyard management practices could also result in several environmental benefits, such as a reduction in the delivery of unwanted herbicides and pesticides to local water bodies. Overall, these technologies have the potential to significantly increase vineyard productivity, production efficiency, and profitability over the long term while minimizing the impact of farm operations on the natural environment.[37]

Notes

1 F. Bodin and R. Morlat, "Characterization of Viticultural Terroirs Using a Simple Field Model Based on Soil Depth I. Validation of the Water Supply Regime, Phenology and Vine Vigour, in the Anjou Vineyard (France)," *Plant and Soil*, no. 281 (2006): 37–54; E. Vaudour, "The Quality of Grapes and Wine in Relation to Geography: Notions of *Terroir* at Various Scales," *Journal of Wine Research* 13, no. 2 (2002): 117–41.

2 F. Meggio, P.J. Zarco-Tejada, L.C. Núñez, G. Sepulcre-Cantó, M.R. González, and P. Martín, "Grape Quality Assessment in Vineyards Affected by Iron Deficiency Chlorosis Using Narrow-Band Physiological Remote Sensing Indices," *Remote Sensing of Environment*, no. 114 (2010): 1968–86; R. Bramley, "Understanding Variability in Winegrape Production Systems 2: Within Vineyard Variation in Quality Over Several Vintages," *Australian Journal of Grape and Wine Research*, no. 11 (2005): 33–42; M. Gishen, P. Iland, R. Dambergs, M. Esler, I. Francis, K. Kambouris, R. Johnstone, and P. Hoj, "Objective Measures of Grape and Wine Quality," in *Proceedings of 11th Australian Wine Industry Technical Conference*, ed. R. Blair, P. Williams, and P. Hoj, 188–94 (Adelaide: AWITC, 2001); M. Krstic, K. Leamon, K. DeGaris, J. Withing, M. McCarthy, and P. Clingeleffer, "Sampling for Wine Grape Quality Parameters in the Vineyard: Variability and Post-Harvest Issues," in *Proceedings of 11th Australian Wine Industry Technical Conference*, ed. R. Blair, P. Williams, and P. Hoj, 87–90 (Adelaide: AWITC, 2001).

3 F. Meggio et al., "Grape Quality Assessment in Vineyards Affected by Iron Deficiency Chlorosis Using Narrow-Band Physiological Remote Sensing Indices"; E. Vaudour, V.A. Carey, and J.M. Gilliot, "Digital Zoning of South African Viticultural Terroirs Using Bootstrapped Decision Trees on Morphometric Data and Multitemporal SPOT Images," *Remote Sensing of Environment*, no. 114 (2010): 2940–50; A. Hall, J.P. Louis, and D.W. Lamb, "Low-Resolution Remotely Sensed Images of Winegrape Vineyards Map Spatial Variability in Planimetric Canopy Area Instead of Leaf Area Index," *Australian Journal of Grape and Wine Research*, no. 14 (2008): 9–17. P. Bowen, C. Bogdanoff, B. Estergaard, S. Marsh, K. Usher, C. Smith, and G. Frank, "Use of Geographic Information System Technology to Assess Viticulture Performance in the Okanagan and Similkameen Valleys, British Columbia," in *Fine Wine and Terroir: The Geoscience Perspective*, ed. R. Macqueen and L. Meinert, 137–52 (St. John's, NL: Geological Association of Canada, 2006); Bramley, "Understanding Variability in Winegrape Production Systems 2"; D.W. Lamb, M.M. Weedon, and R.G.V. Bramley, "Using Remote Sensing to Predict Grape Phenolics and Colour at Harvest in a Cabernet Sauvignon Vineyard: Timing Observations Against Grapevine Phenology and Optimizing Image Resolution," *Australian Journal of Grape and Wine Research*, no. 10 (2004): 45–54. L.F. Johnson, D.E. Roczen, S.K. Youkhana, R.R. Nemani, and D.F. Bosch, "Mapping Vineyard Leaf Area with Multispectral Satellite Imagery," *Computers and Electronics in Agriculture*, no. 38 (2003): 33–44.

4 R. Bramley and R. Hamilton, "Understanding Variability in Winegrape Production Systems 1: Within Vineyard Variation in Yield over Several Vintages," *Australian Journal of Grape and Wine Research*, no. 10 (2004): 32–45.

5 A. Hall, B. Lamb, B. Holzapfel, and J. Louis, "Optical Remote Sensing Applications in Viticulture – A Review," *Australian Journal of Grape and Wine Research*, no. 8 (2002): 36–47.

6　*The Oxford Companion to Wine,* 3rd ed., ed. J. Robinson (New York: Oxford University Press, 2006).

7　Hall et al., "Optical Remote Sensing Applications in Viticulture – A Review."

8　A. Reynolds, I. Senchuk, C. van der Reest, and C. de Savigny, "Use of GPS and GIS for Elucidation of the Basis for Terroir: Spatial Variation in an Ontario Riesling Vineyard," *American Journal of Enology and Viticulture* 58, no. 2 (2007): 145–62.

9　Bramley, "Understanding Variability in Winegrape Production Systems 2."

10　Monique Beech, "Mapping out the Future of Niagara's Grape Industry," Vines and Wine, St. Catharines *Standard,* November 26, 2010.

11　T. Proffitt, R. Bramley, D. Lamb, and E. Winter, *Precision Viticulture: A New Era in Vineyard Management and Wine Production* (Ashford: Winetitles, 2006), 8.

12　R.R. Nemani, L.F. Johnson, and M.A. White, "Application of Remote Sensing and Ecosystem Modeling in Vineyard Management," in *Handbook of Precision Agriculture: Principles and Applications,* ed. A. Srinivasan, 413–29 (New York: Haworth Press, 2006).

13　A. Hall, J. Louis, and D. Lamb, "Characterizing and Mapping Vineyard Canopy Using High-Spatial-Resolution Aerial Multispectral Images," *Computers and Geosciences,* no. 29 (2003): 813–22.

14　Bramley and Hamilton, "Understanding Variability in Winegrape Production Systems 1"; Hall et al., "Characterizing and Mapping Vineyard Canopy Using High-Spatial-Resolution Aerial Multispectral Images."

15　C. Delenne, S. Durrieu, G. Rabatel, and M. Deshayes, "From Pixel to Vine Parcel: A Complete Methodology for Vineyard Delineation and Characterization Using Remote-Sensing Data," *Computers and Electronics in Agriculture,* no. 70 (2010): 78–83. Vaudour et al., "Digital Zoning of South African Viticultural Terroirs Using Bootstrapped Decision Trees on Morphometric Data and Multitemporal SPOT Images."

16　Vaudour et al., "Digital Zoning of South African Viticultural Terroirs Using Bootstrapped Decision Trees on Morphometric Data and Multitemporal SPOT Images"; Robinson, *The Oxford Companion to Wine.*

17　J.H. Rezaei and A.G. Reynolds, "Characterization of Niagara Peninsula Cabernet Franc Wines by Sensory Analysis," *American Journal of Enology and Viticulture,* no. 61 (2010): 1–14.

18　Beech, "Mapping out the Future of Niagara's Grape Industry."

19　Proffitt et al., *Precision Viticulture,* 18.

20　Reynolds et al., "Use of GPS and GIS for Elucidation of the Basis for Terroir."

21　F. Morari, A. Castrignano, and C. Pagliarin, "Application of Multivariate Geostatistics in Delineating Management Zones within a Gravelly Vineyard Using Geo-Electrical Sensors," *Computers and Electronics in Agriculture,* no. 68 (2009): 97–107. Proffitt et al., *Precision Viticulture.*

22　Nemani et al., "Application of Remote Sensing and Ecosystem Modeling in Vineyard Management"; Hall et al., "Characterizing and Mapping Vineyard Canopy Using High-Spatial-Resolution Aerial Multispectral Images"; Hall et al., "Optical Remote Sensing Applications in Viticulture – A Review."

23　L. Johnson, B. Lobitz, R. Armstrong, R. Baldy, E. Weber, J. DeBenedictis, and D. Bosch, "Airborne Imaging Aids Vineyard Canopy Evaluation," *California Agriculture,* no. 50 (1996): 14–18.

24　Nemani et al., "Application of Remote Sensing and Ecosystem Modeling in Vineyard Management"; Hall et al., "Optical Remote Sensing Applications in Viticulture – A Review."

25　Hall et al., "Characterizing and Mapping Vineyard Canopy Using High-Spatial-Resolution Aerial Multispectral Images."

26 Meggio et al., "Grape Quality Assessment in Vineyards Affected by Iron Deficiency Chlorosis Using Narrow-Band Physiological Remote Sensing Indices."
27 Nemani et al., "Application of Remote Sensing and Ecosystem Modeling in Vineyard Management."
28 Nemani et al., "Application of Remote Sensing and Ecosystem Modeling in Vineyard Management"; L.F. Johnson, D.E. Roczen, S.K. Youkhana, R.R., Nemani, and D.F. Bosch, "Mapping Vineyard Leaf Area with Multispectral Satellite Imagery," *Computers and Electronics in Agriculture*, no. 38 (2003): 33–44.
29 Meggio et al., "Grape Quality Assessment in Vineyards Affected by Iron Deficiency Chlorosis using Narrow-Band Physiological Remote Sensing Indices"; Johnson et al., "Mapping Vineyard Leaf Area with Multispectral Satellite Imagery"; Hall et al., "Optical Remote Sensing Applications in Viticulture – A Review."
30 P.R. Clingeleffer and K.J. Sommer, "Vine Development and Vigour Control," in *Canopy Management*, ed. P.F. Hayes (Adelaide: Australian Society of Viticulture and Oenology, 1995), 7–17.
31 Robinson, *The Oxford Companion to Wine*.
32 Hall et al., "Characterizing and Mapping Vineyard Canopy Using High-Spatial-Resolution Aerial Multispectral Images."
33 P. Rodrigues Da Silva and J.R. Ducati, "Spectral Features of Vineyards in South Brazil from ASTER Imaging," *International Journal of Remote Sensing* 30, no. 23 (2009): 6085–98.
34 C. Acevedo-Opazo, B. Tisseyre, H. Ojeda, and S. Guillaume, "Spatial Extrapolation of the Vine (*Vitus vinifera* L.) Water Status: A First Step towards a Spatial Predication Model," *Irrigation Science*, no. 28 (2010): 143–55.
35 Johnson et al., "Airborne Imaging Aids Vineyard Canopy Evaluation."
36 Delenne et al., "From Pixel to Vine Parcel."
37 Some of the data presented in this chapter were acquired through a research grant to Dr. Marilyne Jollineau from the Ontario Centers of Excellence Centre for Earth and Environmental Technologies (OCE- CEET). Victoria Fast prepared the cartographic illustrations included in this chapter.

Nine

The Niagara Peninsula Appellation: A Climatic Analysis of Canada's Largest Wine Region

Anthony B. Shaw

Introduction

G rape production has always been recognized as an important economic activity in the Niagara Region, but only in recent years has this industry gained national and international recognition as a producer of quality wines. A distinctive feature of the Niagara viticulture region is the mild climate that favours the extensive cultivation of a wide range of grape varieties. For Ontario as a whole, total output from some 17.1 million vines covering approximately 6,870 hectares averaged about 56,000 tons annually, with an average farmgate value of about $78 million during the period 2007 to 2011. To date, there are approximately 500 registered grape growers and 123 estate wineries. Ontario wines originating principally from the Niagara Region and the smaller wine-producing regions of Pelee Island, Lake Erie North Shore, Prince Edward County, and the emerging South Coast region account for close to 39 percent of wine sales in Ontario between 2007 and 2011.[1] (See Figure 9.1.) Its location just north of the 43rd parallel would normally endow this region with a continental climate characterized by hot summers and cold winters. However, its position between the cooler waters of Lake Ontario to the north and the eastern end of Lake Erie to the south exposes the region to strong lake breezes that help to cool the summer temperatures. Continental Polar air masses that frequently invade the Niagara Region from the north in the winter are moderated by the relatively warm surface of Lake Ontario before arriving on its south shore. In the spring and summer, the prevailing southwesterly winds traverse the full length of Lake Erie and moderate temperatures in the southern half of the region. This study examines the topographic and climatic attributes of this cool climate wine region and provides a comparative analysis of growing season conditions with other established wine regions in Europe and North America.

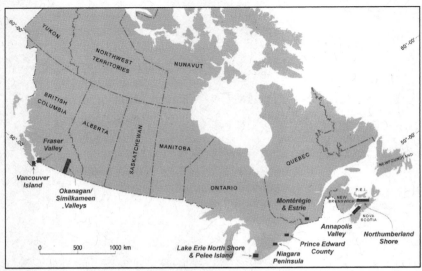

Figure 9.1 The main viticulture and fruit-growing areas of Canada
Source: Anthony B. Shaw, Department of Geography, Brock University.

Physiography and Soils

The Niagara Peninsula can be divided into three broad physiographic areas: the Lake Iroquois Plain, the Niagara Escarpment, and the Haldimand Clay Plain (Figure 9.2). These areas correspond to three broad zones in terms of their suitability for grape production. The Lake Iroquois Plain extends for a distance of 306 kilometres around the western part of Lake Ontario from the Niagara River to the Trent River. The Region's portion of this plain, commonly known as the Niagara Fruit Belt, lies between Lake Ontario and the Niagara Escarpment and includes adjacent terraces to the Escarpment collectively known as Lake Iroquois Bench. Halton Clay Till of variable thickness overlies Queenston shale bedrock in most places, while relatively deep lacustrine fine sandy loam and very fine sandy loam overlie clay till in areas adjacent to Lake Ontario known as the Tender Fruit Belt. The soils that developed over this plain range from imperfectly drained silty clay to moderately well-drained sandy loam. They possess moderately high water-holding capacities, which is a definite advantage for young vines in the drier months of July and August.

From its southern limit along the base of the Niagara Escarpment, the area declines gradually toward the Lake Ontario shoreline. Approximately 90 percent of this area has long, gentle north-facing slopes with gradients less than 3 percent; consequently, the entire area receives uninterrupted sunlight throughout the growing season. Several streams that cross this area to Lake Ontario enhance

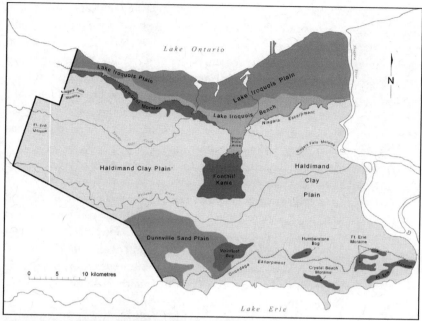

Figure 9.2 Physiographic areas of the Niagara Region
Source: Anthony B. Shaw.

the drainage of excess moisture from adjacent vineyards. During the early spring, they flow vigorously, but in the summer most become dry beds. (See Figure 9.2.)

Over 80 percent of the Niagara Region's *Vitis labrusca*, French–American hybrid, and cold-sensitive *Vitis vinifera* grape varieties are produced on this plain, owing to its relatively favourable soils and climatic conditions. The climate of this area is not homogeneous. Such factors as distance from the lake, slope, elevation, and air flow patterns have created a range of mesoclimates with varying degrees of risk and growing conditions with respect to grape production. Selection of cold-hardy varieties and suitable mesoclimates have enabled growers to reduce the incidence of freeze damage caused by late spring and early fall frosts and low winter temperatures. Nonetheless, the risk of freeze injury still remains high in the vineyards with weak slopes that are located beyond the full moderating influence of Lake Ontario, especially in the eastern section of this plain. For the plain as a whole, freeze injury to the open buds of *vinifera* vines associated with damaging temperatures below minus 2° Celsius in late April and in the first half of May can be expected at least twice yearly, while damaging temperatures below minus 20° Celsius in December, January, and February can be expected at least once yearly.

The Niagara Escarpment, the most prominent topographical feature in the Niagara Peninsula, separates the Iroquois Plain in the north from the Haldimand Plain in the south. This rock-hewn feature sits between 30 to 50 metres above the Iroquois Plain and contains predominantly north-facing slopes that range from those that are too steep for cultivation to those with less steep gradients suitable only for manual labour. The exposed areas of the brow display hard dolomite limestone formations in the upper slopes and soft red shale along the lower slopes. A number of streams descend the Escarpment through precipitous valleys and wind their way to Lake Ontario.

For some distance from the brow back, the dip-slope of the Escarpment has been stripped of its soil and overburden in some areas, thus exposing the dolostone bedrock. The soil textures include well-drained but shallow, droughty loam that requires supplemental irrigation for vines. This area also contains silty and clay loams that are poorly drained and easily compacted. However, the dominant physiographic feature between the top of the Escarpment and the Haldimand Plain is the Vinemount Moraine, a gentle rolling continuous area running parallel to the Escarpment and composed of predominantly Halton Clay Till. On nights with radiation frost, these mainly south-facing slopes provide drainage of cold air for vineyard crops that are more cold-hardy than those grown on the Lake Iroquois Plain. However, in spite of the beneficial effects of these slopes, there is a high risk of freeze injury to *Vitis vinifera* vines in those areas that are located at greater distances from Lake Ontario. In past years, growers have avoided the areas on top of the Escarpment for the production of *Vitis vinifera* varieties and instead prefer mostly the cold-hardy French–American hybrids and *Vitis labrusca* varieties. However, rising land prices due to urbanization and other competing agricultural uses such as greenhouse production on the Lake Iroquois Plain make the Escarpment lands especially attractive for new vineyards.

The Niagara Escarpment also influences the winds and temperature on the Lake Ontario Plain on a daily basis. Under prevailing winds from the southwest, the Escarpment acts as a shelterbelt, decreasing the winds on the leeward side and creating a protected zone that may vary from about 0.3 kilometres to 2.5 kilometres from the base of the Escarpment outward. The width of this protected zone is directly related to the strength of the winds and the angle at which they traverse the Escarpment. Lower wind speeds and moderate temperatures characterize this protected zone. Two terraces commonly known as the Beamsville and St. David's benches, comprised of glacial till and lacustrine sediments and collectively called the Lake Iroquois Bench are located on the face of the Escarpment and enjoy the greatest degree of protection from strong prevailing winds in winter.[2] Moreover, the moderately steep slopes of these terraces also drain the cold air with potentially damaging temperatures away from vineyards on clear, calm winter nights. In summer, the relatively high elevation combined with cool breezes moderate temperatures over this area in the warm months of July and August. Altogether, these climatic and topographic conditions make these bench lands ideally suited to the cultivation of cold-sensitive *Vitis vinifera* varieties and the production of wines that are characteristic of a cool mid-latitude climate.

The third important physiographic area occupying close to three-quarters of the Niagara Peninsula and lying between the Niagara Escarpment and Lake Erie is the Haldimand Clay Plain. This area reaches its highest elevation in the northern part closer to the Niagara Escarpment, in the vicinity of Fonthill and the Short Hills. Good air drainage attributed to the relatively steep slopes of the Fonthill area, together with well-drained, light-textured soils, make this area uniquely suited for specialty crops, such as peaches, sour cherries, apples, and *Vitis labrusca* grapes. The Plain then dips southward toward Lake Erie, becoming mostly flat; closer to the Lake Erie shore, a number of low morainic ridges and the worn-down, low, north-facing scarp of the Onondaga Escarpment are the only break in this flat plain. Extensive field crops such as corn, soya beans, and hay are cultivated on the mostly heavy, fine-textured, and poorly drained soils.

Winters are more severe and the probability of winter injury to vine crops is greater, except on the Fonthill Kame. The latter area experiences damaging temperatures below minus 20° Celsius at least twice annually, at least four times annually in the middle of the Plain, and about twice annually in the area within approximately 5 kilometres of the Lake Erie shoreline. Unlike Lake Ontario, the surface of Lake Erie remains mostly frozen in the months of January, February, and March, so it has no moderating influence on land temperatures in the coldest period of the year. Spring frosts are also a major obstacle to the production of cold-sensitive varieties that experience early bud burst, since there is a tendency for the Haldimand Clay Plain to warm up much earlier than the rest of the Peninsula. For the Plain as a whole, late spring frost in May with temperatures below minus 2° Celsius can be expected at least twice yearly. Consequently, this area has the highest potential for use of wind machines for expanding the area under cold-sensitive crops.

Cultural Practices and Grape Varieties

Growers may choose to harvest by hand or machines. Following the harvest, a common practice is to hill up (cover the graft union with earth in the late fall) around the lower trunks of *Vitis vinifera* varieties to protect them against winter damage. Dormant cane pruning is usually carried out in the months beginning around December, and may extend occasionally through to April before bud break. The occurrence of damaging temperatures in the winter and late spring frosts often requires growers to adjust or delay pruning in order to prevent significant crop loss. Moreover, growers apply several cultural methods to minimize freeze damage, such as site selection, delayed pruning, canopy management, multiple trunks, reduced crop load, and wind machines.

Vitis riparia is the chief indigenous species of *Vitis* found in the Niagara Region. It is also the most widely distributed of any North American species of grape, from east of the Rockies to the Atlantic. Its habitat extends into northern Quebec and New Brunswick, Canada, and south in the United States to northern Texas, Colorado, and Utah. *Vitis riparia* can be found growing in the wild, usually moist, and shady environments along the numerous creeks and riverbanks in the Niagara Region. A vigorous climbing vine, this species is well adapted to cold continental climates and can withstand temperatures to minus 60° Celsius. It is an early bloomer with a short growing season and is also resistant to excess water and to phylloxera. The foliage is rarely attacked by mildews or by black rot. Since it can be easily propagated from cuttings, it was originally used as a rootstock in rebuilding the European vineyards destroyed by phylloxera. In particular, the French grape growers found that good wine could be made from some rootstock varieties (Riparia Gloire, Riparia Glabre, Riparia Scribner, and Riparia Martin) on which *vinifera* grapes were grafted. This variety was never commercialized for winegrape production in the Niagara Region. Early settlers gathered the ripe berries mainly for domestic use. Those who infrequently made wines from this grape found it too sour and harsh for consumption. However, the fruit does develop considerable sweetness and is best for wine when left on the vine until after the first frost.

Unlike the wild *Vitis riparia* grapes, the grapes of *Vitis labrusca* are larger in bunch and berry size. The berry colour can be black, red, or white, and the vines can grow in the wild or be cultivated under a wide range of environmental conditions. Several growers planted a limited number of both the wild and the cultivated varieties of the *Vitis labrusca* grape and made wine on a trial basis prior to the 1840s. However, commercial grape production and winemaking began around the 1850s and grew quickly during the 1860s with the importation of the cuttings of the pure *Vitis labrusca* and *labrusca hybrids* grape varieties, such as Catawba, Isabella, Concord, Delaware, Hartford Prolific, and Diana from the Eastern United States.[3] In the 1880s, the Niagara variety, a *labrusca* hybrid developed in Lockport, New York, during the 1860s, was introduced into the Niagara Region and quickly became the premier white variety. Today, it is still the most widely planted white *labrusca* variety for the table grape and white juice markets and is only second in importance to the Concord variety. In the year 2002, these two varieties accounted for about 89 percent of the vines planted under *Vitis labrusca*. If carefully managed, these cold-hardy varieties can tolerate winter temperatures down to minus 28° Celsius. Concord is the most widely grown of all the *Vitis labrusca* grapes. This hardy, blue-skinned grape was bred in the 1840s in Concord, Massachusetts. From its introduction in the 1860s to the present, the variety has remained the dominant *Vitis labrusca* variety in the Niagara Region. It is well adapted to the soils and climate of the region, is resistant to fungi and pests, matures early, yields well, and seldom experiences freeze injury in the winter or spring. Unlike the port and sherry made from this grape, the wines have a "foxy" taste and aroma. Nowadays, only a small number of acreage under *Vitis labrusca* remains and the crop is processed mainly for jams, jellies, and juice. Other varieties of lesser importance that are also processed for jams and juice include Fredonia, Elvia, Sovereign Coronation, President, Van Buren, Dutchess, Veeblanc, and Ventura.

A shift toward French *Vitis vinifera* and *vinifera* hybrids from the hardier *Vitis labrusca* varieties began to take place on an experimental basis in the 1930s at the Horticultural Research Station in Vineland, Ontario, but it was only after the Second World War that serious attempts were made to import cuttings from France to make dry, European-style wines. Of the several Seibel hybrids (De Chaunac, Chelois, Verdelet, Chancellor) and hardy *vinifera* (Chardonnay, Johannisberg Riesling, Pinot noir) imported from France, only De Chaunac (previously known as Seibel 9495) met with immediate success; later, it was the one to be the most successfully commercialized. This variety became the most widely planted red wine variety in the Niagara Region and in the Eastern United States by the 1970s.[4] However, by the 1990s, it has gradually declined in relative importance in preference for the reds of the French *Vitis vinifera* varieties. Other French hybrids that were later commercialized and are still being cultivated include the red varieties, such as Maréchal Foch, Baco Noir, Chambourcin, and Villard Noir, while the whites include such varieties as Seyval Blanc, Vidal Blanc, and Geisenheim 318. According to the 2002 Grape Vine Census, French hybrids account for about 22 percent of the total grapevines planted in the Niagara Region. The four dominant varieties in terms of the number of vines planted are Vidal, Baco Noir, Seyval Blanc, and Maréchal Foch.[5]

Pure *Vitis vinifera* varieties imported in the 1940s and 1950s met with limited success for various reasons related to such problems as winter injury, diseases, and poor rootstock selection. Of the *Vitis vinifera* varieties imported during this period, only Chardonnay was propagated extensively. By the early 1970s, several European *Vitis vinifera,* such as Johannisberg Riesling, Gamay, and Gewürtztraminer were planted on an experimental basis at the Vineland Research Station, Ontario, and were later commercialized, but on a limited basis. However, it was the emergence of estate wineries, beginning with Inniskillin in 1975, that contributed to the rapid expansion in the acreage under French *Vitis vinifera.* This fact, coupled with the eagerness of the growers and winemakers to match the Region's wide range of soils and micro-climates with the most suitable grape varieties, has not only helped to diversify the varietal selection, but has also pushed their production into other areas of the region previously considered climatically marginal. By the year 2004, the number of estate wineries totalled about fifty-six, while the number of *vinifera* vines amounted to over nine million, just over 58 percent of the total number of vines planted in the region. The dominant white varieties include Chardonnay, Johannisberg Riesling, Sauvignon Blanc, Gewürztraminer, Pinot Gris, Auxerrois, and Geisenheim clones, while the reds include Cabernet Franc, Merlot, Cabernet Sauvignon, Pinot noir, Gamay Noir, and Zweigeltrebe.

Regional Climate

As is characteristic of the mid-latitude climate, the weather in the Niagara Region is highly variable. The area is located in the middle zone of the North American continent where tropical and polar air masses, along with high- and low-pressure systems, govern the daily weather. Beginning in the fall, polar and Arctic air masses sweep into the Great Lakes Region and may bring a diversity of weather conditions as they cross the Niagara Peninsula. All the growing conditions that may culminate in producing a vintage year can deteriorate quickly in September owing to cool, wet weather during the harvest. With the onset of fall, the frequency of moving cyclones increases as they push northeast out of the American Midwest and east from Alberta. The Niagara Region lies directly under a major storm track running northeast from the Midwest, up through the Ohio Valley, over the Great Lakes Region, and then through the St. Lawrence Valley. Occasionally, during the winter, warm, moist, and unstable subtropical air streams may be swept northeast by cyclonic storms that move out of the Midwest. This is common around the third week of January and is often referred to as the January thaw. Along the warm front and north of the centre of low pressure, a mixture of weather conditions, ranging from fog through freezing rain to snow, prevail. Cold outbreaks associated with Arctic air masses often follow on the heels of these warm spells and may result in major damage to the dormant buds of *Vitis vinifera* vines, as happened in January 2003 and 2004. Precipitation in the form of snow can be enhanced considerably as cold, dry winds traverse the Great Lakes. With respect to the Niagara Region, the configuration of topography and the orientation of the Lakes to the prevailing winds are such that the resulting snowfall tends to be of light to moderate intensity, bordering narrow areas along the Lake Erie and Lake Ontario shores.

The spring weather is highly variable and very turbulent, with high wind speeds, frequent cyclonic systems, and alternating warm and cold fronts marking the transition from winter to spring. Cold spells interspersed with warm spells are especially common in the months of March and April as migratory cyclones move through the region in response to large-scale temperature differences between the tropical and polar regions. Occasionally, high-pressure systems accompanied by clear, calm conditions and the possibility of radiation frosts are common in early May. Consequently, these are often the most dangerous periods for the dormant buds of early-season *Vitis vinifera* vines, as they are likely to open in response to warm temperatures after having completed their endo-dormancy or rest period.

Summer weather is dominated by frequent high-pressure systems and maritime tropical air masses from the Gulf of Mexico bringing warm, sunny days, but often humid conditions. Isolated convective systems caused by strong solar heating of the land mass produce most of the precipitation in the months of July and August. Occasionally, cold fronts move through the area, causing heavy and widespread precipitation, but leave behind much-needed moisture for the vineyards. The cool, refreshing dry air keeps humidity levels low and fungal diseases in check. As the warmest month, July's mean daily temperatures range from 20.9° Celsius to 22.2° Celsius and the mean daily maximum varies between 25° Celsius and 27° Celsius over the region.

Climatic Limitations

Notwithstanding the certainty of favourable winter temperatures below minus 10° Celsius that enable the production of its internationally recognized Icewine, a major challenge to commercial viticulture involving *Vitis vinifera* varieties in this region is the high risk of freeze injury resulting from late spring and early fall frosts together with the occurrence of extreme minimum temperatures in the winter. Even the cold-hardy varieties such as Cabernet Franc, Chardonnay, and Riesling are likely to suffer damage to the primary buds when temperatures fall below minus 23° Celsius, while early budding *Vitis vinifera* varieties such as Merlot and Pinot noir are especially susceptible to late spring frosts in the latter half of April and early May. Furthermore, two characteristics of this climate place viticulture at risk. The first relates to the high degree of variability in the daily weather, especially in the months of March and April, marking the transition from winter to spring, and from the summer to autumn in September and October. In these periods, warm spells followed quickly by cold spells can be quite common, subjecting the vines to great physiological stress, including physical damage to the primary buds and canes. In winter, rapid temperature fluctuations and the occurrence of extreme temperatures often below minus 20° Celsius are major causes of freeze damage to the primary buds of *Vitis vinifera* varieties and often lead to splitting in the woody tissue and subsequent infection by "crown gall" (*Agrobacterium tumefaciens*). A second feature of this climate that affects grape yield and wine quality and subjects the industry to a high degree of uncertainty is the year-to-year variability in the growing season conditions.

In some years (e.g., 1995, 1998, 2001, and 2002), the summer and fall temperatures are warmer than normal, allowing the late-season varieties to reach their full ripening potential, producing wines of superior quality. In other years (e.g., 1996, 1997, 2002, and 2003), the growing season can be short and cool, with just enough heat units to ripen the early season varieties, or spring and summer growing conditions that may lead up to a vintage year are often dashed by cool, damp weather during the fall harvest.

The Influence of the Lakes

Lake Ontario

The Niagara Peninsula is bounded by the deep waters of the Niagara River in the east, Lake Ontario in the north, and Lake Erie in the south. What should have been a mid-latitude continental climate with hot summers and cold winters located in the heart of the Great Lakes Basin is instead a semi-maritime climate moderated by these deep and expansive bodies of water. Although both lakes temper the land temperature year round, it is the open waters of Lake Ontario in the winter that enable viticulture and tender-fruit production on its southern shore to be successful. Lake Ontario has the smallest surface area of any of the Great Lakes (19,000 kilometres²), but its average depth (86 metres) ranks second only to Lake Superior. The surface area and depth give the lake a large heat storage capacity that responds slowly to winter air temperatures. By comparison, it has the smallest percentage of ice cover of any of the Great Lakes. The extent of ice cover is a good indication of the potential for freeze injury of grape vines. The area covered by ice can vary yearly. In a normal year, ice cover is confined mostly to the northeastern area of the lake and occupies a total of about 24 percent of the lake area by the first half of February, principally in bays and harbours. Break-up normally starts in late February, with the lake becoming generally open water early in April. Mean surface water temperature hovers at just above 0° Celsius. In a mild winter, ice coverage is about 10 percent, while in a severe winter it can increase to 95 percent. Only three times in the last hundred years has Lake Ontario approached a nearly complete ice cover.[6] Loss of open water, as in 1979, means that the lake behaves like a land surface over which cold and dry Arctic air, often with temperatures below minus 30° Celsius, invades the Niagara Region, with disastrous consequences for the vines.

The lake enters its unstable season when the surface water temperature is higher than the surrounding land and air temperatures. This may typically begin in October, with a surface temperature of about 12° Celsius and cooling to just above 0° Celsius by the end of March. In its unstable season, a warmer lake will delay the onset of autumn frost and extend the growing season of the areas close to it. During this period, with potentially damaging winter temperatures of below minus 20° Celsius, Lake Ontario can also moderate the temperatures in the Niagara viticulture area in three principal ways. First, when winds are out of the north or the northeast, the relatively warmer temperature of the water surface can raise the air temperature several degrees above the damaging threshold temperatures for *Vitis vinifera* vines before finally reaching the south shore of lake. On these occasions, the moderating influence of the lake is felt over a greater area of the region's climate. Second, on clear, calm nights, temperature differentials between the warmer air over the lake and the cold air over the land set up weak circulations that draw the warmer air inland. However, the full beneficial effects of these closed circulation systems tend to be restricted to zones seldom more than a few kilometres from the shoreline. Third, the large band of cloud cover that frequently develops along the lakeshore zone due to the evaporation of moisture from the warm lake may keep the daytime temperature cool but prevents night-time temperatures from falling to the danger zone for vines. Precipitation in the form of lake-effect snow may also help to raise the minimum temperature and insulate the lower trunks of vines from damaging temperatures.

From about early April to the end of September, the surface temperature of the lake is cooler than the surrounding land temperature. During this period, the lake is said to be in its stable season. The average surface temperature rises gradually from about 2° Celsius in April to about 18° Celsius by the end of September. In its stable stage, cooler winds off the lake will delay the warming in vineyards located in the lakeshore zone and allow buds to burst after the danger of late spring frost is past. As the growing season progresses toward the warmer and drier months of July and August, the land mass around the lake becomes a major source of sensible heat that rises in large convectional currents. This process lets cooler and drier air from the lake to reach vineyards inland, slows down the ripening process and lowers the humidity level in the canopy as well as the incidence of fungal diseases.

Lake Erie

Lake Erie is the shallowest, most southerly and warmest of the Canadian Great Lakes. Bordered by an area of low relief, this body of water is the chief catalyst for the fledgling wine industry in southwestern Ontario. Its maximum depth is 64 metres, with an average depth of 19 metres. The greatest width measures 80.5 kilometres and has a surface area of about 26,667 kilometres^2. Lake Erie's expansive surface and its huge reservoir of heat derived from the warm summer help to prolong the growing season into the late fall for grape-growing areas located on its Canadian north and US south shores.

The area covered by ice can vary yearly, depending on the severity of the winters. Ice formation usually begins in the shallow western end of the lake and in Long Point Bay during the third week of December. However, ice has formed as early as the last week of November and as late as the third week of January. Lake Erie is normally ice covered by late January until early March. In a mild year such as 1998, ice will cover approximately 25 percent of the surface, while during a severe winter, 100 percent coverage can occur.[7]

Break-up normally begins close to early March in the western end. By the third week of April, the lake becomes mostly open water. The deeper eastern end of the lake, which borders on the Niagara Peninsula viticulture area, is usually the last area to be ice free, while at the western end, the process is accelerated by strong southwesterly winds that drive the melting ice floes eastward. By the end of May, the mean surface temperature in the western section doubles to about 12° Celsius, twice the level found in the eastern section. Progressive warming occurs throughout the summer months, with temperatures finally peaking at about 24° Celsius in the eastern section, while the rest of the lake's mean surface temperature averages about 22° Celsius.[8]

From about the end of March to the end of August, the lake temperature is generally cooler than the surrounding land temperature. During this period, the prevailing warm winds from the southwest moving over the cooler lake surface tend to moderate temperatures in the southern half of the Niagara Region, especially in the near-shore area.

The lake enters its unstable season when the surface water temperature is warmer than land and air temperatures. This period begins from about September and extends to about late January, by which time the lake surface completely freezes over in most years. In the western section, the process begins earlier and ends earlier, while in the eastern section it is delayed. Mean surface water temperature declines gradually from about 20° Celsius at the end of September to 14° Celsius at the end of October and rapidly declines from about 10° Celsius in November to about 4° Celsius in December. In its unstable season, a warmer lake will delay the onset of fall frost and extend the growing season of the areas close to it. In early winter, as cooler or cold air moves over the warmer lake surface, it gains energy and moisture by evaporation. The greater the distance traversed by the air, the more heat and moisture it is likely to gain. On reaching the northern and eastern shore areas, topographic uplift of this moisture-laden air results in heavy precipitation in the form of lake-effect snow.

Growing Season Conditions

Frost-Free Days

The beginning of the frost-free period for grape production is typically defined as the number of consecutive days when the critical minimum temperature is above minus 2° Celsius at the bud burst stage. Fully emerged shoots or actively growing grapevine tissues will rarely withstand temperatures below this level for 0.5 hours or more. The frost-free period ends in autumn with the first occurrence of critical minimum temperatures at or below minus 2° Celsius, which could result in freeze damage to mature leaves and consequently terminate photosynthetic activity. According to Howell, it is at these temperatures that extra cellular freezing is more likely to result in damage to actively growing tissues caused by dehydration of the cell and the loss of membrane integrity.[9] It should be noted that these temperature thresholds are markedly different from those that are normally defined for other agricultural crops, for which the beginning of the frost-free period is normally described as occurrence of temperature below 0° Celsius and ending with temperature at or below 0° Celsius. Published climatic data on frost-free days based on the latter threshold values are likely to underestimate the length of the frost period for grape production. The term "killing frost" is often used when the low temperature results in freeze damage to grape vine tissue in order to distinguish it from a mere frost that has no noticeable adverse effect on the vegetation and the amount of the yield or fruit quality. Therefore, a more realistic way of defining the end of the growing season should include both the temperature value and the evidence of freeze damage to the canopy to the extent of terminating all photosynthetic activity. Experience has shown that freezing temperatures at minus 2° Celsius for a few hours in the fall, as was the case on October 8, 2001, are not always lethal to the grapevine. Climate data that describe the frost-free days, especially in newly emerging viticulture areas, should be used cautiously when applied to the growing season potential for grape production. A frost-free period of less than 180 days is preferable, especially for late-season varieties such as Riesling, Cabernet Sauvignon, and Cabernet Franc. Early-season varieties (Merlot, Pinot noir, and Chardonnay) could reach full maturity in a season of 160 to 170 frost-free days.

Table 9.1a shows the number of frost-free days calculated from thirty years of daily climatic data for four representative sites in the Niagara Region. Two threshold values are used to provide a comparison of the number of frost-free days and the dates of the first and last occurrences of killing frosts. With respect to general agricultural risk, the average length of the frost-free period based on a threshold value of 0° or lower Celsius ranged from 166 days at Welland to 182 days at Port Colborne. In 90 percent of the years, the last day of frost ended on dates ranging from May 1 at Port Colborne to May 19 at Welland, while in 90 percent of the years, the first fall frost dates ranged from October 30 at Welland to November 11 at Port Colborne. Early to mid-season *Vitis vinifera* and French–American hybrid varieties should achieve maturity within these limits, provided daytime temperatures and sunshine hours are adequate for photosynthesis.

With respect to viticultural risk, the estimated average length of the frost-free period using a threshold value of minus 2° Celsius or lower ranged from 185 to

Table 9.1a Dates of the last spring frost in a thirty-year period

Stations	Mean Frost-Free Period <-1°C	Shortest <-1°C	Longest <-1°C	Earliest Last Spring Frost <-1°C	Latest Last Spring Frost <-1°C	50% >-1°C	90% >-1°C	Mean Frost-Free Period <-2°C	Shortest <-2°C	Longest <-2°C	Earliest Last Spring Frost <-2°C	Latest Last Spring Frost <-2°C	50% <-2°C	90% <-2°C
Vineland	176	147	196	Apr 12 (1991)	May 20 (2001)	Apr 24	May 7	194	152	220	Apr 4 (1991)	May 8 (1974)	Apr 17	Apr 25
Niagara District Airport	179	154	213	Apr 11 (1991)	May 20 (2002)	Apr 24	May 8	198	154	238	Mar 31 (1988)	May 7 (1974)	Apr 15	Apr 25
Welland	166	134	203	Apr 11 (1984)	May 26 (1998)	May 2	May 19	185	149	216	Apr 2 (1991)	May 18 (1997)	Apr 23	May 15
Port Colborne	182	152	207	Apr 17 (1995)	May 10 (1983)	Apr 25	May 1	201	169	225	Mar 25 (1998)	May 1 (1978)	Apr 15	Apr 22

Source: Anthony B. Shaw, Department of Geography, Brock University.

Table 9.1b Dates of the first fall frost in a thirty-year period

Stations	Earliest First Fall Frost >-1°C	Latest First Fall Frost >-1°C	50% >-1°C	90% >-1°C	Earliest First Fall Frost <-2°C	Latest First Fall Frost <-2°C	50% <-2°C	90% <-2°C
Vineland	Sept 23 (1974)	Nov 6 (1973)	Oct 21	Nov 1	Oct 4 (1974)	Nov 15 (1978)	Oct 28	Nov 7
Niagara District Airport	Oct 5 (1981)	Nov 19 (1988)	Oct 20	Nov 7	Oct 4 (1987)	Nov 24 (1988)	Oct 28	Nov 12
Welland	Sept 23 (1974)	Nov 5 (1982)	Oct 14	Oct 30	Oct 3 (1999)	Nov 15 (1975)	Oct 26	Nov 9
Port Colborne	Oct 4 (1987)	Nov 12 (1985)	Oct 25	Nov 11	Oct 3 (1999)	Nov 25 (1985)	Nov 3	Nov 16

Source: Anthony B. Shaw, Department of Geography, Brock University.

201 days for the representative stations. The shortest period ranged from 149 to 169 days and the longest period ranged from 216 to 238 days. In 90 percent of the years, the last spring frost dates ended on April 22 at Port Colborne and May 15 at Welland, while the last dates of fall frost ranged from November 7 at Vineland to November 16 at Port Colborne (Table 9.1b).

In recent years, wind machines have succeeded in extending the frost-free season at certain locations by as much as two weeks. Under clear, calm conditions, wind machines can protect the vine and crop from freeze damage that would normally be caused by a late radiation frost in latter half of April and early May or an early fall frost in the latter half of September and early October. The tower-mounted fans offer excellent advantage for cold protection by mixing the cold air layer at the canopy level with warmer air drawn from an upper inversion layer on a night with radiation frost. Shaw found that close to 90 percent of the frosts that occur in the late spring and early fall are of the radiation type and can be prevented by the use of wind machines.[10]

Heat Units

In addition to freeze injury risks, the accumulated total heat units and its distribution during the seven-month growing season (April 1 to October 31) are also major influential factors on the growth potential of wine grapes in the climatic regions that are characterized by cold winters and relatively short warm summers. Although April 1 normally marks the beginning of the growing season, heat units or biologically effective day degrees above a base of 10° Celsius begin to accumulate only in the latter half of this month (Table 9.2). April's daily weather is highly variable, marked by frequent passages of warm and cold fronts. Open buds of early-season varieties are most susceptible to freeze injury during this month, especially when a cold spell follows a prolonged period of warm temperatures, as in April 27, 2002. At locations away from Lake Erie and Lake Ontario, the middle of May marks a steep rise in heat units, peaking in July, declining gradually in the latter half of August, and followed by a sharp decrease by the middle of September. Although ideally suited for early season varieties, the risk of late spring frost remains high. Vineyards that lie within close proximity of the lakes experience a gradual rise in heat units noticeable toward the end of May, delaying bud break by as much as a week due to the cooler temperatures. Heat units peak in July but decline gradually to mid-October, allowing mid-and late-season varieties to achieve a greater ripening potential.

As is typical of a middle-latitude climate, the photosynthetic active temperature and the daily accumulation of heat units in the spring (April–May) are highly variable due to passage of frequent cyclonic systems and their associated warm and cold fronts. In summer (June, July, and August), numerous high-pressure systems accompanied by weak winds and bright sunshine dominate the

**Table 9.2. Growing season conditions for the
Niagara Peninsula viticultural area based on normalized data**

Month	Vineland Temp C°	Growing degree days (GDD)	Niagara District Airport Temp C°	GDD	Welland Temp C°	GDD	Port Colborne Temp C°	GDD	Precip (mm)	Sunshine (Hours)
Jan.	−4.0	0.2	−4.1	0.5	−4.7	0.2	−4.2	0.0	63.8	88.9
Feb.	−3.3	0.1	−3.6	0.1	−4.3	0.0	−3.8	0.0	55.7	97.3
Mar.	1.1	5.5	1.1	5.8	0.7	4.6	0.7	1.3	70.7	144.8
Apr.	7.1	23.1	7.2	24.7	6.9	25.3	6.3	16.9	74.6	180.6
May	13.4	121.7	13.6	126.0	13.4	123.3	12.9	108.1	74.7	229.7
June	18.8	264.3	18.8	264.4	18.6	258.3	18.3	250.0	80.6	263.9
July	21.9	372.3	21.8	365.6	21.4	354.6	21.6	358.4	79.7	286.4
Aug.	21.0	338.7	20.8	335.3	20.6	327.7	21.1	344.8	74.2	246.1
Sept.	16.9	208.1	16.6	199.9	16.3	190.5	17.0	212.3	88.8	176.6
Oct.	10.6	63.9	10.3	59.2	10.0	55.1	10.9	69.8	70.1	143.1
Nov.	4.9	10.8	4.6	10.6	4.3	9.8	5.0	10.6	79.3	83.3
Dec.	−0.8	1.1	−1.1	1.3	−1.5	1.2	−0.8	0.8	74.5	64.2
Year	9.0	408.9	8.8	1393.4	8.5	1350.6	8.8	1372.1	886.6	2004.9
*		1390		1375		1334		1360		

*April–October

Source: Anthony B. Shaw, Department of Geography, Brock University.

summer weather. Like the spring weather, the growing conditions in the months of September and October can also vary on a daily basis due to the movements of warm and cold fronts under the influence of the mid-latitude jet stream. In a year when the jet stream stays north of the Great Lakes, the entire region enjoys warmer than normal weather, allowing the mid- and late-season varieties to reach their full ripening potential. In other years, early movement of the jet stream south into the Great Lakes brings cooler, wet weather requiring an early harvest in order to prevent degradation of the grape quality. Consequently, it is the transitional periods during the spring and fall that can create a vintage year or a year with mediocre wines. In some years, spring can be warm and dry, favouring early bud burst (1998, 1999, and 2002), and cool and wet in others (1996, 1997, and 2003), while summer conditions can be warm and dry in some years (1999, 2001, and 2002), and cool and wet in other years (1996, 2003, and 2004).

Heat units or biologically effective day degrees for *Vitis vinifera* are normally calculated using a mean daily temperature above a base of 10° Celsius, below which photosynthesis and vine growth are minimal.[11] Estimated heat units for the seven-month growing season based on mean temperatures for the period 1971 to 2002 for four representative locations in the Niagara Region range from 1350 at Welland to 1392 for Vineland (Table 9.2). In 1995, the Grape Growers of Ontario established twelve climatic stations at representative mesoclimates within the main viticultural area situated between the Niagara Escarpment and Lake Ontario. Mean values at these locations for the period 1995 to 2004 ranged from 1588 to 1366 growing degree days (GDD). In the warm years (1999 and 2002), values reached a high of 1700 GDD at some locations, while in the cool years (1997 and 2000), values fell below 1300 GDD. Given an accumulated average total heat units of about 1400 GDD and a Jackson and Cherry latitude-temperature index (LTI) of 362, the Niagara Peninsula can be characterized as a cool-climate wine region.[12] As a general rule, cool climate districts are those that ripen grapes with LTI below 380. Overall, total heat values suggest that the entire viticulture area has the potential to ripen a wide range of European cool- to warm-climate varieties, including the early- to late-season reds (Pinot noir, Gamay Noir, Merlot, Cabernet Sauvignon, and Cabernet Franc) and whites (Chardonnay, Sauvignon Blanc, Chenin Blanc, Gewürztraminer, and Riesling). While total heat units are adequate to ripen most French whites on a consistent basis, only in the warm years are the noble red varieties such as Cabernet Sauvignon and Cabernet Franc likely to achieve their full ripening potential.

Notwithstanding its status as a cool-climate wine region, daily maximum temperatures in the months of July and August can often exceed 30° Celsius, as exemplified in the years 1999, 2002, and 2005. While total heat units may appear impressively high, during these periods with extreme temperatures, photosynthetic activity is typically greatly reduced and vines may require supplemental irrigation in order to avoid moisture stress. Practical experience seems to point to an optimum temperature in the vicinity of 20 to 22° Celsius for most wine grapes, but a range of 23 to 25° Celsius gives potentially the fastest growth.[13]

A comparison of total heat units with those of established wine regions in Europe and the US regions (Table 9.3) shows values comparable to those of Bordeaux and Burgundy. The main difference relates to the distribution of the heat units during the growing season. Although the two Great Lakes (Lake Ontario and Lake Erie) moderate growing season temperatures in the Niagara Region, the climate remains semi-continental, characterized by temperature extremes and year-to-year variability. In the case of Bordeaux, the Atlantic Ocean—in particular, the Gulf Stream—exerts a more pronounced influence on growing-season temperatures of this area. The Gulf Stream ushers in a warmer spring but tempers summer temperatures and delays cooling in the fall. The Niagara Region has warm summers (June, July, and August) in comparison to Prosser (US), cool springs comparable to Summerland (British Columbia, Canada) and Reims (France), and mild falls comparable to Dijon (France).

Table 9.3. A comparison of growing season conditions for the Niagara Peninsula viticultural area with established wine regions

Average Daily Temperature (C°)

Locations	April	May	June	July	August	September	October	April– October
Niagara	7.2	13.6	18.8	21.8	20.8	16.6	10.3	15.9
Geisenheim	9.7	14.2	17.2	18.6	18.1	15	9.7	14.6
Bolzano	13.1	16.6	20.9	22.4	21.4	18.6	12.8	18
Bordeaux	11.7	15.4	18.3	20.5	20.5	18.3	13.8	16.9
Dijon	10.3	14.5	17.6	19.6	19	16.1	10.9	15.4
Reims	10.1	13.4	16.9	18.8	18.2	15.2	10.4	14.7
Colmar	10.4	15	18.2	20.1	19.2	15.7	10.8	15.6
Prosser	10.7	14.8	18.2	21.6	20.7	16.9	11.2	16.3
Napa	13.1	15.2	18.1	19.1	18.9	18.4	16.3	17
Summerland	9	13.6	17.4	20.5	20.2	15	8.8	14.9

Average Growing Degree Days

Locations	April	May	June	July	August	September	October	April– October
Niagara	23	122	264	372	339	208	64	1392
Geisenheim	13	129	217	267	250	150	16	1042
Bolzano	92	203	270	279	279	258	87	1468
Bordeaux	51	167	249	279	279	249	118	1392
Dijon	20	140	228	279	279	183	35	1164
Reims	14	105	207	273	254	156	22	1031
Colmar	24	155	246	279	279	171	32	1186
Prosser	27	149	254	279	279	207	46	1241
Napa	93	161	243	279	276	252	195	1499
Summerland	27	117	221	324	316	153	30	1188

Average Bright Sunshine Hours

Locations	April	May	June	July	August	September	October	Total
Niagara	181	230	264	286	246	177	143	1527
Geisenheim	177	226	228	233	214	159	96	1333
Bolzano	192	177	213	236	211	183	149	1361
Bordeaux	185	205	227	257	250	205	143	1472
Dijon	175	212	241	258	242	192	129	1449
Reims	175	217	224	228	197	178	118	1337
Colmar	168	206	230	242	226	179	118	1369
Prosser	204	236	255	298	267	177	121	1558
Napa	270	310	339	347	319	270	263	2118
Summerland	204	248	263	312	281	219	155	1682

Table 9.3. A comparison of growing season conditions for the Niagara Peninsula viticultural area with established wine regions

Average Precipitation (mm)

Locations	April	May	June	July	August	September	October	Total
Niagara	68	74	81	80	74	89	70	536
Geisenheim	36	41	53	53	53	46	51	333
Bolzano	60	76	68	84	77	72	65	502
Bordeaux	67	65	60	52	47	55	81	427
Dijon	50	55	69	62	61	54	78	429
Reims	47	54	53	67	58	42	67	388
Colmar	39	52	53	60	53	50	48	355
Prosser	14	15	17	4	5	9	19	83
Napa	41	15	5	1	2	8	37	109
Summerland	26	36	36	30	31	20	18	197

Source: Anthony B. Shaw, Department of Geography, Brock University.

Precipitation and Soil Moisture

Classified as a humid continental climate, this viticulture area has abundant precipitation that is almost evenly distributed throughout the year. Annual totals for stations in the Niagara Region range from 807 millimetres to 995 millimetres. The totals for the seven-month growing season range from 534 to 578 millimetres and are considered adequate for this cool-climate wine region. Winter snow and spring rains ensure that the soils are at field capacity at the beginning of the growing season in April. Monthly precipitation during the growing season averages about 77 millimetres, with the maximum value occurring in September and the minimum value in October. The area receives precipitation from several convective and frontal systems throughout the growing season. However, isolated convective systems produce most of the moisture during the warm months of July and August. Consequently, moisture tends to be unevenly distributed over this region during these months. The upper soil layer can become droughty in the warm months of July and August due to high evapotranspiration rates and lower rainfall amounts. Young vines are likely to experience moisture stress during this time and may require supplemental irrigation. With the onset of fall, frontal systems move through the region with greater frequency, recharging soil moisture and providing an equal distribution of moisture over the viticulture area by the end of September. However, precipitation dips in October to its lowest level in the growing season, and often summer-like conditions may prevail owing to the dominance of high-pressure systems that bring clear, calm, sunny days and cool, dry nights. Vineyards that are drenched by September rains dry off in this month before the final harvest.

Snowfall in the Niagara Region can be described as light to moderate. It helps to recharge soil moisture and insulates the newly planted vines and the lower trunk of mature vines from potentially damaging temperatures. Total seasonal accumulations average about 144 centimetres and range from a low of 100 centimetres for inland locations to a high of 177 centimetres for locations adjacent to and downwind from Lake Erie. The relatively high amounts in the south and southwestern portions of the region are due primarily to lake effect. Snowfall produced by lake effect is most frequent in early winter, when the lake surface temperature is relatively mild. The Lake may remain active as long as the greater portion of its surface remains ice free. The Niagara Region has an average of thirty-seven days with measurable snowfall. The highest number of days with snow occurs in January (eleven), followed by February (nine), December (eight) and March (six). Light snowfall is recorded in the latter half of November and the first half of April, averaging three days and one day respectively.

The area below the Niagara Escarpment with the largest area under grape production has a lower snowfall accumulation total in comparison to the area above the Escarpment. This difference is attributed partly to the location of the Niagara Peninsula downwind from Lake Erie. The entire region is under the influence of the prevailing southwesterly winds, which on crossing Lake Erie in the winter months would supplement the snow produced by frontal uplift. This lake-effect snow diminishes northward as the winds cross the Niagara Peninsula. Snowfall induced by Lake Ontario on its southern shore occurs less frequently in the winter. However, heavy snowfall can be produced at times by cold, dry Arctic winds advecting from the northwest and occasionally from the north and northeast across the lake surface. Of the two lakes, Lake Ontario has the greater potential to induce snowfall because of the greater reservoir of heat stored in it and a larger surface that remains ice free throughout the winter. Yet the actual lake-effect snow in the northern areas of the Niagara Peninsula remains low because the optimum over-water trajectory conditions seldom occur.

Conclusions

Having the largest area under viticulture in Canada, a semi-continental climate moderated by the Great Lakes, along with a wide range of mesoclimates, topographies, and soils, are some the important distinctive characteristics of the Niagara Peninsula viticultural area. Although grape production goes as far back as the 1840s, starting with *Vitis labrusca*, it is only in the last thirty years that growers have begun to switch on a commercial scale to the cultivation of the *Vitis vinifera* and French–American hybrid varieties for table wines. Today, approximately 80 percent of the vines planted in the region consist of European and French–American hybrids. The grapes are vinified by approximately fifty-six estate wineries in a wide range of styles for the national and international markets. The grapevine is cultivated across the region over a variety soils and mesoclimates, ranging from the most suitable to marginal. By tradition, the preferred areas for the production of the cold-sensitive *Vitis vinifera* varieties are the Lake Iroquois Bench and the Lake Iroquois Plain, situated below the Niagara Escarpment. Here, the year-round open water of Lake Ontario moderates land temperatures, creating a favourable climate for the production of the several cool-climate grape varieties. However, recurrent freeze injury to *Vitis vinifera* vines from late spring frosts—and, especially, extreme winter temperatures—remain the major climatic limitations. In addition to site selection and matching the variety to the site climate in order to reduce freeze injury, some growers are now resorting to the use of wind machines for freeze protection. This method not only permits the cultivation of a wider range of cold-sensitive varieties, but also allows production in other mesoclimates previously considered marginal.

With respect to the growing season conditions, the region has an average of approximately 1400 growing degree days, 536 millimetres of precipitation, 1527 hours of bright sunshine, and a mean July temperature of 21.8° Celsius. These values are broadly comparable to those of France's Bordeaux and Burgundy regions. However, the choice of grape varieties and wine styles, and the cultural practices of winegrowers, are more reflective of the Niagara Region's unique topographic and climatic characteristics.

Notes

1 Grape Growers of Ontario, *64th Annual Report and Financial Statements for the year ending January 31st, 2012* (Vineland Station, ON: Grape Growers of Ontario, 2004).

2 S.J. Haynes, "Geology and Wine 2: A Geological Foundation for Terroirs and Potential Sub-Appellations of Niagara Peninsula Wines, Ontario, Canada," *Geoscience Canada* 27 (2000): 67–87.

3 W.F. Rannie, *Wines of Ontario* (Lincoln, ON: W.F. Rannie, 1978).

4 T. Aspler, *Vintage Canada* (Scarborough, ON: Prentice-Hall, 1978).

5 Grape Growers of Ontario, *56th Annual Report and Financial Statements for the Year Ending January 31st, 2004.*

6 Canadian Ice Service, *Lake Ice Climatic Atlas, Great Lakes, 1973–2002* (Ottawa: Ministry of Public Works and Government Services of Canada, 2004).

7 Canadian Ice Service, *Lake Ice Climatic Atlas, Great Lakes, 1973–2002.*

8 A. Saulesleja, *Great Lakes Climatological Atlas* (Unpublished report) (Ottawa: Meteorological Service of Canada, 1986).

9 G.S. Howell, "Grapevine Cold Hardiness: Mechanism of Cold Acclimation, Mid-Winter Hardiness Maintenance, and Spring Deacclimation," in *Proceedings of the American Society for Enology and Viticulture, 35–48*, ed. J.M. Rantz (Seattle, WA: American Society for Enology and Viticulture).

10 A.B. Shaw, "A Climatic Assessment of the Niagara Peninsula Viticulture Area of Ontario for the Application of Wind Machines," *Journal of Wine Research* 13 (2002): 143–64.

11 A.J. Winkler, J.A. Cook, W.M. Kliwer, and L.A. Lider, *General Viticulture* (Berkeley, CA: University of California Press, 1974), 19.

12 D.J. Jackson and N.J. Cherry, "Prediction of a District's Grape-Ripening Capacity Using a Temperature-Latitude Index (LTI)," *American Journal of Enology and Viticulture* 3 (1988): 19–26.

13 J. Gladstones, *Viticulture and Environment* (Adelaide: Winetitles, 1992).

Ten

The Wine-Producing Soils of Niagara

Daryl Dagesse

Introduction

O ne would intuitively assume that the nature and properties of the soil, as the growing medium and thus the source of water and nutrients for the vines from which a wine is made, would be of paramount importance. The French have historically argued the importance of soil through the term "terroir." Although commonly defined simply as "soil," in the context of oenology and viticulture, terroir is a much broader term, encompassing all the physical elements of the vineyard habitat, including geological, physiographic, pedological, meteorological, and viticultural processes, acting both singly and together.[1] The role of soil alone and its effect on wine quality remains a contentious issue. Correlation between soil physical and chemical characteristics remains circumstantial, controversial, and anecdotal, yet it remains central to the definition of wine appellations.[2]

This chapter considers the pedological, or soil, aspects of the wine-growing areas within Niagara. As with the more general term "terroir," there are many misconceptions regarding the soil as it pertains to oenology and viticulture in this area. Although the underlying bedrock geology has undeniably played a role in the formation of the soils within the Niagara Peninsula, the events of the much more recent glaciation or, more correctly, deglaciation, of southern Ontario played a much greater role in defining the wine-growing soils of Niagara. As Haynes[3] suggested and will be demonstrated, these fine-grained, clay-rich soils represent an anomaly in the context of the soils of the traditionally considered fine-wine-producing regions of the world.

A Primer on Soils and Their Classification

Soil is arguably the most complex of the natural environments, as it includes physical, chemical, and biological processes occurring simultaneously within a three-dimensional heterogeneous matrix of solid, liquid, and gaseous phases of matter. As such, soil is not a static entity, but rather a continuum with both spatially and temporally varying properties.

Soil classification is therefore pragmatic in nature, in which taxa are based on generalized inherent properties. Although the factors of soil development (climate, organisms, relief, parent material, and time) and processes leading to soil development (additions, removals, transformations, translocations, and mixing) are not directly considered when classifying a soil, they are indirectly included in the final classification through the presence, or absence, of certain diagnostic features and properties present in the vertical horizons of the soil profile. Soil survey reports attempt to capture and categorize these properties via a hierarchical classification scheme as developed by the Canada Soil Survey Committee of the Research Branch of Agriculture Canada.[4] The Canadian System of Soil Classification (cssc) is hierarchical in nature and is organized as follows: Order, Great Group, Subgroup, Family, and Series. It is, in a way, unfortunate that official Soil Surveys map the soils of an area at the Series scale. At this level of detail, soils from different widely separated areas, but classified as the same Series, exhibit similar horizon development, including properties of colour, texture, structure, consistence, thickness, and chemical reaction.[5] This is often the result of soil formation on similar glacially derived parent materials from different geographical areas. Although at times this level of detail may be crucial, in the context of soils within the typical wine-producing area of Niagara, such detail generally confounds the issue. This is well illustrated when viewing vineyard locations on soils maps as compared to maps of soil parent material of glacial origin (Figures 10.1a and 10.1b), where the preferential location of vineyards is much more evident on particular parent materials than on specific soil series. A simpler view is that only two soil Orders are generally represented within the wine-growing areas of Niagara: Luvisolic and Gleysolic.

Luvisols typically develop in mild to very cold, sub-humid to humid environments where the downward flow of water flushes fine clay particles from the surface horizon to deeper levels in the soil. This translocation of finer soil material results in coarser than expected texture in the surface horizons, but concomitantly heavier textures in the lower horizons. This order characterizes the relatively better drained soils of Niagara, although the lower horizons can become waterlogged. If this waterlogging becomes excessive, the soil may be classified as a Gleysol. These soils are primarily distinguished from Luvisols by the presence of gleying, or the mottled appearance their colours display. The bright red, orange, yellow, and blue colours typically found in the lower horizons of these soils result from the chemical reactions of oxidation and reduction as the soil alternately wets and dries throughout the year. This order therefore characterizes the more poorly drained cousins of the Luvisols. The existence, therefore, of a Gleysol adjacent to a Luvisol, while sounding like a drastic difference, as they

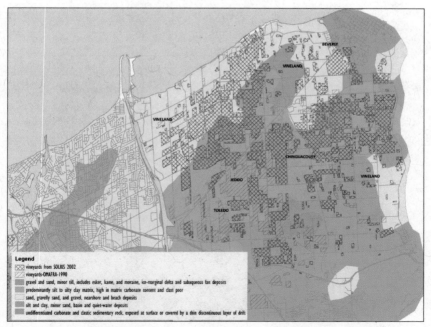

Figure 10.1a Vineyards as located on a soil map (top) and parent material map (bottom), Niagara-on-the-Lake
Source: See note 20.

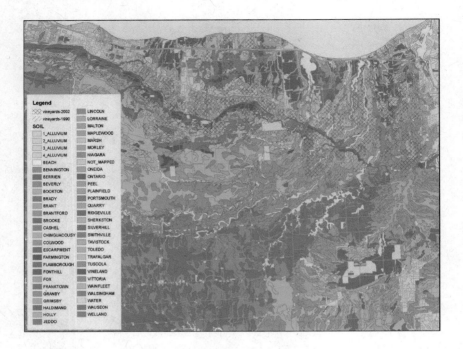

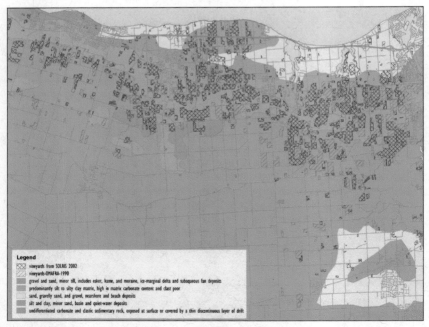

Figure 10.1b Vineyards as located on a soil map (top) and parent material map (bottom), Western Peninsula

Source: See note 21.

represent two different soil Orders, may simply be the result of the location of the former in a poorly drained depression, while the latter is located on a slight topographic high point. This point is highlighted by the predominance of Niagara soils classified as Gleyed Luvisols, which, although dominantly displaying the characteristics and properties of a Luvisol, also secondarily exhibit the mottled colours of the wetter gleysols. Extension of the classification as far as the Series level suggests that these two adjacent soils are completely different, while in fact the only difference in these two end members of the soil continuum may be the effects of slope on water drainage through the profile.

The Role of Soils in Viticulture

Grapes are unusual in that they grow in a wide variety of soils, but typically thrive in marginal, well-drained soils with loamy to sandy and gravelly surface layers over moderately permeable clay subsoils.[6] This fact is borne out through a survey of the great wine-producing regions of the world, suggesting that vines may have initially been cultivated in soils that were so poor as to be unable to reliably produce anything else (Table 10.1). What better way to drown one's sorrows over the poor quality of their farmland than with a glass of wine?

Table 10.1 Traditional wine-producing regions of the world and their associated soil types

Wine-producing Region	Typical Soil Properties
Burgundy, Champagne, and the Loire Valley, France	chalky soil containing at least some limestone clay
Mosel, Germany	steep slate slopes
Bordeaux, France	clay soils, rocky gravel soil, arid sandy, and gravelly soil
Rutherford, Napa Valley	gravelly to sandy to loamy, shattered sandstone and limestone base
Alsace, France	sandstone of Kitterlé, limestone of Zinnkoepflé, granite of Brand, volcanic soils of Rangen
Châteauneuf-du-Pape, France	rocky soils comprised of smooth, oval-shaped stones
Tuscany, Italy	range from tufa and volcanic soil to sandstone and limestone clay
Priorat, Spain	rocky, porous, and free-draining soil comprised of a mix of dark slate and quartzite
Coonawarra, Australia	thin limestone soil over top of a sandstone base
Mendoza, Argentina	combination of sand, granite, schist, and alluvial deposits

Source: Adapted from S. Brown, "The World's Top 10 Wine Soils," http://www.winegeeks.com/articles/139.

Classic compatibilities between grape variety and the type of soils they thrive in are generally accepted (Table 10.2). The heavier textured and poorly drained, although generally fertile, agricultural soils of Niagara are therefore a substantial anomaly in terms of great wine-producing regions of the world.[7] All of this is relative, however, as previous agricultural land use and soil surveys, when there was much less competition with non-agricultural land uses within the traditional fruit belt, revealed that vineyards were preferentially located on soils found less suitable for tender-fruit crops, including peaches, cherries, pears, and plums.[8]

Soil fertility in the form of mineral composition and nutrient availability seems to be of less importance in terms of ultimate wine quality than would be intuitively assumed.[9] Nitrogen, phosphorus, potassium, and other nutrient levels need to be sufficient to promote healthy vine growth, but, if achieved, little benefit is realized from higher levels of most nutrients. Requirements also vary between vines for red wine production versus those for white wine. This, in combination with the complex interrelationships between soil nutrient levels and other factors, including pH, soil temperature, water content, organic matter content, and microbiological populations, preclude simple statements regarding specific nutrient levels for wine production.

It is generally agreed that most crops, including grapes, do best on well-drained, light-textured soils comprised of high percentages of sand. These soils have the advantage of being easily worked, well drained, and aerated, and thus allowing for deep root penetration. This is particularly true in the case of *vinifera* grapes, the saying being that they hate "wet feet."[10] The ease with which water flows through the soil, however, also leads to rapid drying, resulting in a soil often described as droughty. In the Niagara Peninsula, however, even when the surface

Table 10.2. Grape varieties and their preferred soil characteristics	
Grape Variety	**Soil Type**
Chardonnay	Well-drained calcareous limestone
Pinot noir	Well-drained calcareous limestone
Pinot blanc	Well-drained stony or calcareous limestone
Pinot gris	Most types of well-drained soils
Cabernet Sauvingnon, Cabernet Franc	Suits most soils, with a preference for gravelly textures
Sauvingnon Blanc	Sandy or gravelly loams, not excessively dry or calcareous
Riesling	Lighter, well-drained warmer soils
Gewürztraminer (Sauvignon Rosé)	Deep and fertile
Merlot	Suits most soils, with a preference for deeper, warmer soils

Source: Adapted from P.S.W. Laurita, "Wine Grapes and Soil," http://www.bellaonline.com/articles/art36255.asp and Jackson and Schuster (1994).

soil appears light textured, heavy clay lies unseen not far below the surface, often impeding downward drainage. The heavy-textured clay soils more commonly occurring in the Peninsula are generally viewed as being less agriculturally desirable. However, a tile-drained heavy-textured clay-rich soil displaying good soil structure resulting from careful management can exhibit the best properties of both textural extremes in terms of relatively easy workability, acceptable water infiltration rates, and good root penetration, with the added benefit of high water-holding capacities.

An abundant year-round soil pore water supply may actually act as a detrimental factor, as wines produced from vines that experience a water deficit stress during the growing season, particularly during fruit maturation, are generally of higher quality. Good wine-producing soils have typically been considered to be coarse textured and therefore well drained, as this combination of soil properties generally satisfies the vine's requirement of preventing "wet feet." At the same time, however, this drainage introduces the water deficit stress that is desirable for high-quality grapes, particularly in cool climates such as Niagara's, through limiting vine vigour during grape ripening toward the end of the growing season. This has been described as the "Goldilocks approach: just the right amount of water, not too much, not too little."[11]

Niagara's predominantly clay-rich soils promote water deficit stress in the same manner as coarser-textured soils, although through a different mechanism. Soil is a porous media: the voids occurring between the solid particles represent the only space where water can reside within the matrix. The size of the voids between soil particles is generally proportional to the size of the particles

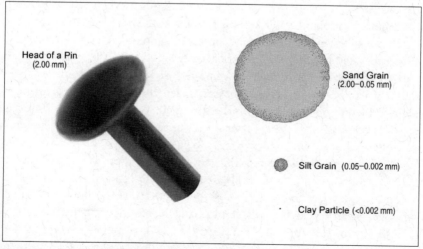

Figure 10.2 The relative sizes of sand, silt, and clay particles
Source: Daryl Dagesse.

themselves. The enormous size difference between sand, silt, and clay particles (Figure 10.2) therefore results in an equally wide range of pore spaces. Coarse-textured soils, such as sands comprised of comparatively large soil particles, contain relatively large void spaces or macropores. Although these large voids are capable of holding large volumes of water, they also drain very easily, resulting in little storage of water within the pore network available for root uptake.

During the summer and early autumn, when water availability through precipitation is generally low, this low soil water storage capacity often results in a desirable water deficit stress to the vines. Conversely, the fine texture of clay soils results in very small pore spaces. Although smaller in size, they are greater in number, thus constituting a total void space or porosity at least comparable to, and often greater than, coarser texture soils. The major difference between the macropores of the coarser-textured soils and the micropores of the finer-textured soils is that the very small micropores hold water tenaciously, in a manner similar to a bath towel. So tightly held is this pore water that the roots of the vines are unable to extract sufficient quantities even when the water content of the soil is apparently still quite high. Sand's large voids may contain much water, but the ease of drainage results in little being retained and therefore little available for plant growth. Conversely, the numerous small pores of clay may also hold large quantities of water, but the tenacity with which it is held similarly results in little being available to satisfy plant demand (Figure 10.3). Therefore, toward the end of the growing season, despite the fine-textured soil's higher water storage capacity, little is actually available, thus triggering a water deficit stress in a manner similar to the coarser-textured soils. Niagara's clay soils thus exhibit soil water retention properties comparable to the coarser-textured sandy or even gravelly soils of some of the great wine-producing regions of the world.

Excessive springtime soil wetness through both snow melt and precipitation has resulted in tile drainage becoming a common management practice to reduce waterlogging of Niagara's heavier-textured soils. The inherently slow transport of pore water through these fine-textured soils to the drains results in soils that, while not waterlogged for excessive lengths of time, still remain fairly wet and therefore warm slowly in the spring. This acts to delay the onset of growth, also serving as a hedge against early season frost damage—a phenomenon more often attributed to proximity to Lake Ontario.

Tile drainage also serves to artificially create the favourable conditions of a shallow rooting depth, as the roots are less likely to penetrate into the wetter conditions below the drains. At these deeper depths, Niagara's heavier-textured clay soils also possess higher-bulk densities and therefore fewer pores that can be exploited by root penetration.[12] For any roots that do penetrate to these greater depths, the underlying clay may therefore serve as more of an advantage than a disadvantage, as it has been observed that the roots of older, well-established

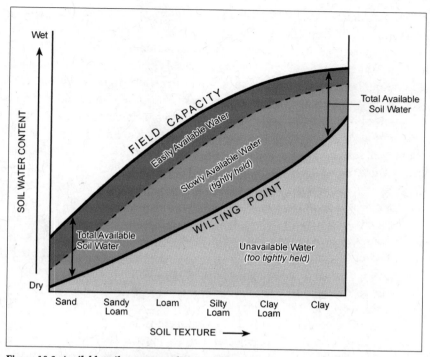

Figure 10.3 Available soil water according to soil texture
Source: Adapted from A. Strahler and O. W. Archibold, *Physical Geography: Science and Systems of the Human Environment*, 5th Canadian Edition (Toronto: John Wiley and Sons Canada, 2011).

vines may be deep enough to penetrate this material to exploit the clay soil's greater soil-water-holding capacity during dry periods.

The Geological Setting

Textbooks covering the development of soil commonly include a diagram showing the underlying bedrock weathering through physical, chemical, and biological processes to produce a regolith of loose material. Through the further action of these weathering processes, notably temperature and moisture as defined by the climate, a soil is ultimately formed on this weathered material. While this process is dominant in many parts of the world, soils in Canada generally, and the Niagara Peninsula specifically, result largely from these weathering processes on materials of much different origin.[13]

The unconsolidated material upon which Niagara's soils have formed is the direct result of the action of the continental glaciers, which had fully retreated from this area only about thirteen thousand years ago. Glacier movement incorporates pre-existing surficial soil and rock materials and grinds away the underlying bedrock to produce material that is later deposited in various forms through different geomorphological processes. While some material can be transported great distances through glacial action, as evident by pieces of rock from the Canadian Shield present as glacial erratics within both the soils and various landforms in Niagara, most glacial material is transported relatively short distances. Thus, the underlying bedrock geology that was ground up and transported by the glacial ice is still of fundamental importance in terms of Niagara soils, as it still, albeit indirectly, forms the parent material upon which the present soils have formed. The reworking and redistribution of this glacially derived material via the processes occurring during the deglaciation of southern Ontario thus resulted in a major disconnect between the underlying bedrock and the adjacent overlying sediments within which the soils actually formed.

Geology is undeniably a dominant factor in the landscape of the Niagara Peninsula, as evident by the Niagara Escarpment. The Niagara Peninsula comprises bedrock formed during the Paleozoic Era (approximately 570 to 225 million years ago). Rock strata from the Ordovician Period (approximately 500 to 430 million years ago) through the Silurian Period (approximately 430 to 395 million years ago) to the Devonian Period (approximately 395 to 345 million years ago) are represented across the Niagara Peninsula.

The oldest rocks of the Ordovician Period underlie the northern part of the Peninsula, along Lake Ontario. The youngest rocks of the Devonian Period underlie the southern Peninsula along the Lake Erie shoreline. Separating these two are the rocks of the Silurian Period. These associations are illustrated in Figure 10.4. The sequence of rocks from the Silurian Period is particularly well illustrated in the Niagara Gorge at the Whirlpool Corner (Figure 10.5). The Niagara Escarpment was formed within this sequence through erosion by a large

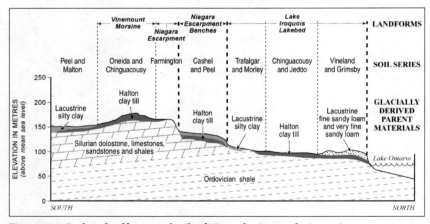

Figure 10.4 Geology, landforms, and soils of Niagara's wine-producing region
Source: Daryl Dagesse.

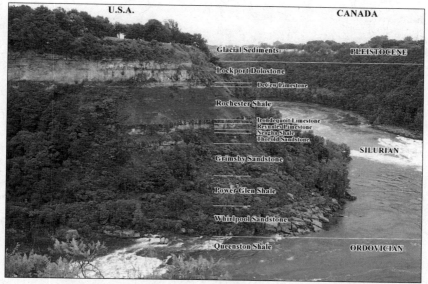

Figure 10.5 **The Silurian sequence of bedrock at Whirlpool Corner in the Niagara Gorge**
Photo: D. Dagesse.

river system that subsequent glaciations, over the past millions of years, have transformed into the present Great Lakes.

Quite contrary to the simplified illustrations often seen in textbooks, the escarpment is a complex landform reflecting the varying hardness of the sequence of rocks comprising it. Rather than a simple steep slope of bedrock, the profile displays several steps or benches controlled by the underlying bedrock (Figure 10.6). The resistant dolomitic limestone of the Lockport Formation forms the topmost scarp face, while the Irondequoit Limestone of the Clinton Group, separated from the Lockport by the softer and therefore more easily erodible Decew Limestone and Rochester Shale, forms a lower scarp face. This dual escarpment form is best exemplified by the two waterfalls comprising Ball's Falls. The Lockport and Irondequoit scarp faces are often separated by a broad bench, up to a kilometre wide. These are the basis of the sub-appellations identified as the Beamsville, Twenty Mile, and Short Hills benches. Although these have been termed "Lake Iroquois benches," the control of the underlying bedrock associated with the form of the Niagara escarpment suggests that "Niagara Escarpment benches" is more appropriate terminology.

Above the escarpment, the regional dip of the bedrock surface to the south has resulted in a generally flat surface descending gently toward Lake Erie. Within a few kilometres of the Lake Erie shoreline, particularly in the southeastern reaches of the peninsula, this surface is sporadically broken by the presence of the Devonian rocks of the Onondaga Escarpment.

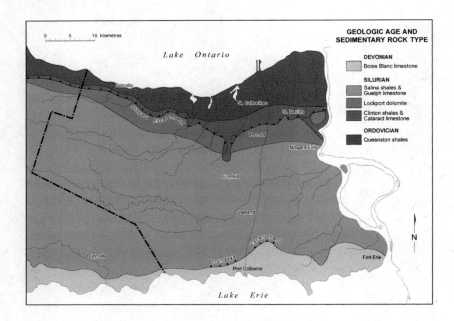

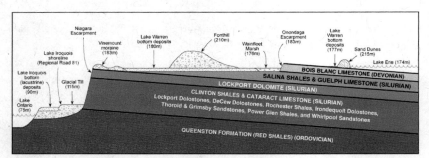

Figure 10.6 Bedrock geology of the Niagara Peninsula in both plan and profile view
Source: Daryl Dagesse.

To the north of the Niagara Escarpment, Ordovician rocks, represented by the Queenston shale, become the underlying bedrock. That this rock unit underlies the Silurian rocks of the Niagara Escarpment is evidence that the escarpment is the result of erosional processes rather than tectonic forces.

Rather than these underlying bedrock surfaces, it was the most recent glacial event that had the most profound effect in terms of the creation of the parent materials upon which Niagara's soils are forming. The actions of glaciation ground up pre-existing surficial soil and the local bedrock and deposited it over the bedrock surface, including the Niagara Escarpment, in the form of the Halton Clay Till.

As the glacier retreated from southern Ontario, and while the ice still occupied the present Lake Ontario basin, the ponding of meltwater in front of the glacier over time formed a series of proglacial lakes. Within these lakes, sediments from the melting glacier were deposited as light-textured lacustrine material thinly covering the underlying Halton Clay Till. As glacial retreat continued, the Niagara Peninsula became free of ice, while the eastern end of the present Lake Ontario basin was still blocked by the glacier. The well-developed proglacial Lake Iroquois formed within the current Lake Ontario basin approximately twelve thousand years ago. This had the effect of eroding the blanket of Halton Clay Till overlying the bedrock surface, in some isolated cases down to the underlying Ordovician Queenston shale bedrock. In other areas, deposition of sandier material occurred on the lake bed in the form of sediment bars as north-flowing streams deposited sediment within Lake Iroquois. The shoreline itself, generally followed by Regional Road 81 from Queenston to Hamilton, comprises features ranging from bluffs cut into the Halton Clay Till to reworked lacustrine deposits from previously higher water levels and beaches and bars similar to those of the current Lake Ontario shoreline.

The parent materials for the soils of the wine-growing area of the Niagara Peninsula are therefore of a heterogeneous background, little of which has to do directly with the underlying bedrock. The exceptions to this are the few areas where near-shore wave action within Lake Iroquois exposed the underlying Queenston shale bedrock through erosion of the overlying glacially derived material.

The Niagara Peninsula has previously been divided into as few as three sub-appellations[14] and as many as ten.[15] These have been delineated not only by soil type, but also on factors including topography, climate, and geology. The following is not an attempt to generate yet another delineation of sub-appellations, but rather to categorize the typical wine-growing soils of Niagara, from strictly a parent material point of view, into three regions, from north to south: the Lake Iroquois lake bed, the Niagara Escarpment benches, and the Niagara Escarpment itself.

The Lake Iroquois Lake Bed

This area has previously been identified as the "Lake Iroquois Plain," "Lakeshore," and "Lakeshore Plain," but the dual effect of both the erosion of the pre-existing blanket of Halton Till overlying the Queenston shale bedrock and the deposition of sediments by virtue of the processes related to coastal erosion and sedimentation suggests that "Lake Iroquois Lake Bed" is more appropriate.

A visit to Lake Ontario during stormy conditions gives an appreciation of the erosive power of wave action against the shoreline bluffs. Waves scour the lake bed and erode the shoreline bluffs, with eroded material moving either further offshore for deposition into deeper water, or, more commonly, along the shoreline to be deposited as sediment bars in the near-shore zone. These same processes, occurring in Lake Iroquois, resulted in three distinctly different zones of parent material upon which the Lake Iroquois Lake Bed soils formed. These zones are, unfortunately, not well reflected in the current sub-appellation delineation, which is based not so much on soil type as on climate as controlled by distance from the moderating effects of Lake Ontario.

In the northernmost zone, furthest from the Lake Iroquois shoreline, at distances ranging up to approximately 4 kilometres from Lake Ontario, fine and very fine sandy materials were deposited as deltaic sediments on the lake bed of eroded Halton Clay Till. These soils are typically represented by the Grimsby (Luvisol) and the wetter Vineland (Gleyed Luvisol) soil series, as they are morphologically similar to those soils formed under like conditions further west in the peninsula. The surface horizon of these soils appear light textured as a result of this sandy material, but the much heavier-textured Halton Clay Till is usually in the order of only a few metres at most below the surface. The droughtiness of these sandy soils is therefore exacerbated as groundwater is not easily drawn upward across this textural boundary. Conversely, the proximity of clay to the surface ironically results in imperfect drainage of these highly permeable soils during times of abundant moisture, as downward percolation is inhibited by this same textural boundary. Although historically used sparingly, they are coming more into use for grape production as the wine industry expands, particularly in the Niagara-on-the-Lake area. These limitations are typically addressed through management practices involving irrigation in the former case and tile drainage in the latter.

A zone typifying the erosive near-shore wave action of Lake Iroquois lies to the south of these lighter-textured soils. The Chinguacousy (Gleyed Luvisol) and the wetter Jeddo (Gleysol) soils are characterized by the heavier clay textures of the Halton Till upon which they are formed. There are some lighter-textured areas resulting from localized deposition of sandier sediments, although these are usually very shallow, typically less than 50 centimetres in depth. The heavy textures are responsible for these soils being poorly drained and slowly permeable, with tile drainage often being used to control excess soil moisture. In spite of these apparent limitations, these soils are used extensively for grape production.

The southernmost zone of the Lake Iroquois Lake Bed represents a complex situation. Both the underlying Queenston shale, lying very close to the surface, and sediments resulting from wave action along the shoreline create a mix of soils. The Trafalgar (Gleyed Luvisol) soil series occurs where the Queenston shale bedrock typically lies less than one metre below the surface. This is perhaps the only instance in the area of the textbook case where the soil actually formed as a result of the weathering of the underlying bedrock. The resultant silty clay textures are responsible for the slowly permeable, imperfectly drained soils, although the greater problem is often the shallow rooting zone resulting from the shallow depth to the underlying bedrock. The associated Morley (Gleysol) soils are very similar, although generally wetter. The major difference is that while the Trafalgar soils are typically reddish in colour, resulting from the oxidized iron of the underlying Queenston shale, the Morley soils often display the green colour of the reduced form of iron typifying the imperfect drainage of these soils.

The Niagara Escarpment Benches

This zone, topographically above the well-developed shoreline of Lake Iroquois to the north but below the Escarpment face to the south, represents an area unaffected by the erosional and depositional actions of Lake Iroquois. The bedrock is no longer the Ordovician Queenston shale underlying the Lake Iroquois Lake Bed, but rather the stratigraphically higher Silurian dolostones, limestones, sandstones, and shales. As on the Lake Iroquois Lake Bed, however, bedrock represents a minor influence, as the soils in this zone developed in the fine-grained lacustrine sediments of proglacial lakes that predated Lake Iroquois. These parent materials give the surface horizons slightly lighter silty clay loam or silty clay textures. However, drainage is often poor, as a result of the compacted Halton Clay Till lying usually less than 1 metre below the surface. The prevalence of different soil types is strongly controlled by local topography, particularly where streams flowing toward Lake Ontario have eroded deep valleys.

The soil series present in these bench areas seem rather diverse, and exhibit a change as one moves east to west through this zone. The Beverly, Cashel, and Peel soil series dominate the benches to the east (i.e., St. Davids Bench), while the Cashel and Peel predominate in the central area (i.e., Short Hills and Twenty Mile benches) and the Chinguacousy and Oneida soil series are common in the west (i.e., Grimsby Bench). Functionally, however, there is much less diversity in the soil than this list would suggest, as they are all members of the Luvisolic order.

The Beverly and Peel soils typically exhibit silty clay loam textures, although certain phases of these soils may also display lighter textures. They tend to be wetter, less well-drained cousins to the Cashel soils. Although classified as Luvisols, their generally poor drainage, resulting from the underlying lacustrine clays and clay till, gives them Gleysolic properties. The Chinguacousy and Oneida soils both tend to exhibit clay loam textures, although, as noted earlier, certain phases of the Chinguacousy soils may also display lighter textures. These soils are also formed over heavy lacustrine clays or the Halton Clay Till, thus leading in places to poor drainage, although they are both generally classified as Luvisols.

The Niagara Escarpment

The Niagara Escarpment reflects the underlying geology of the peninsula as the hard dolostones of the Lockport Formation form a resistant caprock protecting a succession of relatively softer underlying Silurian rocks. Although this bedrock is close to the surface, it is generally buried by an overlying glacial moraine. As is true in other areas within the peninsula, the underlying bedrock has little direct effect as the parent materials of these soils. Rather, it is the properties of the Vinemount Moraine, paralleling the brow of the escarpment, that have the greater effect.

The Vinemount Moraine is characterized as a long, narrow ridge, and likely represents a recessional stage of the glacial ice. The similarity of this Halton Clay Till parent material with that found on the Niagara Escarpment benches below the brow of the escarpment suggests a relatively uniform plastering of this material on the surface as a result of glacial movement. However, the uncharacteristically subdued nature of this moraine suggests that proglacial Lake Warren may have played a role in its formation through facilitating sedimentation into the water or slightly lifting the glacier from its bed as it pushed up and over the escarpment during the deposition of this moraine. In any case, subsequent wave action within Lake Warren would have both smoothed the feature and had the combined effect of washing out some of the finer clay material and depositing capping layers of silty sediments over the heavier clays in a manner similar to the lighter-textured soils on the Lake Iroquois Lake Bed below the escarpment. The Oneida, Jeddo, and Chinguacousy soil series are dominant within this area. Beverly and Toledo soils are also found, but are less likely to host vineyards.

The fine-grained texture of the Oneida soils results in relatively high water-holding capacities, yet they are moderately well drained, more likely due to the rolling topography associated with the Vinemount Moraine than soil texture. These soils are frequently found in association with the more poorly drained Jeddo and Chinguacousy soils, as is the case on the Lake Iroquois Lake Bed. These soils predominantly reflect the heavy-textured characteristics of the Halton Clay Till, upon which they formed with slight modification, in the case of the Oneida soils, by wave action within Lake Warren. This range of drainage characteristics, from poorly drained to well drained, results in the soils being classified as Gleysolic (Jeddo), Gleyed Luvisolic (Chinguacousy), and Luvisolic (Oneida).

The Beverly and Toledo soils are commonly found in association above the escarpment. They are more the result of soil formation on the Halton Clay Till than the Vinemount Moraine, as was the case for the Oneida, Jeddo, and Chinguacousy soils. The Beverly soils are generally imperfectly drained as a result of their heavier texture and compacted subsoils. The Toledo soils are similar in nature but even less well drained and, as such, are typically classified as Gleysolic, in contrast to the slightly better drained Beverly soils (Luvisols).

Niagara's Soils as a Vine-Growing Medium

Although grapes may be grown in a wide variety of soils, a general preference is for warm, well-drained soils with sufficient rooting depth, with good water-holding capacity, and of acceptable pH range with moderate fertility. Niagara's soils naturally possess appropriate pH and fertility levels as a result of the calcareous nature of the parent material and the high cation exchange properties of the clay, so the soils generally need minimal attention from a management perspective. These soil properties are nevertheless usually monitored from a maintenance perspective in order to maintain moderate yields of higher-quality grapes for wine production. The generally poorly drained nature of Niagara's clay-rich soils represent the biggest challenge, as they are an anomaly in terms of the soils of the great wine-producing regions of the world.

Sandy soils, with their large, void spaces, typically exhibit excellent water infiltration rates, good drainage and root aeration, and concomitant good root penetration, while clay soils, with their very small pores, typically exhibit slow infiltration, high water-holding capacities, but potentially poor root aeration and shallow root penetration. The fine-grained texture of Niagara's soils cannot be changed in an economical manner, but the physical properties associated with that texture may be altered via management practices. If fine-grained soils

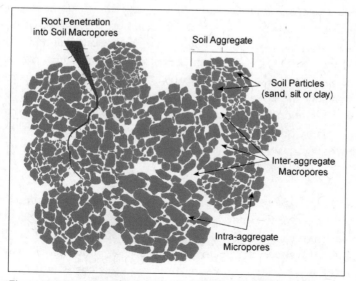

Figure 10.7 Pore space associated with a well-structured clay soil
Source: Daryl Dagesse.

possess good soil structure within the rooting zone, the best properties of a range of textures are realized as the clay particles bind together with organic matter to produce soil aggregates. Large inter-aggregate macropores permit acceptable water infiltration and drainage, while small intra-aggregate micropores hold substantial quantities of water to satisfy water demands during dry periods. The larger inter-aggregate pore spaces also permit acceptable root penetration (Figure 10.7). Maintaining acceptable organic matter contents in fine-grained soils is an important management goal in maintaining adequate levels of soil structure.

The growing of annual field crops permits a more active hand in soil management through seasonal tillage. The adoption of no till management has reduced this once common activity in the production of field crops, although observation of long-term compaction of fields under this practice has resulted in Niagara's clay soils once again being plowed in the fall. The perennial nature of grape growing tends to preclude this practice, although it is not uncommon for vineyards to be subjected to inter-row ripping in an effort to maintain acceptable levels of soil structure.

In the case of the generally poorly drained clay-rich soils of Niagara, the most common practice is that of tile drainage.[16] The grape-producing soils of Niagara typically possess only poor to fair suitability for the production of *vinifera* grapes, yet uniformly display a significant improvement when drained.[17] Tile drainage does not reduce the water-holding capacity of the clay soils, which benefits the vines in the dry summer when drought stress could become a problem, but serves to remove excess soil water from the rooting zone, thus maintaining adequate root zone aeration. This benefit is realized even more so for older vines having established deep root systems through any coarser-textured surficial material and into the underlying clay.[18] Drainage represents a substantial start-up cost in the establishment of a new vineyard, yet the benefits are substantial. Added benefits include the extensive disturbance to the vineyard site to depths in the order of 1 metre or more, thus breaking up any pre-existing plow pans and promoting good initial structural characteristics and the promotion of earlier warming of the soils than may otherwise occur.

Conclusion

The traditional view of soil is often one of a static medium formed through the deterioration of the underlying bedrock. In spite of the obvious geologic importance to the topography of the area through the presence of the Niagara Escarpment, this view is far from the case in the wine-producing soils of Niagara, which owe their origins more to the relatively recent actions of the last glaciation. The resultant fine-grained and generally poorly drained soils have nonetheless been effectively managed so as to produce world-class wines. In an effort to define the terroir of Niagara through the soil, considerable confusion has arisen from the extremely complex classification down to the Series level, when in reality the wine-producing soils of Niagara are of only two closely related Orders and display remarkably similar characteristics across the entire region.[19]

NOTES

1 G. Seguin, "'Terroirs' and Pedology of Vinegrowing," *Experientia* 42 (1986): 861–73; S.J. Haynes, "Geology and Wine 1: Concept of Terroir and the Role of Geology," *Geoscience Canada* 26, no. 4 (1999): 190–94; R.W. MacQueen, "Geology and Wine Retrospect and Prospect," in *Fine Wine and Terroir: The Geoscience Perspective*, ed. R.W. MacQueen and L.D. Meinert, Geoscience Canada Reprint Series 9 (St. John's, NL: Geological Association of Canada, 2006), 71–72.

2 H. Sayed, *Vineyard Site Suitability in Ontario*, Horticultural Research Institute of Ontario (Vineland Station, ON: Ontario Ministry of Agriculture, 1992), 23 pp.

3 S.J. Haynes, "Geology and Wine 2: A Geological Foundation for Terroirs and Potential Sub-Appellations of Niagara Peninsula Wines, Ontario, Canada," in MacQueen and Meinert, *Fine Wine and Terroir*, 9–29.

4 Canada Soil Survey Committee, Subcommittee on Soil Classification, *The Canadian System of Soil Classification* (Ottawa: Supply and Services Canada, Canadian Department of Agriculture Publication 1646, 1978), 164 pp.

5 Haynes, "Geology and Wine 2."

6 R.E. Wicklund and B.C. Matthews, *The Soil Survey of Lincoln County* (Ottawa: Research Branch, Canada Department of Agriculture and the Ontario Agricultural College, Report No. 34 of the Ontario Soil Survey, 1963), 48 pp.; W.J. Dillon, Grape Production in the Niagara Peninsula: Production Costs, Returns and Management Practices, 1959–1962 (Toronto: Farm Economics, Co-operatives and Statistics Branch, Ontario Dept. of Agriculture and Food, 1968), 34 pp.; K.H. Fisher, O.A. Bradt, and R.A. Cline, *The Grape in Ontario* (Toronto: Ontario Ministry of Agriculture and Food, 1982), 50 pp.

7 Haynes, "Geology and Wine 2."

8 Ontario Ministry of Agriculture and Food, Factors Affecting Land Use in a Selected Area in Southern Ontario: A Land Use and Geographic Survey of Louth Township in Lincoln County (Toronto: Ontario Department of Agriculture in consultation with Dept. of Geography, University of Western Ontario, Ontario Dept. of Planning and Development, Ontario Dept. of Municipal Affairs, Ontario Dept. of Highways, compiled by R.M. Irving, 1957), 148 pp.; R.R. Krueger, "Changing Land-Use Patterns in the Niagara Fruit Belt," *Transactions of the Royal Canadian Institute*, V. XXXII, Part II, No. 67 (1959), 39–140; E.W. Ellis, *Grape Distribution in a Selected Area of the Niagara Peninsula* (Unpublished Bachelor of Arts [Geography] thesis, Brock University, 1969), 128 pp.

9 L.D. Meinert, "Introduction," in MacQueen and Meinert, *Fine Wine and Terroir*, ix–xiv.

10 Haynes, "Geology and Wine 2."

11 Meinert, "Introduction."

12 R. Morlat and A. Jacquet, The Soils Effects on the Grapevine Root System in Several Vineyards in the Loire Valley (France), *Vitis* 32 (1993): 35–42.

13 L.R. Webber and D.W. Hoffman, *Origin, Classification and Use of Ontario Soils* (Toronto: Ontario Department of Agriculture and Food, Publication 51, 1967), 58 pp.; L.J. Chapman and D.F Putnam, *The Physiography of Southern Ontario*, 3rd ed. (Ontario Ministry of Natural Resources, 1984), 244 pp.

14 Sayed, "Vineyard Site Suitability in Ontario"; J. Wiebe and E.T. Anderson, Map, Site Selection for Grapes in the Niagara Peninsula (Vineland Station, ON: Ontario Ministry of Agriculture, Horticultural Research Institute of Ontario, 1977).

15 A.B. Shaw, Delimiting Sub-Appellations within the Niagara Peninsula and Lake Erie North-Shore Viticulture Areas: Report prepared on behalf of the Vintners Quality Alliance Ontario (2004), p. 55.

16 Haynes, "Geology and Wine 2."

17 M.S. Kingston and E.W. Presant, The Soils of the Regional Municipality of Niagara (Guelph, ON: Ontario Institute of Pedology, Report No. 60, 1989), V. 1 & 2.

18 Haynes, "Geology and Wine 1."

19 E.g., D. Douglas, M.A. Cliff, and A.G. Reynolds, "Canadian Terroir: Characterization of Riesling Wines from the Niagara Peninsula," Food Research International 34 (2001), 559–63; D. Kontkanen, A.G. Reynolds, M.A. Cliff, and M. King, "Canadian Terroir: Sensory Characterization of Bordeaux-Style Red Wine Varieties in the Niagara Peninsula, Food Research International 38 (2005): 417–25; J. Schlosser, A.G. Reynolds, M. King, and M.A. Cliff, "Canadian Terroir: Sensory Characterization of Chardonnay in the Niagara Peninsula," Food Research International 38 (2005): 11–18.

20 Southern Ontario Land Resource Information System [computer file]. Regional Municipality of Niagara, ON: Ministry of Natural Resources, 2002. Available: Brock University Map Library Controlled Access S:\MapLibrary\DATA\MNR\SOLRIS\; Agriculture Land Use of Southern Ontario [computer file]. Guelph, Ontario: Ontario Ministry of Agriculture and Food, 1990. Available: Brock University Map Library Controlled Access S:\MapLibrary\DATA\OMAFRA\NIAGARA\NIAGARA\ NIAGARA; Soils of Southern Ontario [computer file]. Guelph, Ontario: Ontario Ministry of Agriculture and Food, 1990. Available: Brock University Map Library Controlled Access S:\MapLibrary\DATA\OMAFRA\NIAGARA\NIAGARA\ NIAGARA; Southern Ontario Land Resource Information System [computer file]. Regional Municipality of Niagara, ON: Ministry of Natural Resources, 2002. Available: Brock University Map Library Controlled Access S:\MapLibrary\DATA\MNR\ SOLRIS\.S:\MapLibrary\DATA\OMAFRA\NIAGARA\NIAGARA\NIAGARA; Ontario Geological Survey 1997. Quaternary Geology of Ontario—Seamless Coverage of the Province of Ontario, Ontario Geological Survey, ERLIS Data Set 14, 1997.

21 Southern Ontario Land Resource Information System [computer file], 2002; Agriculture Land Use of Southern Ontario [computer file], 1990; Soils of Southern Ontario [computer file], 1990; Southern Ontario Land Resource Information System [computer file], 2002; Ontario Geological Survey, 1997.

Eleven

Viticultural Practices and Their Effects on Grape and Wine Quality

Andrew G. Reynolds

Introduction

Wine quality is the result of a complex set of interactions, which include geological and soil variables, climate, and many viticultural decisions (Figure 11.1). Taken as a whole, they may be described as the terroir effect. The components normally associated with the terroir effect, such as geology, site factors, and soil fertility, depth, and moisture, have a significant impact upon the microclimate of the vine, particularly fruit exposure and leaf exposure. The myriad cultural practices such as vine spacing, training system, and what is collectively known as "canopy management" also impact canopy microclimate. Thus, both traditional terroir factors and canopy management practices exert their influences through their effects upon vine vigour (Figure 11.1). This chapter will focus primarily upon those viticultural practices beyond soil (see Chapter 10) and climate (see Chapter 9) that impact grape composition and wine quality, with a particular emphasis on wine aroma compounds and phenolic analytes.

Wine growers have likely recognized for centuries that specific viticultural practices result in better fruit composition and, hence, better wines. Specific practices such as the trimming of vine canopies to minimize shading of fruit; removal of leaves in the fruit zone, again for purposes of fruit exposure; adaptation of particular training systems to optimize fruit composition; and reduction of crop levels to increase rate of fruit maturation are but four specific practices that have been used by wine growers before the advent of scientific research in this field. A specific common goal among many of these viticultural practices is that of fruit exposure. It is one of four "pillars" of the Cool Climate Paradigm, which is a set of general principles common to most premium wine-growing regions of the world:

1. Keep the fruit warm, presumably by either natural cluster exposure provided by an appropriate training system or by canopy management practices such as basal leaf removal;
2. Keep the leaves exposed to sunlight, again, by viticultural practices such as appropriate training, shoot density, and row and vine spacing;
3. Achieve vine balance between vegetative growth and crop size; and
4. Avoid water stress.[1]

These pillars will be briefly addressed throughout this chapter as they pertain to various cultural practices.

Research into viticultural practices that impact wine quality did not appear in the literature until the mid-1980s.[2] Part of the reason for this was the development during that time of sensory descriptive analysis, which allowed for viticulturists and sensory scientists to collaborate on field-based projects with

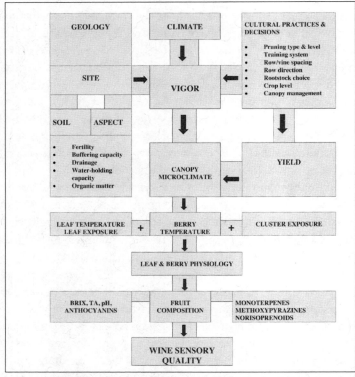

Figure 11.1 A conceptual view of factors contributing to vine balance and vine quality
Source: From: A.G. Reynolds, "Impact of Trellis/Training Systems and Cultural Practices on Production Efficiency, Fruit Composition, and Vine Balance," *American Journal of Enology and Viticulture* 51: 309–17. (Originally published figure was modified from R.E. Smart, J.B. Robinson, G. Due, and C.J. Brien, "Canopy Microclimate Modification for the Cultivar Shiraz. I. Definition of canopy microclimate," *Vitis* [1985], 24, 17–24.)

an ultimate goal of microvinification of fruit from individual field treatments and subsequent sensory evaluation of those wines. Among the first of these studies were those of Smart et al.[3] in Australia, clearly showing the positive impact of alleviating canopy shade upon fruit composition and wine quality in Shiraz through accommodation of vine vigour by canopy division. In France, Carbonneau et al.[4] similarly introduced the Lyre system for *Vitis vinifera*. Later work that followed this theme included evaluation of divided canopies in California,[5] British Columbia,[6] and New Zealand.[7] The original basis for these then-radical ideas was the work of Shaulis, Amberg, and Crowe in Concord, who clearly showed that optimization of cluster and leaf light environment by the Geneva double curtain divided canopy system led to substantial improvements in fruitfulness, yield, and fruit composition.[8] This work was verified in Australia using Thompson Seedless.[9] It led to a revolution in the viticultural world that might be described by the Small Vine (Old World) philosophy of narrow rows, close vine spacing, and narrow, vertical trellises, versus the Large Vine (New World) concept that might include divided canopies, wide rows, and wide vine spacing as a combined utility for accommodation of vine vigour. However, we are now realizing that both approaches are easily reconciled. Whether one searches for premium terroirs, or uses a battery of viticultural practices, the end result encompasses the four pillars of the Cool Climate Paradigm.

Impact of Vineyard Site and the Terroir Effect

Impact of Vineyard Site on Monoterpenes

Before discussing impacts of grapevine cultural practices, one must address the effects of vineyard site. The influence of site on fruit composition is difficult to define objectively, when site-based differences in canopy density, phenology, soil type, and cultural practices are simultaneously involved. Only minor differences were found in monoterpene concentrations (compounds responsible for aromas in Muscat and most aromatic white wine grapes) between Moscato Bianco and Moscato Giallo grapes grown on several sites in the Piemonte and Val d'Aosta regions in northern Italy.[10] Subsequent work showed that higher terpene concentrations were associated with warm sites.[11] Likewise, few differences were found between Chardonnay wines whose origins included Monterey (region I), Oakville (Napa County; region III), and Livermore (Alameda County; region III).[12] Multivariate analysis was used to distinguish between wine-growing regions in Catalunya, Spain.[13] A cool, high elevation site (High Eden, South Australia) produced Riesling fruit with highest terpene concentration, but terpene concentrations could not be linked to wine scores.[14] Thus, although great volumes of anecdotal evidence exist for differentiating sites, very few objective studies have been carried out to quantify these differences.

Work in British Columbia attempted to distinguish between sites based on monoterpene concentrations by locating vineyards of similar soil type and vine vigour and by maintaining the vines using identical cultural practices. Fruit maturation proceeded faster at warmer sites on a daily basis, and free volatile terpenes (FVT) and potentially volatile terpenes (PVT) were therefore usually higher in berries from these sites on any given sampling day.[15] Cooler sites matured their fruit more quickly when expressed on a per growing degree day (GDD) basis. Musts from warm sites tended to be higher in FVT and PVT, although harvested at similar titratable acidity (TA) and pH. Tasters distinguished between wines from the warm and cool sites on the basis of aroma for only one of the four cultivars, but the sites could be distinguished on the basis of flavour for three of the four cultivars.[16] For Bacchus and Schönburger, the warm sites were clearly identified as having the more intense Muscat flavour.

In a three-site experiment, no clear pattern emerged regarding the relationship between site and FVT, but berries from both an Oliver, British Columbia, and Kelowna site were highest in PVT in two of five years.[17] Must FVT and PVT were highest from the Kelowna site. The young wines from the Oliver and Kelowna sites were identified as most spicy. Aged wines from the Oliver and Kelowna sites were high in citrus aroma, those from a third Kaleden site were primarily vegetative, acidic, and astringent, while wines from the Oliver site were characterized by apricot, butter, cedar, and Muscat flavours, high astringency, aftertaste, and body. These results corresponded with the PVT concentrations measured in the berry samples taken at harvest.

Application of Geomatics to the Study of Sites: The Terroir Effect

European viticulturists have maintained that soil is a primary determinant of wine quality, whereas a prevalent philosophy in New World viticulture suggests that soil merely mediates vine growth and vigour, and that the skill by which this vigour is accommodated determines wine quality.[18] Since 1998, our research group has focused upon use of geomatic techniques such as global positioning systems (GPS) and geographic information systems (GIS) (see Chapter 8) to define spatial variability in vineyard blocks. Use of GIS also allows the user to elucidate spatial correlations between important variables.

Chardonnay Terroir Study in Ontario

An example of the use of geomatic tools to study the terroir effect is derived from a study of five Chardonnay sites in Ontario between 1998 and 2002.[19] Specifically, it was hypothesized at the outset of this study that soil type would play a minor role in the determination of aroma compound concentration and wine sensory attributes, and that vine vigour, crop load, and fruit environment would play the major roles. To test this hypothesis, trials were established to study concurrently the impact upon flavour compounds and wine sensory attributes of Chardonnay, whereby soil type varied within vineyard blocks while mesoclimate was kept constant, and the comparative magnitude of the effects of soil texture (sand vs. clay), vine vigour (high vs. low), and crop size (high vs. low) were studied by mapping these variables using GPS and GIS. Five Niagara Peninsula Chardonnay vineyards located on sites with heterogeneous soil types were chosen for study. Multiple-hectare blocks were delineated using GPS, and a series of 75 to 200 data ("sentinel") vines per block were geolocated within a sampling grid imposed on each block. Data were collected on soil texture, soil composition, tissue elemental composition (1998 only), vine performance (yield components and weight of cane prunings ["vine size"]), and fruit composition. These variables were mapped

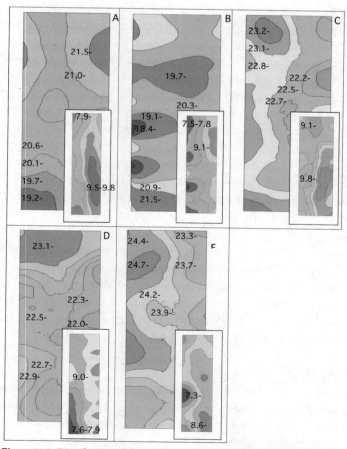

Figure 11.2 Distribution of Brix 1998 to 2002 in a Chardonnay block in one Niagara Peninsula vineyard
Source: Andrew G. Reynolds.

using GIS, and relationships between them were elucidated (Figure 11.2). Wines were made from the various vine size/yield × soil texture combinations and assessed by a trained sensory panel. Soil texture and composition were occasionally correlated to yield components and fruit composition, but often these relationships were site specific.

Sensory descriptive analysis of wines produced from individual soil × vine size categories displayed little differences between sites, but soil texture and vine size had independent effects. When sensory data were analyzed by site, a handful of differences between the wines were apparent. High vine size was associated with greater sweetness and a longer/more intense finish (two sites); higher fruity flavour, more body, and more colour (one site); and less colour (one site). Vines located in sandy portions of the vineyards led to wines with higher fruity and apple aromas, less cedar aroma, higher fruity and apple flavours, and less cedar flavour; however, one site produced wines with less fruity flavour from sandy portions of the vineyard block. Two sites displayed an association between sand and reduced sweetness in the wines, while one site showed an opposite trend. Similar opposing trends were observed for astringency and body. One site also had wines from sandy portions of the site associated with higher perception of acidity, while another had wines from sandy soils with longer finish and greater colour. Several soil texture × vine size interactions were also observed for two sites: Buis (a Lakeshore site) and Château des Charmes (St. Davids Bench). At the Buis site, cedar aroma and mushroom aroma and flavour were decreased in sandy regions where vine size was low, but increased when vine size was higher. Small vines on clay had lowest cedar aroma and mushroom flavour. Fruity flavour increased in sandy regions of the vineyard as vine size increased, but decreased in intensity with respect to increased vine size on clay. At Château des Charmes, wine colour increased with vine size in sandy areas of the block, but was unresponsive to vine size in wines from clay sections. Apple and fruity flavours increased and cedar flavour decreased with vine size in sand portions of the vineyard, but the opposite response was found from wines produced from grapes growing in high clay portions of the vineyard. The take-home message appears to be that there are no patterns with respect to soil texture or vine size vs. wine sensory variables that can be anticipated with certainty.

Riesling Terroir Study in Ontario

To elucidate the potential basis for terroir, a number of factors were measured in an Ontario Riesling vineyard in 1998–2002: vine size and soil texture effects on yield components; soil, vine tissue, and fruit composition; and wine sensory attributes.[20] "Sentinel" vines were geolocated using GPS. GIS was used to delineate spatial variation in soil texture and soil vine tissue composition (1998) and in yield components, berry composition, and weight of cane prunings (vine size) over four years from each sentinel vine. Much like the Chardonnay study, vines were classified as "large" or "small," based upon the previous season's vine size within each of two soil texture classes (clay and sand); fruit from these four categories was separated for winemaking; berry, must, and wine chemical compositional data were determined; and sensory descriptive analysis was undertaken on the wines. Correlations were observed between soil texture and composition vs. berry weight and potentially volatile terpenes (PVT). GIS-derived maps indicated that free volatile terpenes (FVT) and PVT in 1998 and FVT in 1999 were associated areas in the vineyard containing high percentages of sand; however, the pattern for PVT in 1999 was reversed relative to that for 1998, perhaps due to vineyard renovation on the west side of the block during that summer (Figure 11.3). There were no consistent soil texture or vine-size effects on berry, must, or wine composition. High vine size increased berry titratable acidity (TA) and PVT; decreased must pH; and increased wine FVT in individual seasons. Sandy soil (vs. clay soil) reduced wine TA and must PVT, and increased berry TA and must soluble solids (Brix) in individual seasons. Vine size and soil texture did not consistently affect wine sensory attributes across vintages. High vine vigour decreased mineral aroma and citrus flavour, and increased apple attributes in individual vintages. Clay soil increased mineral aroma and citrus attributes in

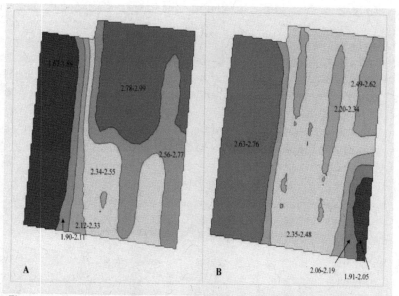

Figure 11.3 Spatial distribution of berry potentially volatile terpenes (PVT, in mg/kg berry), St. Urban Vineyard, Vineland Estates Winery, Vineland, ON (A: 1998; B: 1999)

individual vintages, but decreased apple aroma. Citrus aroma and petrol flavour increased in some vine size × soil texture combinations in one season. Vintage and wine age had greater impacts on wine sensory attributes than vine size or soil texture. Once again, it appeared that although vine size and soil texture had some effects upon berry composition and wine sensory attributes, these effects could be overshadowed by vintage effects.

A similar example of the application of spatial variability of vine vigour is a study that explored the relationship between vigour and phenolic analytes.[21] Fruit exposure effects on Pinot noir anthocyanins were examined from a standpoint of vigour-induced fruit exposure effects. High vigour zones in two Oregon vineyards had lower Brix and higher TA, and there was a trend toward lower anthocyanin concentration in the high vigour zones. In one year, there was a higher proportion of malvidin-3-O-glucoside and lower proportions of four other anthocyanins (delphinidin-, cyanidin-, petunidin-, and peonidin-3-O-glucosides) commonly found in Pinot noir. In both years studied, one site had proportionally higher peonidin-3-O-glucoside and lower malvidin-3-O-glucoside than the other site. Authors opined that some of these differences might have been related to the higher exposure and temperatures found in site B compared to site A, which were found also in the low vigour zones.

Riesling Terroir: The Impact of Vine Water Status in Ontario

Research in the Bordeaux region has strongly suggested that soil itself is not a determinant of the terroir effect, but that the impact of soil upon vine water status is of crucial importance.[22] Perhaps the first published use of geomatic tools to map vine water status showed some clear spatial correlations between berry sugar carbon isotope ($6^{13}C$) and stem water potential (ψ).[23] This supported data showing relationships between predawn leaf ψ and berry $6^{13}C$.[24] In 2005 to 2007, ten Riesling vineyards with heterogeneous soil types were selected from throughout the Niagara Peninsula to test if soil and vine water status were important determinants of terroir, particularly monoterpene concentration and wine sensory attributes.[25] The objectives were

1. To ascertain the impact of vine and soil water status on FVT/PVT and wine sensory attributes;
2. To enumerate the comparative magnitude of effects of soil texture, water status, and vine vigour; and
3. To elucidate relationships between these variables and wine sensory quality.

Therefore, the major focus of this research was to explain terroir effects that might impact wine varietal character. We sought to elucidate potential determinants of terroir by choosing vine water status as a major factor of the terroir effect. One hypothesis was that consistent water status zones could be identified within vineyard sites and that differences in wine sensory attributes could be related to vine water status. To test this hypothesis, ten commercial Riesling vineyards representative of each sub-appellation created by the Vintners Quality Alliance of Ontario were selected within the Niagara Peninsula. These vineyards were delineated using GPS and 75 to 80 sentinel vines were georeferenced within a sampling grid for data collection. GIS-generated maps were analyzed by spatial correlation analysis. During the 2005 to 2007 growing seasons, vine water status measurements (midday leaf ψ) were collected biweekly from a subset of these sentinel vines. Data were collected on soil texture and composition, soil moisture, vine performance (yield components, vine size) and fruit composition. These variables were mapped using GIS software and relationships between them were elucidated. Vines were categorized into "low" and "high" water status regions within each vineyard block through the use of these geospatial maps, and replicate wines were made from each region. Many geospatial patterns and relationships were found to be spatially and temporally stable within vineyards. Vine water status was found to be temporally stable within vineyards despite different weather conditions during each growing season. Generally, spatial relationships between vine water status, soil moisture, vine size, berry weight, and yield were also stable from year to year. Vine water status also had some impact on the fruit composition in several vineyards. GIS-generated maps were analyzed by spatial correlation analysis. In some instances, FVT /PVT were correlated with leaf ψ or soil moisture or both, suggesting that mild water stress may be beneficial for wine flavour. Sand and clay content of the soils were usually inversely correlated. Soil moisture content was usually higher in clay-dominated areas of the vineyards. Vine water status (leaf ψ) was higher in clay soils. Leaf ψ was often inversely correlated with vine size; i.e., vine water status was improved in low vine-size areas. Berry weight and Brix were both positively correlated with vine water status, while TA was inversely correlated. Spatial relationships in vine water status appeared to be temporally stable, and patterns observed in 2005 appeared similar in 2006 and 2007 despite different weather conditions. In addition to these reasonably good spatial correlations between soil moisture and leaf ψ, there were strong spatial relationships between leaf ψ and both vine size and soil texture. These observations suggest indirectly that vine size and soil texture, in addition to water status, are major contributors to the terroir effect. Particularly compelling were the facts that vine water status was temporally stable, and that temporally stable relationships were elucidated between variables such as leaf ψ, berry weight, Brix, TA, and monoterpenes (Figure 11.4).

Wines from the 2005 and 2006 vintages were subjected to both sorting and descriptive sensory analysis. Initially, a sorting task (multidimensional scaling) was performed on wines from the 2005 and 2006 vintages; wines of similar water status had similar sensory properties. Descriptive analysis using a trained

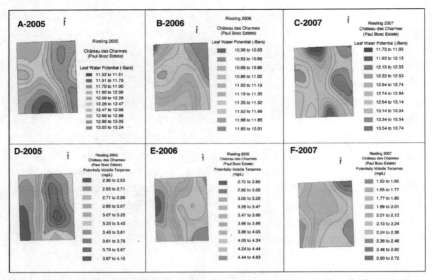

Figure 11.4 Spatial variation in leaf water potential (A–C) and potentially volatile monoterpenes (D–F; mg/L; high and low ranges only) in Riesling, Château des Charmes, St. Davids, Ontario, 2005–7
Source: A.G. Reynolds, C. de Savigny, and J.J. Willwerth, "Riesling Terroir in Ontario Vineyards: The Roles of Soil Texture, Vine Size and Vine Water Status," *Progrès Agricole et Viticole* 127, no. 10 (2010): 212–22.

panel further indicated that water status had an effect on wine sensory profiles. Similar attributes were different for wines from different water status zones. Through multivariate analyses, specific sensory attributes, viticultural variables, and chemical variables were associated with wines of different water status. Vine water status was a major contributor to the terroir effect, as it had a major impact on vine size, berry weight, fruit composition, and wine sensory characteristics. Partial least squares (PLS) analysis in fact showed that the greatest percentage of sensory attributes were associated with leaf ψ, soil moisture, and vine size.

At the outset of this work, we hypothesized, consistent with Seguin, that the main factors driving the terroir effect might be soil texture based.[26] We also hypothesized, consistent with Van Leeuwen,[27] Van Leeuwen and Seguin,[28] and Van Leeuwen et al.,[29] that the terroir effect would be strongly based upon soil moisture, vine water status, or both. These hypotheses were, for the most part, proven. In the majority of situations, distinct spatial patterns in soil texture, soil moisture, and leaf ψ were demonstrated. Moreover, the spatial patterns in soil moisture and leaf ψ were in most cases temporally stable, and any temporal variations in their spatial patterns were likely influenced by the volatile precipitation patterns that are typical of the region. Finally, there were clear spatial correlations between soil moisture, leaf ψ, and many soil physical and composition variables, including soil texture (percentage of sand and clay), soil pH, organic matter (OM), cation-exchange capacity (CEC), base saturation (BS), phosphorus (P), potassium (K), calcium (Ca), and magnesium (Mg).

Riesling Terroir Defined by Airborne Imaging in Ontario

A ten-hectare Riesling site in Beamsville, Ontario (Thirty Bench Wine Makers), was chosen for this study. The block had heretofore been divided into three sub-blocks for small lot winemaking, and at the outset of the study, we sub-divided two of these into high and low vigour zones based upon low elevation (1000 metres) imaging, and delineated an additional sixth sub-block. This study ultimately demonstrated that remote sensing using high-definition multispectral airborne imaging can be used to identify useful differences within a vineyard.[30] Ground data for berry composition from 2006 to 2008 were used to show that there were significant differences between the six vineyard blocks, and that sub-division of some blocks between low and high vigour zones was appropriate. Subdivision within blocks into three vine water status zones was also shown to be appropriate. Values for a normalized difference vegetation index (NDVI), calculated from leaf reflectance data from a ground-based spectrometer, matched the pattern of NDVI values calculated from high-definition multispectral airborne imaging. NDVI correlated well with vine performance measurements such as yield and vine size, as well as leaf ψ and soil moisture (Figure 11.5). Although correlations of vegetation indices with berry and wine composition variables were not as strong as for vine performance metrics, values for red edge inflection point, another index, were used to predict the location of vineyard areas that produced high PVT concentrations. Study of the same vineyard in the 2007 and 2008 growing seasons verified the patterns of variation between blocks and sub-zones. Hence, the spatial variability was temporally stable, and the relationships between the remotely sensed leaf reflectance data and the ground data remained the same. We were able to tentatively conclude that airborne imaging might be used someday to help determine the vineyard segregation boundaries for small-lot wine production.

Cabernet Franc Terroir: The Impact of Vine Water Status in Ontario

Much like the Riesling study discussed previously, the aim of this study was to look at spatial variation in soil texture and composition, yield components, and berry composition, and to see whether these variables were spatially correlated.[31] Moreover, an objective was also to examine the impact of vine water status on sensory and chemical characteristics of Cabernet Franc wines on non-irrigated sites in the Niagara Peninsula to assess whether vine water status might be a key factor in the determination of so-called terroir effects. The effects of vine water

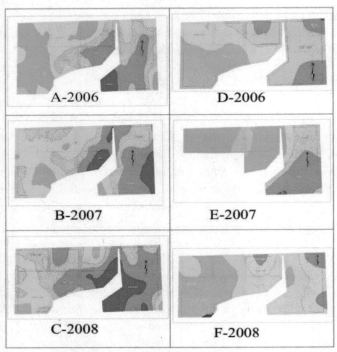

Figure 11.5 Map of soil moisture (% v/v) and leaf water potential across six vineyard sub-blocks at Thirty Bench vineyards in Beamsville, Ontario, 2006 to 2008. North is oriented to the top. A to C: soil moisture 2006, 2007, and 2008, respectively; D to F: leaf water potential, 2006, 2007, and 2008, respectively. Orange-shaded areas near the right of each map are those with lowest values; blue zones near the left are highest in value.
Source: A.G. Reynolds, M. Marciniak, R.B. Brown, L. Tremblay, and L. Baissas, "Using GPS, GIS and Airborne Imaging to Understand Niagara Terroir," *Progrès Agricole et Viticole* 127, no. 12 (2010): 259–74.

status on wine sensory characteristics were thereby studied in Cabernet Franc in the Niagara Peninsula in the 2005 and 2006 vintages. Vine water status was monitored throughout the growing season in ten vineyard blocks using midday leaf ψ values.

Areas of low water status were positively correlated with areas of high Brix, A520, A420+520, total anthocyanins, and total phenols, and negatively correlated with berry weight and TA (Figure 11.6).

In most vineyards, areas of high and low colour intensity were positively correlated with areas of high and low anthocyanins and total phenols. Chemical and descriptive sensory analyses were performed on experimental wines to elucidate differences between wines from high and low water status zones in each vineyard. Judges evaluated six aroma and six flavour (red fruit, black cherry,

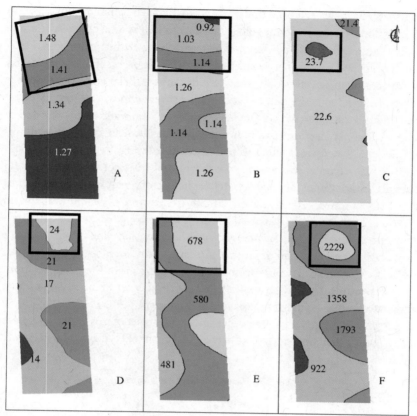

Figure 11.6 Spatial variation in A: Leaf water potential (-bars); B: Berry weight (g); C: Berry Brix; D: Berry color absorbance (A420 + A520); E: Berry total anthocyanins (mg/L); F: Berry total phenols (mg/kg), Château des Charmes, St. Davids, ON, 2005.
Zones marked by black polygons are those with lowest leaf water potential and berry weight, and highest Brix, absorbance, total anthocyanins and total phenols. Values represent the lower limit within each zone.
Source: Andrew G. Reynolds. Derived from the Ph.D. thesis of Javad Hakimi Rezaei (Brock University, 2009): "Delineation of Within-Site Terroir Effects Using Soil and Vine Water Measurement: Investigation of Cabernet Franc."

blackcurrant, black pepper, bell pepper, and green bean) and three mouthfeel/taste (astringency, bitterness, and acidity) sensory attributes, as well as colour intensity. Leaf ψ varied within and between vineyards in both years. In 2005, low water status wines had higher colour intensity (four sites), black cherry flavour (one site), and red fruit aroma and flavour (two sites). Similar trends were observed in the 2006 vintage. No temporal differences were found between the wines produced from the same vineyard, indicating that the attributes of these wines were consistent despite markedly different conditions in the 2005 and 2006 vintages. PLS analysis showed that leaf ψ was associated with red fruit and berry aromas and flavours, and wine colour intensity, as well as total phenols, Brix, and anthocyanins, while soil moisture was associated with acidity, green bean aroma and flavour, and bell pepper aroma and flavour. Measurement of midday leaf ψ was successful in detecting differences among vine water status levels throughout the growing season. The range of leaf ψ values were almost consistent at most sites in both 2005 and 2006 years. Differences in vine water status resulted in wines with different composition, aroma, flavour, and colour intensity. At almost all sites, low water status wines were associated with high red fruit aroma and flavour, black fruit aroma and flavour, berry and wine colour intensity, total phenols, anthocyanins, and berry pH. Despite two different vintages of hot and dry (2005) and wet (2006) seasons, similar trends were observed in high and low water status wines. PLS analysis illustrated that leaf ψ was positively correlated with red fruit aroma/flavour, berry colour intensity, wine colour intensity, total phenols and Brix, and negatively correlated with soil moisture, green bean aroma/flavour, and bell pepper aroma/flavour (Figure 11.7). The strong relationships between leaf ψ and sensory attributes of Cabernet Franc suggested that vine water status was a major basis for the terroir effect.

As with Riesling, we had hypothesized at the outset of this work, consistent with Seguin,[32] that the main factors driving the terroir effect might be soil texture

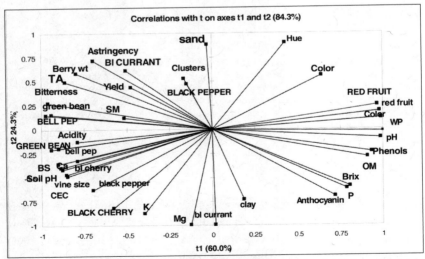

Figure 11.7 Partial least squares analysis of field and sensory data for nine Cabernet Franc wines from Niagara Peninsula, ON, 2005. Aroma attributes are represented in lowercase and flavour attributes are represented in uppercase.
Source: J. Hakimi Rezaei and A.G. Reynolds, "Impact of Vine Water Status on Sensory Evaluation of Cabernet Franc Wines in the Niagara Peninsula of Ontario," *Journal International des Sciences de la Vigne et du Vin* 44 (2010): 61–75.

based. We also hypothesized, consistent with Van Leeuwen,[33] Van Leeuwen and Seguin,[34] and Van Leeuwen et al.[35] that the terroir effect would be strongly based upon soil moisture, vine water status, or both. These hypotheses were, for the most part, proven with Cabernet Franc. In the majority of situations, distinct spatial patterns in soil texture, soil moisture, and leaf ψ were demonstrated. Moreover, the spatial patterns in soil moisture and leaf ψ were in most cases temporally stable, and any temporal variations in their spatial patterns were likely influenced by the volatile precipitation patterns that are typical of the region. Finally, there were clear spatial correlations between soil moisture, leaf ψ, and many soil physical and composition variables, including soil texture (percentage sand and clay), soil pH, organic matter (OM), cation-exchange capacity (CEC), base saturation (BS), phosphorus (P), potassium (K), calcium (Ca), and magnesium (Mg).

As suggested earlier in this chapter, there is a very clear link between understanding terroir and the implementation of canopy management. In situations where leaf ψ does not vary temporally, this knowledge might allow one to implement specific viticultural practices in high water status situations (canopy management, crop reduction) or in drought situations (e.g., deficit irrigation).[36] There is also the potential for establishing temporally stable zones of different flavour potential.[37] In Cabernet Franc, 2-methoxy-3-isobutylpyrazine (IBMP) is ubiquitous worldwide, and a substantial influence is exerted by soil type (less IBMP in gravel soils).[38] The norisoprenoid β-damascenone has a huge impact upon odor activity of wines, and although it has odor impact by itself, it also enhanced fruity notes of ethyl cinnamate and ethyl caproate and suppressed the odor activity of IBMP in red Bordeaux wines including Cabernet Franc; moreover, its concentration varied according to soil type.[39] The cysteine precursors of odor-active thiol compounds were shown to be closely linked to nitrogen status in Sauvignon Blanc, and therefore zones within vineyards with high nitrogen supply can potentially increase varietal typicity in this cultivar.[40]

Effects of Fruit Exposure Effects on Aroma Compounds

A primary pillar of the Cool Climate Paradigm is that fruit must be kept warm. Fruit exposure can be substantially impacted by the texture and depth of the soil. For example, a deep, coarse-textured soil is likely to produce a deep root system, a vigorous vine, and the potential for substantial canopy shade. On the other hand, a shallow, fine-textured soil is very likely to produce a low-vigour vine with a high degree of cluster exposure. In addition to naturally occurring cluster exposure that is based exclusively on vineyard ecology, cluster exposure can also be a factor that is manipulated through cultural practices such as training system or growing season canopy management. In general, exposing fruit to the sun will increase fruit temperature along with the enzymatic activities therein. Consequently, when compared to shaded fruit, exposed fruit will normally contain lower malic acid concentration,[41] lower methoxypyrazines,[42] higher anthocyanins and phenols, higher monoterpenes,[43] and, occasionally, higher Brix.[44] Increased fruit exposure has also been shown to reduce methoxypyrazines[45] and to increase norisoprenoids.[46] However, there are two main components to fruit exposure—light and temperature—and it has been exceedingly difficult to separate the effects of these factors.

Effects of Viticultural Practices on Fruit Composition and Wine Quality

Effects of Growing Season Canopy Management

A second important pillar of the Cool Climate Paradigm is to keep leaves well exposed to sunlight. This may be accomplished in many ways, but ultimately the vigour of the vine must be accommodated such that canopy shade is avoided. Carbonneau et al.[47] and Smart et al.[48] were among the first to take a viticultural experiment "from the field to the glass" to conclusively demonstrate the positive impact of canopy management for minimization of canopy shading—in this case, use of divided canopies. The degree of shade in Shiraz grapevine canopies was varied by four treatments—control; shade; severe hedging; and Geneva double curtain—to create a naturally occurring vigour gradient, hence providing different canopy microclimates.[49] Constraining foliage into a smaller volume ("shade") increased shading over control vines, and Geneva double curtain training and slashing reduced it. A shaded canopy microclimate was associated with increased K^+ concentration in the leaves, petioles, and stems at veraison (the onset of ripening), as well as reduced Brix, and higher malic acid and K^+ concentrations in fruit. Wines from these musts had higher pH, K^+ and reduced proportions of ionised anthocyanins. An eight-character visual scorecard of grapevine canopies was evaluated and used to describe the canopies. A conceptual model was proposed to explain how soil and climatic factors and cultural practices can affect canopy microclimate. High vine vigour had similar effects on must and wine composition as shading treatments.

Hedging the tops and sides of vigorous canopies, and removing leaves in the fruit zone, have likely been traditional techniques in Europe for centuries. Vertically shoot-positioned canopies are typically hedged to eliminate overhanging shoots, reduce canopy width, and generally reduce shade in dense canopies. Normally this is done about two weeks after fruit set, but can be repeated one or more times during the growing season. Basal leaf removal is generally done one week or more after hedging by removing the lower two to four leaves at the base of the shoot to expose the clusters. This can be done on both sides of the canopy, particularly in the case of red wine cultivars and during wet

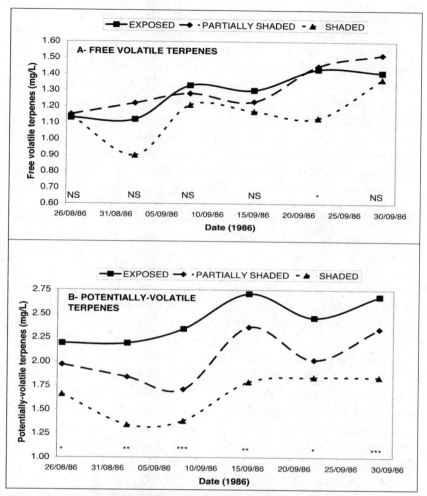

Figure 11.8 Effects of cluster exposure on monoterpene composition of Gewürztraminer, Kaleden, British Columbia, 1986. A: Free volatile terpenes (mg/L); B: Potentially volatile terpenes (mg/L). *,**,***, ns: Significant at $p < 0.05, 0.01, 0.001$, or not significant, respectively. Source: Redrawn from A.G. Reynolds and D.A. Wardle, "Influence of Fruit Microclimate on Monoterpene Levels of Gewürztraminer," *American Journal of Enology and Viticulture* 40 (1989): 149–54.

growing seasons; typically, it is done on the least sunny side (east or north side in the northern hemisphere) to avoid the possibility of sunburning the fruit. These practices have been addressed through research only since the mid-1980s. Those who focused on growing season canopy management include a study that showed that basal leaf removal on Chardonnay vines in New York increased Brix and lowered TA.[50] Hedging of Riesling canopies followed by inhibition of lateral shoot growth enhanced fruit composition,[51] while severe hedging of

De Chaunac vines, particularly when performed late in the season, was detrimental to berry composition.[52] Smith et al. reduced TA using fruit zone leaf removal, and also found substantial increases in several free and glycosylated aroma compounds (including monoterpenes) in Sauvignon Blanc berries from leaf-removed treatments.[53] Leaf removal reduced Brix in some situations;[54] however, TA was reduced and concentrations of free and bound monoterpenes increased in Gewürztraminer[55] and Riesling,[56] in several early-maturing Muscat cultivars,[57] and most recently in Ontario Chardonnay Musqué.[58] Basal leaf removal also increased glycosyl-glucose (GG) concentrations[59] in Riesling[60] and Chardonnay[61] in Virginia. Basal leaf removal also reduces methoxypyrazines, as shown in studies in California[62] and New York.[63]

Effects of Shoot Density and Crop Level

A third essential pillar of the Cool Climate Paradigm is that the vine must be balanced with respect to vine vigour and crop size. If cluster thinning has an impact upon vine size, it may be argued that it is therefore affecting vine balance. However, thinning is frequently carried out at veraison, at which time vegetative growth should have ceased. Fruit composition and wine quality may therefore be impacted, but vine size remains unchanged. Regardless, the more recent studies that have focused upon the effects of crop level upon fruit composition and wine quality are worthy of mention here. Cluster thinning has been associated with increased yield, vine size, vine hardiness, fruit Brix, flavour compounds, anthocyanins, and, sometimes, wine quality. Studies in North America include those with Chardonnay Musqué,[64] Gewürztraminer,[65] Müller-Thurgau,[66] Pinot noir,[67] Riesling,[68] and French–American hybrid cultivars such as De Chaunac[69] and Seyval.[70] Reduced crop level tended to increase monoterpenes in aromatic white wine cultivars.[71] In a trial testing Riesling shoot density × cordon age, increasing volume of "old" wood also increased berry and must FVT and PVT.[72] Crop thinning can also help decrease methoxypyrazine concentration in grapes because excessive crop to leaf area can delay the rate of fruit maturity and therefore the degradation of methoxypyrazines.[73]

Influence of Training Systems

Research on training systems with respect to their impact on wine quality did not begin until the 1980s. Much of this was inspired by the seminal work of Shaulis et al. on Concord in New York.[74] The most noteworthy of those first studies dedicated to training and wine grape quality include the work of Carbonneau et al., who described the beneficial impact of canopy division (the Lyre trellis) on fruit composition and wine quality of Cabernet Sauvignon in Bordeaux.[75] Smart[76] and Smart et al.[77] showed the benefits of divided canopy training on fruit composition and wine quality of Shiraz in Australia. Studies assessing training systems in North America alone include those on the French–American hybrids Seyval,[78] Chancellor,[79] Vignoles,[80] and Vidal,[81] and *Vitis vinifera* cultivars Chardonnay and Cabernet Franc[82] and Riesling.[83] Generally speaking, changes to a training system within a standard vertical shoot–positioned system produces minor differences in fruit composition.[84] Increased trunk height in French–American hybrids has been shown to increase Brix and decrease TA as a result of enhanced fruit exposure,[85] but this is not always the case.[86] Divided canopies can also increase monoterpenes. A divided canopy system (alternate double crossarm, or ADC) produced yields in Riesling in British Columbia as high as 33 tonnes per hectare, along with lower TA, and higher FVT and PVT than standard *pendelbogen* and bilateral cordon systems.[87] In Virginia, divided canopies generally produced wines higher in some terpenes and in overall glucoconjugates.[88] As in the case of fruit exposure and basal leaf removal, certain training systems may enhance fruit environment and potentially lead to changes in methoxypyrazine concentration. Nonetheless, gobelet- and bilateral cordon-trained Cabernet Sauvignon vines did not produce fruit with different berry IBMP concentrations.[89] However, IBMP concentration of the final wines was much higher in the cordon-trained vines.

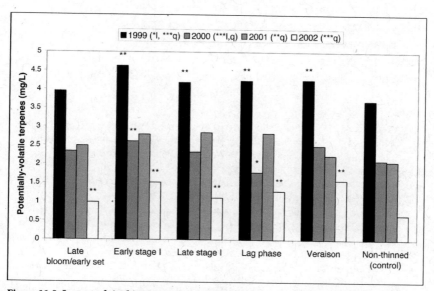

Figure 11.9 Impact of six thinning times on Chardonnay musqué berry potentially volatile terpene concentration, 1999 to 2001. Asterisks indicate significant difference from the control, $p \leq 0.05$, Dunnett's t-test.

Source: A.G. Reynolds, J.W. Schlosser, D. Sorokowsky, R. Roberts, and C. de Savigny, "Magnitude and Interaction of Viticultural and Enological Effects. II. Relative Impacts of Cluster Thinning and Yeast Strain on Composition and Sensory Attributes of Chardonnay Musqué," *American Journal of Enology and Viticulture* 58 (2007): 25–41.

Numerous training systems are used around the world for wine grapes. One of the many objectives of a successful training system affords optimal fruit exposure that is appropriate for the region in which the grapes are grown. A great many studies have compared the effects of training systems on fruit composition and wine quality.

Conclusions

The results of these experiments suggest that fruit exposure, canopy manipulation, pre-fermentation practices, and vineyard site may influence monoterpene concentration of berries and juices of several *Vitis vinifera* cultivars. These differences can sometimes be confirmed organoleptically in wines. A failure to find good agreement between analytical and sensory results may be due to variability among judges, but may also be ascribed in part to the confounding taster response to non-floral monoterpenes such α-terpineol. This underscores the need to follow up work of this nature with gas chromatographic analyses of wines, to overcome problems of this nature. Our work and that of others has demonstrated that vineyard and cellar practices may affect aroma compound concentration in berries and juices, and much of this work has shown that organoleptic evaluation may confirm these analytical results. Our specific conclusions to date are: (1) PVT are more responsive to viticultural and enological practices than FVT; (2) FVT and PVT are rarely correlated with Brix, TA, or pH, and thus cannot be predicted by standard harvest indices; (3) losses in FVT and PVT can occur between the berry and juice stages, hence the desirability of skin contact; (4) FVT and PVT concentrations can, in some cases, be related to wine-tasting results.

Notes

1   A.G. Reynolds, "Viticultural and Vineyard Management Practices and Their Effects on Grape and Wine Quality," in *Managing Wine Quality: Viticulture and Wine Quality*, ed. A.G. Reynolds, Vol. I (Cambridge, UK: Woodhead Publishing, 2010): 365–444.

2   R.E. Smart, "Vine Manipulation to Improve Wine Grape Quality," in *Proceedings of the University of California, Davis, Grape and Wine Centennial Symposium*, ed. A.D. Webb (Berkeley, CA: University of California Press, 1982): 362–75.

3   R.E. Smart, J.B. Robinson, G. Due, and C.J. Brien, "Canopy Microclimate Modification for the Cultivar Shiraz. I. Definition of Canopy Microclimate," *Vitis* 24 (1985): 17–24; R.E. Smart, J.B. Robinson, G. Due, and C.J. Brien, "Canopy Microclimate Modification for the Cultivar Shiraz. II. Effects on Must and Wine Composition," *Vitis* 24: 119–28.

4   A. Carbonneau, P. Casteran, and P. Leclair, "Essai de détermination en biologie de la plante entière, de relations essentielles entre le bioclimat naturel, la physiologie de la vigne et la composition du raisin," *Ann. Amelior. Plantes* 28 (1978): 195–221.

5   W.M. Kliewer, J.J. Marois, A.M. Bledsoe, S.P. Smith, M.J. Benz, and O. Silvestroni, "Relative Effectiveness of Leaf Removal, Shoot Positioning, and Trellising for Improving Winegrape Composition," in *Proceedings of the 2nd International Symposium on Cool Climate Viticulture and Oenology*, ed. R.E. Smart, R.J. Thornton,

S.B. Rodriguez, and J.E. Young (Auckland: New Zealand Society for Viticulture and Oenology, 1988): 123–26.

6 A.G. Reynolds and D.A. Wardle, "Impact of Training System and Vine Spacing on Vine Performance and Berry Composition of Seyval Blanc," *American Journal of Enology and Viticulture*, 45 (1994): 444–51; A.G. Reynolds and D.A. Wardle, "Flavour Development in the Vineyard," *South African Journal for Enology and Viticulture* 18, 3–18; A.G. Reynolds, D.A. Wardle, and A.P. Naylor, "Impact of Training System and Vine Spacing on Vine Performance and Berry Composition of Chancellor," *American Journal of Enology and Viticulture*, 46 (1995): 88–97; A.G. Reynolds, D.A. Wardle, and A.P. Naylor, "Impact of Training System, Vine Spacing, and Basal Leaf Removal on Riesling. Vine Performance, Berry Composition, Canopy Microclimate, and Vineyard Labor Requirements," *American Journal of Enology and Viticulture*, 47 (1996): 63–76.

7 R.E. Smart and S.M. Smith, "Canopy Management: Identifying the Problems and Practical Solutions," in *Proceedings of the 2nd International Symposium on Cool Climate Viticulture and Oenology*, ed. R.E. Smart, R.J. Thornton, S.B. Rodriguez, and J.E. Young (Auckland: New Zealand Society for Viticulture and Oenology, 1988): 109–15; S. Smith, I.C. Codrington, M. Robertson, and R.E. Smart, "Viticultural and Oenological Implications of Leaf Removal for New Zealand Vineyards," in Smart et al., *Proceedings of the 2nd International Symposium on Cool Climate Viticulture and Oenology*, ed. R.E. Smart, R.J. Thornton, S.B. Rodriguez, and J.E. Young (Auckland: New Zealand Society for Viticulture and Oenology, 1988): 127–33.

8 N.J. Shaulis, H. Amberg, and D. Crowe, "Response of Concord Grapes to Light, Exposure, and Geneva Double Curtain Training," *Proceedings American Society of Horticultural Science* 89 [1966], 268–80.

9 N.J. Shaulis and P. May. "Response of 'Sultana' vines to training on a divided canopy and to shoot crowding," *American Journal of Enology and Viticulture* 22, no. 4 (1971): 215–22.

10 R. Di Stefano and L. Corino, "Valutazione comparativa fra Moscato bianco e Moscato giallo con particolare riferimento alla componente terpenica," *Riv. Vitic. Enol. Conegliano*, 37 (1984): 657–70; R. Di Stefano and L. Corino, "Caratteristiche chimiche ed aromatiche di vini secchi prodotti con Moscato Bianco e Giallo di Chambave e con Moscato Bianco di Canelli," *Riv. Vitic. Enol. Conegliano*, 39 (1986): 3–11.

11 L. Corino and R. Di Stefano, Comportamento del vitigno Moscato Bianco in relazione ad ambienti di coltivazione diversi e valutazione di sistemi di allevamento e potatura," *Riv. Vitic. Enol. Conegliano* 41 (1988): 72–85.

12 A.C. Noble, "Evaluation of Chardonnay Wines Obtained from Sites with Different Soil Compositions," *American Journal of Enology and Viticulture* 30 (1979): 214–17.

13 M.S. Larrechi and F.X. Ruiz, "Multivariate Data Analysis Applied to the Definition of Two Catalan Viticultural Regions. I. Cluster analysis. *Z. Lebensm. Unters. Forsch.*, 185 (1987): 181–84; M.S. Larrechi, J. Guasch, and F.X. Ruiz, "The Definition of Two Catalan Viticultural Regions by Classification Methods," *Acta Alimentaria* 17 (1988): 177–82.

14 A.J.W. Ewart, "Influence of Vineyard Site and Grape Maturity on Juice and Wine Quality of *Vitis vinifera* cv. Riesling," in *Proceedings of the 6th Australian Wine Industry Technical Conference*, 14–17 July, 1986, Adelaide, South Australia, ed. T. Lee (Adelaide, S. Australia: Australian Industrial Publishers, 1987): 89–93.

15 A.G. Reynolds, D.A. Wardle, J.W. Hall, and M.J. Dever, "Fruit Maturation in Four *Vitis vinifera* Cultivars in Response to Vineyard Location and Basal Leaf Removal," *American Journal of Enology and Viticulture* 46 (1995): 542–58.

16 Reynolds et al., "Fruit maturation in Four *Vitis vinifera* Cultivars in Response to Vineyard Location and Basal Leaf Removal."

17 A.G. Reynolds, D.A. Wardle, and M.J. Dever, "Vine Performance, Fruit Composition, and Wine Sensory Attributes of Gewürztraminer in Response to Vineyard Location and Canopy Manipulation," *American Journal of Enology and Viticulture* 47 (1996): 77–92.

18 J. Coipel, B. Rodriguez-Lovelle, C. Sipp, and C. Van Leeuwen, "«Terroir» Effect, as a Result of Environmental Stress, Depends More on Soil Depth than on Soil Type (Vitis vinifera L. cv. Grenache noir, Côtes du Rhône, France, 2000). *Journal International des Sciences de la Vigne et du Vin* 40 (2006): 177–85; G. Seguin, "Terroirs and Pedology of Wine Growing," *Experientia,* 42 (1986): 861–73; C. Van Leeuwen, P. Friant, X. Choné, O. Tregoat, S. Koundouras, and D. Dubourdieu, "The Influence of Climate, Soil and Cultivar on Terroir," *American Journal of Enology and Viticulture* 55 (2004): 207–17.

19 A.G. Reynolds and C. De Savigny, "Use of GPS and GIS to determine the basis for terroir." *Proceedings of the Space Age Winegrowing Symposium,* ed. A.G. Reynolds, American Society for Enology and Viticulture, Eastern Section, pp. 79–102. Presented at ASEV/ES meeting, 11–12 July 2001.

20 A.G. Reynolds, I. Senchuk, and C. de Savigny, "Use of GPS and GIS for Elucidation of the Basis for Terroir. Spatial Variation in an Ontario Riesling Vineyard," *American Journal of Enology and Viticulture* 58 (2007): 145–62; A.G. Reynolds, C. de Savigny C, and J.J. Willwerth, "Riesling Terroir in Ontario Vineyards. The Roles of Soil Texture, Vine Size and Vine Water Status," *Progrès Agricole et Viticole* 127, no. 10 (2010): 212–22.

21 J.M. Cortell, M. Halbleib, A.V. Gallagher, T.L. Righetti, and J.A. Kennedy, "Influence of Vine Vigour on Grape (*Vitis vinifera* L. cv. Pinot noir) Anthocyanins. 1. Anthocyanin Concentration and Composition in Fruit," *Journal of Agricultural and Food Chemistry,* 55 (2007): 6575–84.

22 Coipel et al., "«Terroir» Effect, as a Result of Environmental Stress, Depends More on Soil Depth than on Soil Type"; Seguin, "Terroirs and Pedology of Wine Growing"; C. Van Leeuwen, "Terroir: The Effect of the Physical Environment on Vine Growth, Grape Ripening and Wine Sensory Attributes," in *Managing Wine Quality, Volume 1: Viticulture and Wine Quality,* ed. A. Reynolds (Cambridge, UK: Woodhead Publishing, 2010): 273–315; Van Leeuwen et al., "The Influence of Climate, Soil and Cultivar on Terroir"; Van Leeuwen et al., "Vine Water Status Is a Key Factor in Grape Ripening and Vintage Quality for Red Bordeaux Wine."

23 C. Van Leeuwen, J.-P. Goutouly, C. Azais, A.-M. Costa-Ferreira, E. Marguerit, J.-P. Roby, X. Choné, and J.-P. Gaudillère, "Intra-Block Variations of Vine Water Status in Time and Space," VIth International Terroir Congress, 2–7 July 2006, ENITA de Bordeaux – Syndicat Viticole des Coteaux du Languedoc, France, 2006): 64–69.

24 J.-P. Gaudillère, C. Van Leeuwen, and N. Ollat, "Carbon Isotope Composition of Sugars in Grapevine: An Integrated Indicator of Vineyard Water Status," *J. Exp. Bot.* 53 (2002): 757–63.

25 Reynolds et al., "Riesling Terroir in Ontario Vineyards"; J.J. Willwerth, A.G. Reynolds, and L. Lesschaeve, "Terroir Factors: Their Impact in the Vineyard and on the Sensory Profiles of Riesling Wines," *Progrès Agricole et Viticole* 127, no. 8 (2010): 159–68.

26 Seguin, "Terroirs and Pedology of Wine Growing."

27 Van Leeuwen, "Terroir."

28 C. Van Leeuwen and G. Seguin, "Incidences de l'alimentation en eau de la vigne, appréciée par l'état hydrique du feuillage, sur le développement de l'appareil végétatif et la maturation du raisin (Vitis vinifera variété Cabernet franc, Saint-Emilion, 1990)," *Journal International des Sciences de la Vigne et du Vin* 28 (1994): 81–110.

29 Van Leeuwen et al., "The Influence of Climate, Soil and Cultivar on Terroir"; Van Leeuwen et al., "Vine Water Status Is a Key Factor in Grape Ripening and Vintage Quality for Red Bordeaux Wine."

30 A.G. Reynolds, M. Marciniak, R.B Brown, L. Tremblay, and L. Baissas, "Using GPS, GIS and Airborne Imaging to Understand Niagara Terroir," *Progrès Agricole et Viticole* 127, no. 12 (2010): 259–74.

31 J. Hakimi Rezaei and A.G. Reynolds, "Characterization of Niagara Peninsula Cabernet Franc Wines by Sensory Analysis," *American Journal of Enology and Viticulture* 61 (2010): 1–14; J. Hakimi Rezaei and A.G. Reynolds, "Evaluation of Cabernet Franc Wines in the Niagara Peninsula," *Progrès Agricole et Viticole* 127, no. 4 (2010): 87–92; J. Hakimi Rezaei and A.G. Reynolds, "Impact of Vine Water Status on Sensory Evaluation of Cabernet Franc Wines in the Niagara Peninsula of Ontario," *Journal International des Sciences de la Vigne et du Vin* 44 (2010): 61–75.

32 Seguin, "Terroirs and Pedology of Wine Growing."

33 Van Leeuwen, "Terroir."

34 Van Leeuwen and Seguin, "Incidences de l'alimentation en eau de la vigne, appréciée par l'état hydrique du feuillage, sur le développement de l'appareil végétatif et la maturation du raisin (Vitis vinifera variété Cabernet franc, Saint-Emilion, 1990)."

35 Van Leeuwen et al., "The Influence of Climate, Soil and Cultivar on Terroir"; Van Leeuwen et al., "Vine Water Status Is a Key Factor in Grape Ripening and Vintage Quality for Red Bordeaux Wine."

36 Van Leeuwen et al., "Vine Water Status Is a Key Factor in Grape Ripening and Vintage Quality for Red Bordeaux Wine."

37 Willwerth et al., "Terroir factors."

38 C. Peyrot des Gachons, C. Van Leeuwen, T. Tominaga, J.P. Soyer, J.P. Gaudillère, and D. Dubourdieu, "Influence of Water and Nitrogen Deficit on Fruit Ripening and Aroma Potential of *Vitis vinifera* L cv Sauvignon blanc in Field Conditions," *Journal of Agricultural and Food Chemistry*, 85 (2005): 73–85; D. Roujou de Boubée, C. Van Leeuwen, and D. Dubourdieu, "Organoleptic impact of 2-methoxy-3-isobutylpyrazine on red Bordeaux and Loire wines. Effect of environmental conditions on concentrations in grapes during ripening," *Journal of Agricultural and Food Chemistry* 48 (2000): 4830–4.

39 B. Pineau, J.-C. Barbe, C. Van Leeuwen, and D. Dubourdieu, "Which Impact for β-damascenone on Red Wine Aroma?" *Journal of Agricultural and Food Chemistry*, 55 (2007): 4103–8.

40 X. Choné, V. Lavigne-Cruege, T. Tominaga, C. Van Leeuwen, C. Castagnede, C. Saucier, and D. Dubourdieu, "Effect of Vine Nitrogen Status on Grape Aromatic Potential: Flavor Precursors (S-cysteine conjugates), Glutathione and Phenolic Content in *Vitis vinifera* L. cv. Sauvignon Blanc Grape Juice," *Journal International des Sciences de la Vigne et du Vin* 40 (2006): 1–6.

41 A.N. Lakso and K.M. Kliewer, "The Influence of Temperature on Malic Acid Metabolism in Grape Berries. I. Enzyme responses," *Plant Physiology*, 56 (1976): 370–72.

42 M.S. Allen, M.J. Lacey, R.L.N. Harris, and W.V. Brown, "Contribution of Methoxypyrazines to Sauvignon Blanc Wine Aroma," *American Journal of Enology and Viticulture* 42 (1991): 109–12; Y. Kotseridis, A. Anocibar Beloqui, C.L. Bayonove, R.L. Baumes, A. Bertrand, "Effects of Selected Viticultural and Enological Factors on Levels of 2-methoxy-3-isobutylpyrazine in Wines. *Journal International des Sciences de la Vigne et du Vin* 33 (1999): 19–23.

43 A. Belancic, E. Agosin, A. Ibacache, E. Bordeu, R. Baumes, A.J. Razungles, and C.L. Bayonove, "Influence of Sun Exposure on the Aromatic Composition of Chilean Muscat Grape Cultivars Moscatel de Alejandría and Moscatel Rosada," *American Journal of Enology and Viticulture* 48 (1997): 181–86; S.M. Bureau, A.J. Razungles,

and R.L. Baumes, "The Aroma of Muscat of Frontignan Grapes: Effect of the Light Environment of Vine or Bunch on Volatiles and Glycoconjugates," *Journal of Agricultural and Food Chemistry,* 80 (2000): 2012–20; L.E. Macaulay and J.R. Morris, "Influence of Cluster Exposure and Winemaking Processes on Monoterpenes and Wine Olfactory Evaluation of Golden Muscat," *American Journal of Enology and Viticulture* 44 (1993): 198–204; A.G. Reynolds and D.A. Wardle, "Influence of Fruit Microclimate on Monoterpene Levels of Gewürztraminer," *American Journal of Enology and Viticulture* 40 (1989): 149–54 (Fig. 8); Reynolds and Wardle, "Flavour Development in the Vineyard"; P.A. Skinkis, B.P. Bordelon, and E.M. Butz, "Effects of Sunlight Exposure on Berry and Wine Monoterpenes and Sensory Characteristics of Traminette," *American Journal of Enology and Viticulture* 61 (2010): 147–56.

44   W.M. Kliewer and L.A. Lider, "Influence of cluster exposure to the sun on the composition of Thompson Seedless Fruit," *American Journal of Enology and Viticulture* 19 (1968): 175–84; D.D. Crippen and J.C. Morrison, "The Effects of Sun Exposure on the Compositional Development of Cabernet Sauvignon Berries," *American Journal of Enology and Viticulture* 37 (1986): 235–42; A.G. Reynolds, R.M. Pool, and L.R. Mattick, "Influence of Cluster Exposure on Fruit Composition and Wine Quality of Seyval blanc," *Vitis* 25 (1986): 85–95.

45   Roujou de Boubée et al., "Organoleptic Impact of 2-methoxy-3-isobutylpyrazine on Red Bordeaux and Loire Wines"; I. Ryona, B.S. Pan, D.S. Intrigliolo, A.N. Lakso, and G.L. Sacks, "Effects of Cluster Light Exposure on 3-isobutyl-2-methoxypyrazine Accumulation and Degradation Patterns in Red Wine Grapes (*Vitis vinifera* L. cv. Cabernet Franc)," *Journal of Agricultural and Food Chemistry* 56 (2008): 10838–46.

46   S.M. Bureau, R.L. Baumes, and A.J. Razungles, "Effects of Vine or Bunch Shading on the Glycosylated Flavor Precursors in Grapes of Vitis vinifera L. cv. Syrah," *Journal of Agricultural and Food Chemistry* 48 (2000): 1290–97; S.-H. Lee, M.-J. Seo, M. Riu, J.P. Cotta, D.E. Block, N.K. Dokoozlian, and S.E. Ebeler, "Vine Microclimate and Norisoprenoid Concentration in Cabernet Sauvignon Grapes and Wines, *American Journal of Enology and Viticulture* 58 (2007): 291–300; J. Marais, C.J. van Wyk, and A. Rapp, "Effect of Sunlight and Shade on Norisoprenoid Levels in Maturing Weisser Riesling and Chenin Blanc Grapes and Weisser Riesling Wines," *South African Journal for Enology and Viticulture* 13 (1992): 23–32; A.J. Razungles, R.L. Baumes, C. Dufour, C.N. Sznaper, C.L. Bayonove, "Effect of Sun Exposure on Carotenoids and C-13-norisoprenoid Glycosides in Syrah berries (*Vitis vinifera* L.)," *Sciences des Aliments* 18 (1998): 361–73.

47   Carbonneau et al., "Essai de détermination en biologie de la plante entière, de relations essentielles entre le bioclimat naturel, la physiologie de la vigne et la composition du raisin."

48   Smart et al., "Canopy Microclimate Modification for the Cultivar Shiraz. I. Definition of Canopy Microclimate"; Smart et al., "Canopy Microclimate Modification for the Cultivar Shiraz. II. Effects on Must and Wine Composition."

49   Smart et al., "Canopy Microclimate Modification for the Cultivar Shiraz. I. Definition of Canopy Microclimate"; Smart et al., "Canopy Microclimate Modification for the Cultivar Shiraz. II. Effects on Must and Wine Composition."

50   T.K. Wolf, R.M. Pool, and L.R. Mattick, "Responses of Young Chardonnay Grapevines to Shoot Tipping, Ethephon, and Basal Leaf Removal," *American Journal of Enology and Viticulture* 37 (1986): 263–68.

51   A.G. Reynolds, "Inhibition of Lateral Shoot Growth in Summer-Hedged 'Riesling' Grapevines by Paclobutrazol," *HortScience*, 23 (1988): 728–30; A.G. Reynolds, D.A. Wardle, A.C. Cottrell, and A.P. Gaunce, "Advancement of 'Riesling' Fruit Maturity by

Paclobutrazol-Induced Reduction of Lateral Shoot Growth, *Journal of the American Society for Horticultural Science*, 117 (1992): 430–35.

52 A.G. Reynolds and D.A. Wardle, "Effects of Timing and Severity of Summer Hedging on Growth, Yield, Fruit Composition, and Canopy Characteristics of De Chaunac. II. Yield and Fruit Composition," *American Journal of Enology and Viticulture* 40 (1989): 299–308.

53 Smith et al., "Viticultural and Oenological Implications of Leaf Removal for New Zealand Vineyards."

54 A.G. Reynolds and D.A. Wardle, "Impact of Several Canopy Manipulation Practices on Growth, Yield, Fruit Composition, and Wine Quality of Gewürztraminer," *American Journal of Enology and Viticulture* 40 (1989): 121–29; Reynolds et al., "Impact of Training System, Vine Spacing, and Basal Leaf Removal on Riesling."

55 Reynolds et al., "Vine Performance, Fruit Composition, and Wine Sensory Attributes of Gewürztraminer in Response to Vineyard Location and Canopy Manipulation."

56 Reynolds et al., "Impact of Training System, Vine Spacing, and Basal Leaf Removal on Riesling."

57 Reynolds et al., "Fruit Maturation in Four *Vitis vinifera* Cultivars in Response to Vineyard Location and Basal Leaf Removal."

58 A.G. Reynolds, J.W. Schlosser, R. Power, R. Roberts, and C. de Savigny, "Magnitude and interaction of viticultural and enological effects. I. Impact of canopy management and yeast strain on sensory and chemical composition of Chardonnay musqué," *American Journal of Enology and Viticulture* 58 (2007): 12–24.

59 N.A. Abbott, P.J. Williams, and B.G. Coombe (1993): "Measure of Potential Wine Quality by Analysis of Grape Glycosides," in *Proceedings of the Eighth Australian Wine Industry Technical Conference*, eds. C.S. Stockley, R.S. Johnstone, P.A. Leske and T.H. Lee (Adelaide, S. Australia: Winetitles, 72–75).

60 B.W. Zoecklein, T.K. Wolf, J.E. Marcy, and Y. Jasinski, "Effect of Fruit Zone Leaf Thinning on Glycosides and Selected Aglycone Concentrations of Riesling *(Vitis vinifera* L.) Grapes," *American Journal of Enology and Viticulture* 49 (1998): 35–43; B.W. Zoecklein, T.K. Wolf, S.E. Duncan, J.E. Marcy, and Y. Jasinski, "Effect of Fruit Zone Leaf Removal on Total Glucoconjugates and Conjugate Fraction Concentrations of Riesling and Chardonnay *(Vitis vinifera* L.) Grapes," *American Journal of Enology and Viticulture* 49 (1998): 259–65.

61 Zoecklein et al., "Effect of Fruit Zone Leaf Removal on Total Glucoconjugates and Conjugate Fraction Concentrations of Riesling and Chardonnay *(Vitis vinifera* L.) Grapes."

62 R.A. Arnold and A.M. Bledsoe, "The Effect of Various Leaf Removal Treatments on the Aroma and Flavor of Sauvignon Blanc Wine." *American Journal of Enology and Viticulture* 41 (1990): 74–76; A.M. Bledsoe, W.M. Kliewer, and J.J. Marois, "Effects of Timing and Severity of Leaf Removal on Yield and Fruit Composition of Sauvignon Blanc Grapevines," *American Journal of Enology and Viticulture* 39 (1988): 49–54.

63 J.J. Scheiner, G.L. Sacks, B.S. Pan, S. Ennahli, L. Tarlton, A. Wise, S.D. Lerch, and J.E. Vanden Heuvel, "Impact of Severity and Timing of Basal Leaf Removal on 3-isobutyl-2-methoxypyrazine Concentrations in Red Winegrapes," *American Journal of Enology and Viticulture* 61 (2010): 358–64.

64 Reynolds et al., "Magnitude and Interaction of Viticultural and Enological Effects. I. Impact of Canopy Management and Yeast Strain on Sensory and Chemical Composition of Chardonnay Musqué"; A.G. Reynolds, J.W. Schlosser, D. Sorokowsky, R. Roberts, and C. de Savigny, "Magnitude and Interaction of Viticultural and Enological Effects. II. Relative Impacts of Cluster Thinning and Yeast Strain on

Composition and Sensory Attributes of Chardonnay Musqué," *American Journal of Enology and Viticulture* 58 (2007): 25–41 (Fig. 9).

65 Reynolds and Wardle, "Impact of Several Canopy Manipulation Practices on Growth, Yield, Fruit Composition, and Wine Quality of Gewürztraminer."

66 R. Eschenbruch, R.E. Smart, B.M. Fisher, and J.G. Whittles, "Influence of Yield Manipulations on the Terpene Content of Juices and Wines of Müller Thurgau," in *Proceedings of the 6th Australian Wine Industry Technical Conference*, 14–17 July 1986, Adelaide, South Australia, ed. T. Lee (Adelaide, S. Australia: Australian Industrial Publishers, 1987): 89–93.

67 A.G. Reynolds, S.F. Price, D.A. Wardle, and B.T. Watson, "Fruit Environment and Crop Level Effects on Pinot Noir. I. Vine Performance and Fruit Composition," *American Journal of Enology and Viticulture* 45 (1994): 452–59; A.G. Reynolds, S. Yerle, B.T. Watson, S.F. Price, and D.A. Wardle, "Fruit Environment and Crop Level Effects on Pinot Noir. III. Composition and Descriptive Analysis of Oregon and British Columbia Wines," *American Journal of Enology and Viticulture* 47 (1996): 329–39.

68 M.G. McCarthy and B.G. Coombe, "Water Status and Winegrape Quality," *Acta Horticulturae*, 171 (1985): 447–56; A.G. Reynolds, "Riesling Vines Respond to Cluster Thinning and Shoot Density Manipulation. *Journal of the American Society for Horticultural Science* 114 (1989): 264–68; A.G. Reynolds, C.G. Edwards, D.A. Wardle, D.R. Webster, and M.J. Dever, "Shoot Density Affects Riesling Grapevines. I. Vine Performance. *Journal of the American Society for Horticultural Science* 119 (1994): 874–80; A.G. Reynolds, C.G. Edwards, D.A. Wardle, D.R. Webster, and M.J. Dever, "Shoot Density Affects Riesling Grapevines. II. Wine Composition and Sensory Response," *Journal of the American Society for Horticultural Science* 119 (1994): 880–92.

69 K.H. Fisher, O.A. Bradt, J. Wiebe, and V.A. Dirks, Cluster Thinning 'De Chaunac' French Hybrid Grapes Improves Vine Vigour and Fruit Quality in Ontario. *Journal of the American Society for Horticultural Science* 102 (1977): 162–65; N.J. Looney, "Some Growth Regulator and Cluster Thinning Effects on Berry Set and Size, Berry Quality, and Annual Productivity of De Chaunac Grapes," *Vitis* 20 (1981): 22–35.

70 A.G. Reynolds, R.M. Pool, and L.R. Mattick, "Effect of Shoot Density and Crop Control on Growth, Yield, Fruit Composition and Wine Quality of 'Seyval Blanc,'" *Journal of the American Society for Horticultural Science* 111 (1986): 55–63.

71 Eschenbruch et al., "Influence of Yield Manipulations on the Terpene Content of Juices and Wines of Müller Thurgau"; McCarthy and Coombe, "Water Status and Winegrape Quality"; A.G. Reynolds and D.A. Wardle, "Impact of Several Canopy Manipulation Practices on Growth, Yield, Fruit Composition, and Wine Quality of Gewürztraminer," *American Journal of Enology and Viticulture* 40 (1989): 121–29; A.G. Reynolds, C.G. Edwards, D.A. Wardle, D.R. Webster, and M.J. Dever, "Shoot Density Affects Riesling Grapevines. I. Vine Performance," *Journal of the American Society for Horticultural Science* 119 (1994): 874–80; A.G. Reynolds, C.G. Edwards, D.A. Wardle, D.R. Webster, and M.J. Dever, "Shoot Density Affects Riesling Grapevines. II. Wine Composition and Sensory Response," *Journal of the American Society for Horticultural Science* 119 (1994): 880–92; A.G. Reynolds, J.W. Schlosser, R. Power, R. Roberts, and C. de Savigny, "Magnitude and Interaction of Viticultural and Enological Effects. I. Impact of Canopy Management and Yeast Strain on Sensory and Chemical Composition of Chardonnay Musqué," *American Journal of Enology and Viticulture* 58 (2007): 12–24; A.G. Reynolds, J.W. Schlosser, D. Sorokowsky, R. Roberts, and C. de Savigny, "Magnitude and Interaction of Viticultural and Enological Effects. II. Relative Impacts of Cluster Thinning and Yeast Strain on Composition and Sensory Attributes of Chardonnay Musqué," *American Journal of Enology and Viticulture* 58 (2007): 25–41.

72 A.G. Reynolds, D.A. Wardle, and M.J. Dever, "Shoot Density Effects on Riesling Grapevines: Interaction with Cordon Age," *American Journal of Enology and Viticulture* 45 (1994): 435–43.

73 J. Marais, "Sauvignon Blanc Cultivar Aroma – A Review," *South African Journal for Enology and Viticulture,* 15 (1994): 41–45; D.M. Chapman, M.A. Matthews, and J.X. Guinard, "Sensory Attributes of Cabernet Sauvignon Wines Made from Vines with Different Crop Yields," *American Journal of Enology and Viticulture* 55 (2004): 325–34; D.M. Chapman, J.H. Thorngate, M.A. Matthews, J.X. Guinard, and S.E. Ebeler, "Yield Effects on 2-methoxy-3-isobutylpyrazine Concentration in Cabernet Sauvignon Using a Solid Phase Microextraction Gas Chromatography/Mass Spectrometry Method," *Journal of Agricultural and Food Chemistry* 52 (2004): 5431–35.

74 N.J. Shaulis, H. Amberg, and D. Crowe, "Response of Concord Grapes to Light, Exposure, and Geneva Double Curtain Training," in *Proceedings of the American Society of Horticultural Science* 89 (1966): 268–80.

75 Carbonneau, Casteran, and Leclair, "Essai de détermination en biologie de la plante entière."

76 Smart, "Vine Manipulation to Improve Wine Grape Quality."

77 Smart et al., "Canopy Microclimate Modification for the Cultivar Shiraz. I. Definition of Canopy Microclimate"; Smart et al., "Canopy Microclimate Modification for the Cultivar Shiraz. II. Effects on Must and Wine Composition."

78 A.G. Reynolds, R.M. Pool, and L.R. Mattick, "Effect of Training System on Growth, Yield, Fruit Composition and Wine Quality of Seyval Blanc," *American Journal of Enology and Viticulture* 36 (1985): 156–65; Reynolds and Wardle, "Impact of Training System and Vine Spacing on Vine Performance and Berry Composition of Seyval Blanc"; A.G. Reynolds, D.A. Wardle, M.A. Cliff, and M.J. King, "Impact of Training System and Vine Spacing on Vine Performance, Berry Composition, and Wine Sensory Attributes of Seyval and Chancellor," *American Journal of Enology and Viticulture* 55 (2004): 84–95.

79 Reynolds, Wardle, and Naylor, "Impact of Training System and Vine Spacing on Vine Performance and Berry Composition of Chancellor"; Reynolds et al., "Impact of Training System and Vine Spacing on Vine Performance, Berry Composition, and Wine Sensory Attributes of Seyval and Chancellor."

80 G.S. Howell, D.P. Miller, C.E. Edson, and R.K. Striegler, "Influence of Training System and Pruning Severity on Yield, Vine Size, and Fruit Composition of Vignoles Grapevines," *American Journal of Enology and Viticulture* 42 (1991): 191–98.

81 G.S. Howell, T.K. Mansfield, and J.A. Wolpert, "Influence of Training System, Pruning Severity, and Thinning on Yield, Vine Size, and Fruit Quality of Vidal Blanc Grapevines," *American Journal of Enology and Viticulture* 38 (1987): 105–12.

82 J.E. Vanden Heuvel, J.T.A. Proctor, J.A. Sullivan, and K.H. Fisher, "Influence of training/trellising system and rootstock selection on productivity and fruit composition of Chardonnay and Cabernet Franc grapevines in Ontario, Canada," *American Journal of Enology and Viticulture* 55 (2004): 253–64.

83 A.G. Reynolds, "Response of Riesling Vines to Training System and Pruning Strategy," *Vitis* 27 (1988): 229–42; Reynolds, Wardle, and Naylor, "Impact of Training System, Vine Spacing, and Basal Leaf Removal on Riesling"; A.G. Reynolds, D.A. Wardle, M.A. Cliff, and M.J. King, "Impact of Training System and Vine Spacing on Vine Performance, Berry Composition, and Wine Sensory Attributes of Riesling," *American Journal of Enology and Viticulture* 55 (2004): 96–103.

84 Reynolds, "Response of Riesling Vines to Training System and Pruning Strategy."

85 Reynolds, Pool, and Mattick, "Effect of Training System on Growth, Yield, Fruit Composition and Wine Quality of Seyval Blanc."

86 Howell, Mansfield, and Wolpert, "Influence of Training System, Pruning Severity, and Thinning on Yield, Vine Size, and Fruit Quality of Vidal Blanc Grapevines"; Howell et al., "Influence of Training System and Pruning Severity on Yield, Vine Size, and Fruit Composition of Vignoles Grapevines."

87 Reynolds, Wardle, and Naylor, "Impact of Training System, Vine Spacing, and Basal Leaf Removal on Riesling"; Reynolds et al., "Impact of Training System and Vine Spacing on Vine Performance, Berry Composition, and Wine Sensory Attributes of Riesling."

88 B.W. Zoecklein, T.K. Wolf, L. Pélanne, M.K. Miller, and S.S. Birkenmaier, "Effect of Vertical Shoot-Positioned, Smart-Dyson, and Geneva Double-Curtain Training Systems on Viognier Grape and Wine Composition," *American Journal of Enology and Viticulture* 59 (2008): 11–21.

89 C. Sala, O. Busto, J. Guasch, and F. Zamora, "Influence of Vine Training and Sunlight Exposure on the 3-alkyl-2-methoxypyrazines Content in Musts and Wines from the *Vitis vinifera* Variety Cabernet Sauvignon," *Journal of Agricultural and Food Chemistry* 52 (2004): 3492–97.

Twelve

Wine Tasting

Ronald S. Jackson

Introduction

C ritical assessment is essential to genuine wine appreciation. Without serious appraisal, appreciation may become little more than an act of possession or affected self-esteem. Understanding how and where wines are made can enliven intellectual discussion, but what sets wine connoisseurship apart from other forms of artistic appreciation is tasting. Nonetheless, detailed assessment is a two-edged sword. It exposes not only the full range of a wine's qualities, but also its failings. Close scrutiny has its greatest value when sampling superb wines, but herein lies a problem. How does one know in advance that a wine merits detailed analysis? Price, prestige, and provenance are inconsistent and, too frequently, poor predictors of quality. It is only upon opening that one can ascertain a wine's true worth. However, it is partially this uncertainty that gives wine much of its intrigue. Regrettably, variability also permits many a mediocre, but expensive, wine to masquerade under the premium label. Excuses such as a poor vintage, bottle-to-bottle variation, or a lack of personal acuity are all too often invoked to explain away nondescript quality in pricey wines. Vinous quality can be observed only in the glass, preferably in the absence of all potentially biasing information. Tasting should be a search for the truth—personal truth. It should not be an attempt to subjugate personal perception to that of some "authority," nor should it become an act of genuflection to label prestige or implied superiority.

Everyone eventually evolves their own tasting procedure, one that is adjustable to personal preferences and different situations. For example, critical appraisal is usually inappropriate at the dinner table. It diverts attention away from the central functions of the occasion—enjoyment of the dining experience and the company of family or friends. However, when the purpose is to compare wines, every possible sensory attribute should be assessed. These facets

Each sample should be poured into identical, clear, tulip-shaped, wine glasses. They should each be filled (1/4 to 1/3 full) with the same volume of wine.

I. Appearance

1 – View each sample at a 30° to 45° angle against a bright, white background.
2 – Record separately the wine's:
    clarity (absence of haze)
    color hue (shade or tint) and depth (intensity or amount of pigment)
    viscosity (resistance to flow)
    effervescence (notably sparkling wines)

II. Odor "in-glass"

1 – Sniff each sample at the mouth of the glass before swirling.
2 – Study and record the nature and intensity of the fragrance[a] (see Figs 1.3 and 1.4)
3 – Swirl the glass to promote the release of aromatic constituents from the wine.
4 – Smell the wine, initially at the mouth and then deeper in the bowl.
5 – Study and record the nature at intensity of the fragrance.
6 – Proceed to other samples.
7 – Progress to tasting the wines (III)

III. "In-mouth" sensations

(a) Taste and mouth-feel

1 – Take a small (6 to 10 ml) sample into the mouth.
2 – Move the wine in the mouth to coat all surfaces of the tongue, cheeks and palate.
3 – For the various taste sensations (sweet, acid, bitter) note where they are percieved, when first detected, how long they last, and how they change in perception and intensity.
4 – Concentrate on the tactile (mouth-feel) sensations of astringency, prickling, body, temperature, and "heat".
5 – Record these perceptions and how they combine with one another.

(b) Odor

1 – Note the fragrance of the wine at the warmer temperatures of the mouth.
2 – Aspirate the wine by drawing air through the wine to enhance the release of its aromatic constituents.
3 – Concentrate on the nature, development and duration of the fragrance. Note and record any differences between the "in-mouth" and "in-glass" aspects of the fragrance

(c) Aftersmell

1 – Draw air into the lungs that has been aspirated through the wine for 15 to 30 s.
2 – Swallow the wine (or spit it into a cuspidor).
3 – Breath out the warmed vapors through the nose.
4 – Any odor detected in this manner is termed aftersmell; it is usually found only in the finest or most aromatic wines.

IV. Finish

1 – Concentrate on the olfactory and gustatory sensations that linger in the mouth.
2 – Compare these sensations with those previously detected.
3 – Note their character and duration.

V. Repetition of assessment

1 – Reevaluate of the aromatic and sapid sensations of the wines, beginning at II.3—ideally several times over a period of 30 min.
2 – Study the duration and development (change in intensity and quality) of each sample.

Finally, make an overall assessment of the pleasurableness, complexity, subtlety, elegance, power, balance, and memorableness of the wine. With experience, you can begin to make evaluations of its *potential*— the likelihood of the wine improving in its character with additional aging.

[a] Although fragrance is technically divided into the *aroma* (derived from the grapes) and *bouquet* (derived from fermentation, processing and aging), descriptive terms are more informative.

Figure 12.1 Wine-tasting sequence
Source: From R.S. Jackson, *Wine Tasting: A Professional Handbook,* 2nd ed. (London: Academic Press, 2009). © Elsevier. Reproduced by permission.

are necessary to evaluate its quality as well as to divine its varietal, stylistic, or regional attributes. Figure 12.1 outlines a programmed set of steps designed to focus sequentially on every sensory aspect a wine may possess.

One of the most frequently asked questions by those new to wine is whether they can distinguish gradations between ordinary and superior wine. Assuming that they have a normal sense of taste and smell, and are willing to focus on these facets, the answer is definitively "yes." Whether they may appreciate or value these aspects is less certain. Experience, habituation, inclination, maturity, and peer pressure can all play a role in developing a love for what may initially not enamour. The most essential ingredient of a serious taster is the willingness, desire, and ability to direct one's attention to the task. As in other aspects of life, what one gets out of an activity is usually directly proportional to what one puts in.

Detailed Tasting Technique

In critical tastings, the palate is usually cleansed between samples, to avoid sensations from previous samples influencing those subsequently tasted. This is best accomplished with a 1 percent pectin solution.[1] Unsalted crackers are an acceptable, though imperfect, substitute, and far superior to water. Water is ineffectual in removing residual tannic constituents and dilutes the saliva. Cheese should be avoided. It tends to suppress several wine sensations.[2] Suppression of sensory attributes by cheese (and food in general) is only of value if it is specifically desired. For example, the saltiness of hard cheeses, such as Parmigiano-Reggiano, can markedly mollify the bitter/astringent character of tannic wines.

For assessment purposes, the majority of wines are best sampled using clear, tulip-shaped glasses. The International Standards Organization (ISO) wine tasting glass (Figure 12.2) is ideal for the purpose.[3] Glasses of this shape facilitate swirling the wine and concentrate its fragrance in the bowl. This, of course, does not obviate the psychological benefits that may accrue from the use of highly advertised crystal goblets, with specific forms supposedly designed for particular styles. Different glass shapes do subtly affect the dynamics of aromatic release from wine.[4] However, whether one shape is "better" than another depends on the person's mindset, not on physics or chemistry. Regardless of the shape of the glass, it is important that all wines be presented under equivalent conditions: in identical glasses, at the same temperature, and containing equal

Figure 12.2 ISO wine tasting glass: Royal Leerdam Wine Taster #9309RL – 229 ml, 7¾ oz.
Source: Photo courtesy Libbey Inc., Toledo, OH.

volumes. Samples of about 30 to 50 millilitres (one-quarter to one-third full) make it easier to angle the glass (for viewing colour and clarity) and to swirl the wine (enhancing the release of wine aromatics). The only exception is when sampling sparkling wines. Here, tall, flute-shaped glasses facilitate appraisal of the wine's effervescence. Regrettably, their open mouths, and the level to which they are usually filled, do little to facilitate detection of the subtle fragrance of most sparkling wines.

Appearance

In assessing a wine's appearance, colour and clarity are paramount. Colour is not only the first feature typically detected, but it also generates the wine's first esthetic pleasure. The countenance provided by a rich, vivacious red, or the golden opulence of a sweet white wine, act as harbingers of the sapid pleasures to follow. Conversely, an aberrant colour often foretells the presence of an atypical fragrance or off odour. Nonetheless, these clues can bias the perception of even experienced tasters. Thus, visual clues must be regarded with caution to fairly assess the wine's gustatory, tactile, and olfactory sensations. It is for this reason that wines sampled for research purposes may be sampled under dim red lights (to disrupt colour perception) or tasted using black glasses.

To optimally view a wine's colour, tilt the glass at about a 35- to 45-degree angle, against a bright white background. This permits viewing the hue through varying wine depths. For example, a young red wine typically shows a clear rim and a purplish cast. The earliest indication of aging (or undesirable premature oxidation) expresses itself as a brickish coloration at the rim. Viewing the wine at an angle also gives the best indication of its colour intensity (depth).

Tilting the glass also facilitates assessing wine clarity. All commercial wines should be perfectly clear, with no detectable haziness. The sediment that may accumulate at the bottom of some aged red wines is of little concern, if the wine is first slowly decanted to separate the wine from the precipitated material (primarily tannins and tartrate salts). Repeated tipping, during pouring, resuspends the bitterish to clayey-tasting sediment.

Only wines possessing very high sugar, glycerol, and/or alcohol contents may show slight modifications in perceptible viscosity.[5] Thus, viscosity is rarely a property of significant interest. The same also applies to "tears." These form following swirling. Once swirling has ceased, the more rapid evaporation of alcohol from the wine, relative to water, increases the surface tension in the wine adhering to the sides of the glass. This, in turn, leads to the formation of droplets along its rim. As the droplets enlarge, gravity initiates their decent—ipso facto, the formation of tears.

Occasionally, bubbles may form after pouring, usually along the bottom and sides of the glass. This typically results if the wine is bottled before the dissipation of excess carbon dioxide, supersaturated in all newly fermented wines. Pronounced and continuous effervescence is characteristic only of sparkling wines. In the latter instance, the size, number, and duration of the bubbles are valued as important quality features.

Visual features can provide hints as to the wine's age, conformity to accepted norms, and potential faults. Despite their potential diagnostic value, it is generally more productive for consumers to concentrate on the pleasure derived from the wine's entrancing countenance than to focus on its predictive potential. In addition, if misinterpreted, appearance may consciously or subconsciously prejudice appraisal of the wine's other attributes.

Odour Detection

Assessing a wine's fragrance has two distinct aspects—that assessed by smelling (orthonasal) and that detected via the back of the throat (retronasal). Although assessing more or less the same aromatic compounds, their qualitative perceptions can differ markedly. Volatile compounds are thought to be detected in a linear, temporal sequence over the olfactory patches of the nose (similar to a gas chromatograph).[6] This means that aromatics detected via the throat are assessed in the reverse order to those detected via the nostrils. Because odours apparently are recognized as time sequence patterns, when the sequence is reversed, the perceived pattern can differ markedly. Contemplate the difference between the odour and in-mouth perception of Limburger cheese. The disparity is equivalent to playing a familiar musical theme backwards. In addition, the concentration of aromatic compounds received via the throat is usually lower than that received via the nostrils (affecting intensity-sensitive reactions). Furthermore, selective absorption and enzymatic modification in the mouth can change the chemical composition of the wine. The temperature difference between wine in the glass and the mouth can also influence relative volatilization.

To facilitate smelling, the wine is swirled in the glass. This increases the surface area over which wine aromatics can escape and become concentrated in the bowl. Periodic swirling also helps replace aromatics that escape via the mouth of the glass. It is primarily for this reason that glasses should not be more than one-third full. It also helps if the sides of the glass curve inwards (Figure 12.2). This feature also favours the concentration and retention of aromatic compounds.

Whiffs are taken at the rim and then in the bowl. This permits sensation of the fragrance at different concentrations, potentially generating distinct sensations. Where the fragrance is mild, some professionals place a watch glass over the mouth of the glass. This permits vigorous swirling and traps the aromatics released from the wine. Substitutes for watch glasses include the cover of a small petri dish, or an appropriately sized plastic coffee-cup cover.

No special method has been demonstrated to significantly improve effective smelling. Typical sniffing, usually less than a second, and separated sequentially by 20 to 30 seconds, seems fully adequate.[7] This relates to the adaptation of olfactory receptors to odours, and the time taken for them to regain their intrinsic sensitivity.[8] Nonetheless, differences in sniffing procedure might affect the qualitative perception of odours.[9] In addition, prolonged inhalation periods can occasionally be useful with some complexly aromatic wines, such as ports. As particular olfactory receptors become fatigued (fail to respond), the pattern of impulses sent by those still being activated is likely to be interpreted differently. As the effective pattern of olfactory sensations changes, a kaleidoscope of odour perceptions may be activated—the olfactory equivalent of the Dance of the Seven Veils.

Professional tasters are usually trained in odour recognition, using samples specifically designed for the purpose. Reference samples are typically available during tastings. Fragrance and off-odour charts may also aid in training.[10] Precise term use is critical for scientists to define and measure the relative differences among wines. This is less critical for lay use. It is more valuable for consumers to concentrate on memorizing the holistic fragrance characteristics that distinguish particular wines than to name individual aromatic resemblances. It is akin to the greater value in recognizing a person's face than denoting its distinctive features. In addition, overattention to describing a wine's aroma can detract from the actual appreciation of the wine. Is the enjoyment of a Granny Smith apple enhanced by being able to name its individual aromatic constituents, or is the appreciation of a painting dependent on recognizing the artist's distinctive brush strokes?

Both positive and negative impressions should be recorded. Figure 12.3 provides an example of a table for reporting comments. Alternately, a graphic representation, such as illustrated in Figure 12.4, is a quick, succinct, and effective means of recording attributes as well as temporal modifications in flavour intensity and character.

The primary value of recording impressions is the concentration it demands. It also helps develop a wine lexicon. The terms employed tend to strongly reflect a person's familial, cultural, and geographic upbringing. That the language employed may not be as precise, nor as legitimate, as required for scientific investigation is not a serious drawback. The primary purpose of recording impressions is to focus one's attention on the wine's attributes and how they change with time. In addition, using any olfactory vocabulary, both consistently and accurately, is difficult. Typically, people have great difficulty correctly naming even familiar odours (without visual clues). The problem of naming recognizable odours has been aptly dubbed the "tip-of-the-nose" phenomenon.[11] This probably reflects the dominance of visual information in human sensory perception.

Figure 12.3 Wine-tasting sheet for quality assessment

Sample Number: ___	Wine Category: ___	Exceptional	Very Good	Above Average	Average	Below Average	Poor	Faulty	Comments
Visual	Clarity								
Odor (orthonasal)	Intensity*								
	Duration**								
	Quality***								
Flavor (taste, mouth-feel, retronasal odor)	Intensity								
	Duration								
	Quality								
Finish (after-taste and lingering flavor)	Duration								
	Quality								
Conclusion									

*   Intensity: the perceived relative strength of the sensation—too weak or too strong are equally undesirable.
** Duration: the interval over which the wine develops or maintains its sensory impact; long duration is usually a positive feature if not too intense.
*** Quality: the degree to which the feature reflects appropriate and desirable varietal, regional or stylistic features of the wine, plus the pleasure these features give the taster.

Figure 12.3   Wine-tasting sheet for quality assessment
Source: From R.S. Jackson, *Wine Tasting: A Professional Handbook*, 2nd ed. (London: Academic Press, 2009). © Elsevier, reproduced by permission.

Considerable concentration is usually required in recognizing varietal, stylistic, or regional attributes. It often demands repeated attempts, employing both inductive and deductive memory searches. Features both present and absent are often required. Nonetheless, as the primary source of a wine's unique character, the study of fragrance warrants the intense scrutiny it demands.

Oral Assessment

After assessing in-glass fragrance, attention turns to oral-based sensations. For this, sip a small amount of the wine. As far as feasible, the volume of each sample should be kept identical. This permits valid comparison among the wines. Active churning ("chewing") brings the wine into contact with all sensory regions in the mouth.

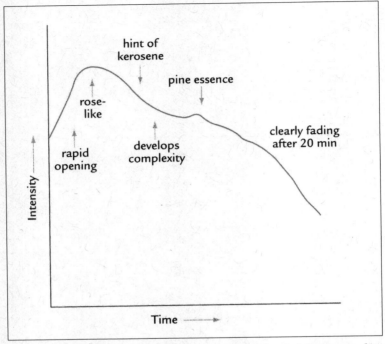

Figure 12.4 Graphic representation of the development of a wine's fragrance and its relative intensity. Specific observations can be written on the graph to represent when it was detected.
Source: From R.S. Jackson, *Wine Tasting: A Professional Handbook,* 2nd ed. (London: Academic Press, 2009). © Elsevier, reproduced by permission.

The sensation most rapidly detected (if present) is sweetness. This is followed by sourness (primarily on the sides of the tongue and cheeks). Bitterness is detected more slowly, and is sensed primarily at the back of the tongue. It can take upward of fifteen seconds before the perception of bitterness reaches its peak.[12] Subsequently, focus shifts to mouth-feel sensations, such as the dry, dust-in-the-mouth, occasionally velvety aspect of astringency, the potential burning sensation of alcohol, or the prickling aspect of carbon dioxide. These and other tactile sensations are dispersed throughout the mouth.

Depending on the intention of the tasting, it may be worth noting differences between the first and subsequent sips. Tannins and acids react with salivary and membrane-bound proteins in the mouth, initially diminishing their respective bitter/astringent and sour/astringent sensations. Thus, the initial sample will often appear smoother and more balanced than subsequent samplings. The first sip approximates the perception generated when wine is taken with food. Proteins in the food react with tannins and acids, reducing their perceived intensity. If assessing how a wine might balance with food, sampling should progress slowly. This provides time for activated saliva production to compensate for its dilution during tasting.

Some vinous sensations, perceived as tastes, may be mistaken aspects of retronasal odour. Classic examples are the "sweet" taste associated with some dry white wines, due to the presence of fruity-smelling aromatics.[13] Another is the peppery aspect of Shiraz, generated by the aromatic compound rotendone. Nonetheless, most peppery sensations in wine are legitimate taste sensations, usually induced by tannins. Another example of sensory misinterpretation is the metallic aspect occasionally detected in some wines. It is probably associated with volatile compounds synthesized by the oxidation of fats[14] and is likely catalyzed by the presence of metal ions in the wine. Fatty acid oxidation by metal ions is also involved in the potential generation of a fishy aftertaste in seafood and wine pairing.[15]

To enhance retronasal odour detection, tasters may aspirate the wine. This involves tightening the jaws, holding the lips slightly ajar (or pursing the lips), and slowly drawing air through the wine. This is analogous to swirling wine in the glass. The volatilized compounds flow up into the nasal passages, especially during swallowing and breathing out. Intentional concentration on slow, deliberate exhalation apparently improves retronasal identification.[16] The olfactory sensations so induced, combined with gustatory sensations, generate the multimodal sensation termed "flavour." The duration of flavour detection, following swallowing or expectoration, is termed the wine's "finish." It is the vinous equivalent of a sunset. The importance of retronasal olfaction to flavour detection is easily demonstrated by clamping the nose. Typically, the longer the finish, the more highly rated the wine (unless it involves some off odour). Fruity-floral essences, associated with refreshing acidity, epitomize most superior white wines, whereas complex berry/jammy fragrances, combined with flavourful velvety tannins, exemplify the best red wines. Fortified wines, possessing more intense flavours, have very long finishes. Oral perceptions should be recorded quickly, as they can be evanescent and can change unpredictably.

Although significant in some critical tastings, noting a wine's individual taste attributes is less important than concentrating on how they integrate to produce holistic perceptions, such as flavour, balance, and body.

When sequentially tasting several wines, expectorating each sample is the norm. This helps minimize an increase in blood alcohol content, and a corresponding decline in sensory acuity and discrimination skill.[17] In social situations, though, expectoration is usually anathema, and frequently a waste of limited amounts of good wine.

Integration, Interpretation, and Quality Assessment

For maximal assessment of a wine's attributes, it is essential to sample the wine periodically over a fifteen- to thirty-minute period. This is required to detect features such as duration and development. Duration relates to how long the wine retains its aromatic character, whereas development refers to qualitative changes over time. Development is comparable to watching a blossom open with time-lapse photography. Duration, development, and a lingering finish are all highly prized features, and are particularly important in premium wines. The higher costs of such wines are justifiable only if accompanied by exceptional sensory endowments.

During tasting, attention finally shifts from observing individual attributes to focusing on their integration and interpretation. This may involve assessing aspects such as conformity with, and distinctiveness within, regional norms, varietal origin, stylistic characteristics, and an appraisal of the wine's overall quality.

A wine's quality (as independent from its geographic, stylistic, or varietal characteristics) is described in terms typical of art appreciation. Complexity refers to the presence of many distinctive aromatic elements, rather than one or a few easily recognizable odours. Balance (harmony) denotes the perceptive equilibrium of all olfactory and oral sensations, where individual perceptions do not dominate. Occasionally, an aspect of imbalance may be appealing, donating an impression of intriguing nervousness. Fine wines should ideally show extended duration, a fascinating development, and a lingering and appealing finish. These aspects both enthrall and sustain interest. Implied, but seldom enunciated, are properties such as power, elegance, and finesse. Without these attributes, attractiveness is shallow. Occasionally, these perceptions are so remarkable that the experience becomes unforgettable—a feature termed "memorableness" by Amerine and Roessler.[18] Although highly appreciated, it must be as unexpected as it is striking (essential for memory branding). Thus, it is regrettably rare, but transcendental when it occurs!

Alternately, quality may be measured relative to features considered traditional for regional appellations, wine styles, grape varieties, or how the wine combines with a meal.

Other Tasting Situations

The previous section has dealt with serious tastings, whereby all of a wine's sensory delights (and faults) are assessed. Nonetheless, under most circumstances, its application is either unnecessary (wines of mediocre character), inappropriate (during a meal), or too prolonged (large stand-up tastings). Under such circumstances, the procedure needs to be truncated. Suggestions for such situations are briefly noted below.

Most wine and cheese parties are really designed for social interaction. The wine and cheese act primarily to occupy hands and lubricate throats. The wines selected are all too frequently as innocuous or as inexpensive as possible in order to avoid interfering with the true function of the gathering, and because few people at such events are really interested in the quality of the wine or cheese. Under such circumstances, there is little value in attempting a critical analysis, unless it is to confirm, if necessary, the nondescript nature of both the wine and the cheese.

Commercial tastings are usually designed to showcase participant wines, to raise funds, or both. In both situations, some of the wines may be of a quality worth assessing, although the setting is usually ill designed for critical appraisal. Conditions are improved if the number of wines is limited and participants are seated during the tasting. Regrettably, the food and cheese accompaniments typically supplied are usually inappropriate, albeit appreciated by most participants. The only exception relates to whether one is more interested in the wine's food-association potential than its inherent quality. As noted previously, food tends to suppress or interfere with the perception of several wine attributes. In addition, the social distractions of the occasion complicate valid (objective) assessment. Despite these limitations, a reasonable evaluation of a select number of wines is possible if considerable concentration is applied to counter the all too abundant distractions. With large stand-up, commercial tasting, sampling is usually limited to a quick swirl, sniff, sip, and spit (or swallow). Because of the number of wines sampled, expectoration is advisable where possible. Conditions for note taking are usually limited and primitive. The tasting conditions also make it impossible to assess important aspects such as wine development. There is simply insufficient time to observe this property, as wines are sampled individually, not together. Thus, it is advisable to carefully select those wines that appear to warrant attention. All one can usually do is evaluate the wine's immediate impression. Although this approach does injustice to most wines of quality, it favours the positive perception of wines that are ready for immediate consumption. This is not a serious problem, as usually these are the only wines presented. Because of these limitations, it is wise to decide in advance on what basis you intend to evaluate the wines. It could be on how well they reflect some geographic norm or varietal expression, but the most practical basis is on a price/quality ratio. Wines that are two or three times as expensive, but appear only marginally better, are not necessarily wise (or at least economically sound) purchases. Nonetheless, if impressing friends or colleagues is the principal concern, then the more expensive are probably preferable.

It is a generally accepted maxim that wine is ideally suited to accompany food. This long-established European view probably reflects wine's antimicrobial properties and its ready availability in many European countries. In reality, there is no inherent compatibility between wine and food. It is a culturally derived tradition, based primarily on habituation. White wines are predominantly acidic and alcoholic, with floral fragrances, whereas reds are relatively bitter and astringent, with fruity to jammy flavors. None of these attributes find equivalents in most foods, although they occasionally do with some condiments. For example, only a few wines possess herbaceous flavors and, if these are marked, it is considered a fault. Correspondingly, the flavours found in meat, poultry, and fish dishes, or the tastes typical of raw or cooked vegetables, find few liaisons with wines. The most evident association between food and wine involves their respective palate-cleansing actions. This permits both the wine and the food to be savoured afresh, if taken in sequence. There is also good evidence suggesting that the premier benefit of cheese and wine combination relates to their mutual suppression of each other's more negative aspects.[19] The result is an enhanced mutual appreciation. This feature also probably explains why some wines appear better (smoother and more balanced) with food. Thus, when attempting to critically assess a wine at the dinner table, it is better to sample the wine(s) in advance of the food's arrival. This avoids any skewing or reduction in the sensory characteristics of the wines. This is especially important for better wines, permitting their attributes to be fully detected. It also provides an opportunity to make comparisons, take notes, and share perceptions with fellow aficionados. Comparisons are more informative (less chance for knowledge-based bias) if the identity of the wines, or at least their order, is withheld during the sampling period. In contrast, if the tasting occurs synchronously with the meal, concentration on the wine's features is complicated, due to competition and interaction with food odours and flavours, and the social context of the meal.

Application to Niagara Wines

Up to now, this chapter has been of a general nature. What follows are a few suggestions on how readers may hone their sensory skills on, and increase their knowledge of, Niagara wines. In this, it is important to reiterate that preferences are a complex mixture of genetic and cultural factors, plus individual experience. Perception is as idiosyncratic as personal appearance. We are all "an island." Wine critics are standard bearers only for those who wish to be led.

I have purposely avoided padding my comments with flavour descriptors. Descriptive terms have a legitimate and essential place in wine sensory evaluation. In wine appreciation courses, descriptors help focus attention on the diversity of a wine's sensory attributes. However, once this has become second nature, there is no real value in constantly searching for verbal descriptors. Wine appreciation should not degenerate into a game of Charades.

For red wines, I recommend those made from French–American hybrids. These are crosses between North American and European grapevines. Not only can varietals such a Maréchal Foch and Baco Noir produce superb wines in Niagara, they should be its signature wines. Chambourcin is another excellent variety, regrettably rarely grown in Niagara. Not only does Niagara do a great job with these varieties, but Europe has turned its back on them. Their mistake is Niagara's gain. That the wines contain both European and North American traits gives them a quality superior to most European varieties. Niagara does not have to kowtow to Eurocentric tastes. I hope we have outgrown our colonial status. All the varieties just noted produce rich berry/jammy favors with smooth textures. They have characters that do not demand imagination to detect, which is too often a feature in some European wines. Former students of mine half-jokingly referred to Baco Noir as "baco-tobacco." This might seem insulting, but only to those unfamiliar with the exotic and seductive aroma of freshly cured tobacco leaves. It was a compliment! These wines also superbly envelop the flavours of the best Niagara cuisine. Furthermore, they are more affordable than their pure *vinifera* counterparts—a feature of no little marketing advantage. Niagara Cabernets and Merlots are fine. Regrettably, they are not clearly superior to similar wines produced under more favourable climates, unless you prefer wine with more bell pepper than redcurrant flavors. Niagara Pinot noir can, as elsewhere, occasionally express those elusive flavours that have made the variety both world famous and infamous. Correctly or not, this feature is frequently described as beetroot. Although I am not a beet lover, this feature, combined with complex and intriguing fruit flavours, can make Pinot noir enthralling.

Of white cultivars, Niagara has done superbly well with the French–American hybrid Vidal. It initially became famous as the cultivar first associated with Icewine production (and with more affordable late-harvest wines). It has many of the attributes of Riesling wines, but more subtle. When it comes to *vinifera* wines, my expectations are a mix of Euro- and Americentric views. Niagara definitely can produce excellent Rieslings, up to the standard of better Prädikats and Spätleses from Germany, the variety's homeland. Because Riesling contains naringin (a bitter compound characteristic of grapefruit), the wine has traditionally been produced retaining some sweetness. Sugar is well known to mask bitterness. Riesling should also be sampled on the cool side. This accentuates its rose-floral aspect and classic hints of pine resin. If you do not detect these attributes, it is not a personal failing, any more than it is a fault that I do. Above all, unto thyself be true. Niagara also can do particularly well with Chardonnay. Depending on the grape's maturity and individual clonal attributes, Chardonnay may express apple through peach notes. The Musqué clone of Chardonnay produces a mild Muscat aroma, familiar to anyone used to Asti Spumante. Typically, flavour of Niagara Chardonnays (as does its Rieslings) benefits with several years of in-bottle aging. The one white variety that is disappointing is Gewürztraminer. The intense, lychee, "spicy" aspects typical of this variety are muted. To confirm its varietal origin, consulting the label is often required. This is a far cry from my detecting Gewürztraminer, as an unlisted component, in an Italian Chardonnay several years ago. Gewürztraminer constituted only 15 percent of that blend. Nonetheless, if your preference is for a mild Gewürztraminer, then Niagara versions are for you.

Finally, I would encourage the reader to sample Niagara sparkling wines. They are favourably distinct from Champagnes. Of those I have tried recently, their fruity to floral flavours were a pleasant change from the nonvarietal, toasty attributes of champagne. This is a positive step. It distinguishes Niagara sparkling wines from the crowd.

A fascinating investigation of these aspects can include a blind tasting of wines from distinct sub-regions (for example, the Bench, Lakeshore Plain, and Lakeshore). Several recent papers have looked into sensory differences in distinctive sites within the Niagara region.[20] They detected significant and repeatable differences in Riesling, Chardonnay, and Cabernet wines. Admittedly, these differences are relatively subtle and can be affected by climatic and winemaking conditions. Thus, do not be overly disappointed if you do not find clear differences among a few samples. Alternately, you may enjoy a comparative tasting of several vintages of a winery's products. To add an element of intrigue, repeat some of the wines more than once, or add a mystery wine. Can people recognize the repeat or mystery wine? These tests can be both fun and a challenge in perfecting your sensory skills and knowledge of Niagara wines.

Postscript

Headache production is one of the occupational hazards of wine tasting. If this is a recurring problem, it can often be avoided by the prior consumption of a small dose of acetylsalicylic acid (e.g., aspirin), acetaminophen (e.g., Tylenol), or ibuprofen (e.g., Advil).[21] They limit the production of prostaglandins in the brain by wine tannins (and its associated blood-vessel dilation). Large polymeric tannins, formed during aging, are less able to traverse the intestinal wall and enter the bloodstream. This probably explains why aged red wines induce fewer headaches than their younger counterparts. Idiosyncrasies in sensitivity to wine-induced headaches may be partially a function of differences in the activity of platelet phenol sulfotransferase. This enzyme plays a critical role in the inactivation and removal of phenolic compounds from the blood.

Notes

1 A.E. Colonna, D.O. Adams, and Ann C. Noble, "Comparison of Procedures for Reducing Astringency Carry-Over Effects in Evaluation of Red Wines," *Australian Journal of Grape and Wine Research* 10 (2004): 26–31; Carolyn F. Ross, Catherine Hinken, and Karen Weller, "Efficacy of Palate Cleansers for Reduction of Astringency

Carryover During Repeated Ingestions of Red Wine," *Journal of Sensory Studies* 22 (2007): 293–312.

2　I.T. Nygren, I.-B. Gustafsson, and L. Johansson, "Perceived Flavour Changes in White Wine after Tasting Blue Mould Cheese," *Food Service Technology* 2 (2002): 163–71.

3　Margaret A. Cliff, "Impact of Wine Glass Shape on Intensity of Perception of Color and Aroma in Wine," *Journal of Wine Research* 12 (2001): 39–46.

4　Jeannine F. Delwiche and Marcia L. Pelchat, "Influence of Glass Shape on Wine Aroma," *Journal of Sensory Studies* 17 (2002): 19–28; T. Hummel, Jeannie F. Delwiche, C. Schmidt, and K.-B. Hüttenbrink, "Effects of the Form of Glasses on the Perception of Wine Flavors: A Study in Untrained Subjects," *Appetite* 41 (2003): 197–202.

5　Gary J. Pickering, David A. Heatherbell, L.P. Vanhaenena, and M.F. Barnes, "The Effect of Ethanol Concentration on the Temporal Perception of Viscosity and Density in White Wine," *American Journal of Enology and Viticulture* 49 (1998): 306–18; S. Yanniotis, G. Kotseridis, A. Orfanidou, and A. Petraki, "Effect of Ethanol, Dry Extract and Glycerol on the Viscosity of Wine," *Journal of Food Engineering* 81 (2007): 399–403.

6　Johannes Frasnelli, Saskia van Ruth, Irina Kriukova, and Thomas Hummel, "Intranasal Concentrations of Orally Administered Flavors," *Chemical Senses* 30 (2005): 575–82; Dana M. Small, Johannes C. Gerber, Y.E. Mak, and Thomas Hummel, "Differential Neural Responses Evoked by Orthonasal Versus Retronasal Odorant Perception in Humans," *Neuron* 47 (2005): 593–605.

7　David G. Laing, "Natural Sniffing Gives Optimum Odour Perception for Humans," *Perception* 12 (1983): 99–117.

8　G. Ekman, B. Berglund, U. Berglund, and T. Lindvall, "Perceived Intensity of Odor as a Function of Time of Adaptation," *Scandinavian Journal of Psychology* 8 (1967): 177–86.

9　Valéry Normand, Shane Avison, and Alan Parker, "Modeling the Kinetics of Flavour Release During Drinking," *Chemical Senses* 29 (2004): 235–45.

10　See Ronald S. Jackson, *Wine Tasting: A Professional Handbook*, 2nd ed. (London: Academic Press, 2009); Ann C. Noble, R.A. Arnold, B.M. Masuda, S.D. Pecore, J.O. Schmidt, and P.M. Stern, "Progress Towards a Standardized System of Wine Aroma Terminology," *American Journal of Enology and Viticulture* 35 (1984): 107–9.

11　Harry Lawless and Trygg Engen, "Associations of Odors: Interference, Mnemonics and Verbal Labeling," *Journal of Experimental Psychology: Human Learning and Memory* 3 (1977): 52–59.

12　Jean-Xavier Guinard, Rosemary M. Pangborn, and Michel J. Lewis, "The Time-Course of Astringency in Wine upon Repeated Ingestion," *American Journal of Enology and Viticulture* 37 (1986): 184–89.

13　Richard J. Stevenson, Robert A. Boakes, and John Prescott, "Changes in Odor Sweetness Resulting from Implicit Learning of a Simultaneous Odor-Sweetness Association: An Example of Learned Synesthesia," *Learning and Motivation* 29 (1998): 113–32.

14　David A. Forss, "Role of Lipids in Flavors," *Journal of Agriculture and Food Chemistry* 17 (1969): 681–85; Harry T. Lawless, Serena Schlake, John Smythe, Juyun Lim, Heidi Yang, Kathryn Chapman, and Bryson Bolton, "Metallic Taste and Retronasal Smell," *Chemical Senses* 29 (2004): 25–33.

15　Takayuki Tamura, Kiyoshi Taniguchi, Yumiko Suzuki, Toshiyuki Okubo, Ryoji Takata, and Tomonori Konno, "Iron is an Essential Cause of Fishy Aftertaste Formation in Wine and Seafood Pairing," *Journal of Agriculture and Food Chemistry* 57 (2009): 8550–56.

16　Joshua Pierce and Bruce P. Halpern, "Orthonasal and Retronasal Odorant Identification Based upon Vapor Phase Input from Common Substances," *Chemical Senses* 21 (1996): 529–43.

17 Paul Scholten, "How Much Do Judges Absorb?" *Wines and Vines* 69 (1987): 23–24.
18 Maynard A. Amerine and Edward B. Roessler, *Wines: Their Sensory Evaluation*, 2nd ed. (San Francisco: Freeman, 1983).
19 Ingemar T. Nygren, Inga-Brit Gustafsson, and Lisbeth Johansson, "Effects of Tasting Technique – Sequential Tasting vs. Mixed Tasting – On Perception of Dry White Wine and Blue Mould Cheese," *Food Service Technology* 3 (2003a): 61–69; Nygren, Gustafsson, and Johansson, "Perceived Flavour Changes in Blue Mould Cheese after Tasting White Wine," *Food Science Technology* 3 (2003b): 143–50; Berenice Madrigal-Galan and Hildegarde Heymann, "Sensory Effects of Consuming Cheese Prior to Evaluating Red Wine Flavor," *American Journal of Enology and Viticulture* 57 (2006): 12–22.
20 David Douglas, Margaret A. Cliff, and Andrew G. Reynolds, "Canadian Terroir: Sensory Characterization of Riesling in the Niagara Peninsula," *Food Research International* 34 (2001): 559–63; Derek Kontkanen, Andrew G. Reynolds, Margaret A. Cliff, and Marjorie King, "Canadian Terroir: Sensory Characterization of Bordeaux-Style Red Wine Varieties in the Niagara Peninsula," *Food Research International* 38 (2005): 417–25; J. Schlosser, Andrew G. Reynolds, Marjorie King, and Margaret Cliff, "Canadian Terroir: Sensory Characterization of Chardonnay in the Niagara Peninsula," *Food Research International* 38 (2005): 11–18; Hakimi J. Rezaei and Andrew G. Reynolds, "Characterization of Niagara Peninsula Cabernet Franc Wine by Sensory Analysis," *American Journal of Enology and Viticulture* 61 (2010): 1–14.
21 Herbert S. Kaufman, "The Red Wine Headache and Prostaglandin Synthetase Inhibitors: A Blind Controlled Study," *Journal of Wine Research* 3 (1992): 43–46.

Thirteen

Vintning on Thin Ice: The Making of Canada's Iconic Dessert Wine

Gary Pickering and Debbie Inglis

C anada is the largest producer by volume of Icewine—one of the world's truly unique beverages. And, without a doubt, that uniqueness comes at the price of pushing winemakers to their wits' end. In fact, some winemakers have termed the gruelling process "Extreme Winemaking." From draping the crop in miles of awkward netting to thwart bird attacks, to picking in the middle of bone-chilling nights (including holidays), to the irksome stickiness on everything it touches in your winery, making Icewine can be a ruthless pursuit. And the sticky stuff doesn't ease up when it comes to fermenting it. This is when yeast are put through the microbial equivalent of Marine Corps basic training. Join us as we unveil the "frozen truth" about Icewine production, what it truly takes to make Canada's ultra-premium dessert wine, and the research from Brock University's Cool Climate Oenology and Viticulture Institute (ccovɪ) that supports this Canadian icon.[1]

Introduction

Icewine is a late-harvest wine produced from the juice of grapes that have naturally frozen on the vine. During pressing, the water in the frozen berries is retained as ice crystals, so the resulting juice is highly concentrated in sugar, acid, and flavour compounds. It is valued for its rarity, sweetness, intensity of aroma and flavour, and—not least of all—its purity. Its origins are equivocal, but a date often cited as the first "genuine" Icewine harvest is 1794, in Germany.[2] For most of its history since that date, it is likely that it was a very irregular occurrence, with one German wine writer referenced in Schreiner calculating that there were only ten Icewine (Eiswein) vintages between 1875 and 1962.

The question of when, where, and by whom the first Icewine was produced in Canada is contentious; this chapter will not fuel the provincial and individual rivalries that flavour that particular debate. Suffice it to say that the answers probably vary depending on whether we are considering hobby, commercial, or research wine, whether authentic or artificial freezing of the grapes was used, and whether the wines were *vinifera* or non-*vinifera* based. Hence, the pride associated (and rightly so!) with the origins of one of the world's most exquisite wine styles.

What is less contentious is that the beginnings of commercial Icewine production, both in Ontario and in British Columbia, owe much to the enterprise of European immigrants who successfully transplanted their experiences from Germany and Austria to the Canadian situation. Production was restricted to only a few wineries during the early to mid-1980s, and the volume produced was relatively small.

A significant catalyst for the rapid growth witnessed in the 1990s was the award of the Grand Prix d'Honneur at Vinexpo in 1991 to Inniskillin's 1989 Vidal Icewine. Considered by many as the most coveted award in the international wine show circuit, it helped to put Canada on the international wine map, as well as give a valuable boost to domestic sales. Canada now dominates the global Icewine scene in terms of both quantity and—arguably—quality.

The hot summers and cold, dependable winters of Canada's main grape-growing regions create the requisite conditions for consistent and reliable production. While many European ice wine producers cannot be assured of sufficiently cold conditions at the correct time and for long enough to harvest the fruit, this is generally not a problem with Canada's long, cold winters. Most of its viticultural regions produce Icewine, and these are spread across a continent—from the Pacific province of British Columbia, to Nova Scotia, on the Atlantic Ocean. It is in Ontario, however, and on the Niagara Peninsula in particular, that the majority (approximately 75 percent) of Icewine is made and where projected growth is greatest.[3] While yields can vary from vintage to vintage, Canada's total Icewine production has increased dramatically over the last decade, and current estimates are approximately one million litres.[4]

"Icewine" is used here to delineate Canadian-produced product (the term is regulated under the Vintners Quality Alliance Act in Ontario), "Eiswein" for this style in Germany and Austria, and "ice wine" as a generic term encompassing both styles as well as those from other nations. Simulated ice wine—which makes use of artificial means to concentrate the grapes or juice—will not be considered here, except for a brief mention on new research initiatives aimed at identifying fraudulent products.

The CCOVI was established in 1997 at Brock University in St. Catharines, in the heart of Niagara's wine region (Figure 13.1). As a partnership between industry, government, and academia, a major part of its mandate is to develop both applied and basic research programs of particular relevance to the demands of cool climate grape growing and winemaking. One of its strengths has been

Figure 13.1 The Cool Climate Oenology and Viticulture Institute at Brock University, St. Catharines, Ontario
Photo: Divino Mucciante.

its integration of research across the value chain, literally from the grape to the consumer, and its multidisciplinary approach to the needs of industry. Icewine has been one of the major foci of the Institute since its inception; we highlight in the text some of the research results from the Institute where they illustrate the uniqueness of Icewine and some of the technical solutions that have helped it to grow to its current status.

How Is Icewine Made?

Pre-Harvest Considerations

Specific climatic conditions must prevail in order to produce Icewine regularly. Sufficient sun and heat during summer and autumn are needed to mature grape flavours, accumulate sugar, and maintain healthy fruit, while reliably cold temperatures during late fall and winter are required to freeze the fruit and concentrate their contents.

The potential for loss of crop from bird predation is considerable given the extended hang time of Icewine grapes and the concurrent loss of other food sources for these avian pests. Starlings are a particular problem, and may devastate a vineyard in a matter of hours (Figure 13.2), particularly later in winter, when numbers can be high and alternate food sources more limited when the ground is covered in snow. Thus, bird netting is a necessity. For Icewine

Figure 13.2 Starlings flock over a Vidal Icewine vineyard in the Niagara Peninsula
Photo: Debbie Inglis.

grapes that are to be machine harvested, a netting mesh size of approximately 2.5 square centimetres is ideal; this is small enough to prevent bird damage, yet large enough to allow for frozen berries to fall through the mesh during harvest. Bird bangers and other devices designed to scare off birds are also employed, but generally seem to be less effective in winter. Adding to the burden of protecting the crop until the late harvest, deer, raccoons, cattle, coyotes, and even bears (in British Columbia) can be significant pests as they hunt for alternate food during the meagre offerings of winter.[5]

As with table wine, sound, ripe grapes are needed to produce quality Icewine. If late summer and autumn are warm and wet—conditions more common in Ontario than British Columbia—the risk of mildews and Botrytis developing will equally affect those grapes intended for Icewine production.[6] This may be further confounded if the winter is unusually warm, as a delay in harvest into February or March increases the risk of rain-related rots and other disease, particularly in those varieties with thinner skins. In contrast with many other dessert wines, Noble rot is not a welcome visitor for grapes intended for Icewine. The dehydration caused by significant infection from Botrytis cinerea concentrates the sugars to the extent that yields would be miniscule by the time further concentration occurs during the winter harvest. Depending on the extent of infection, freezing may not even be possible, due to the high sugar content obtained from this double concentration. In addition, the integrity of the stems is sometimes compromised by Botrytis, which may add to the early loss of fruit from the vine. Also, many producers argue that Botrytis characters detract from the purity of fruit flavours found in Icewines derived from "clean" grapes.[7] Additional protection against Botrytis bunch rot using late-season fungicide sprays are normally practised for grapes destined for Icewine production, with the last pesticide applications done approximately 90 days prior to Icewine harvest. In addition, prior to netting the grapes, any diseased fruit is removed from the vine to reduce further spread of disease.

Cropping levels can influence final Icewine quality, in terms of overall flavour, as studied in Riesling and Vidal Icewines,[8] although less so than with table wines, because of the dominant influence of temperature during harvest and pressing on juice composition. Dr. Karl Kaiser of Inniskillin Wines believes that an ideal Icewine juice yield (based on grape weight as estimated in October) is approximately 150 litres per tonne for Vidal and 125 litres per tonne for Riesling. This compares to approximately 700 litres per tonne of juice if the grapes were pressed during a normal fall harvest. In the vineyard, the Icewine grape yield translates to not more than seven tonnes per acre for Vidal and five tonnes per acre for Riesling (as calculated in October).[9]

Harvest

Grapes must be picked and pressed at minus 8° or lower, with optimal harvest and pressing temperatures between minus 10 to minus 12° Celsius. Although not a necessity, harvest often occurs at night, during a six- to eight-hour period of low temperatures. In Ontario, harvesting at the end of December through to mid-January is common and is generally considered optimal for the final quality of the wine, although on a rare occasion, harvest has been delayed until late March. Anecdotally, it is believed that a number of freeze/thaw events on the vine are necessary prior to harvest in order for the grapes to achieve the complexity of flavour expressed in the best Icewines. This theory was recently tested in the Niagara Peninsula. In general, grapes from later harvest dates (in the month of February) had significantly higher aroma and flavour intensities.[10] In British Columbia, the harvest dates are typically earlier, often in November. Practical limitations affecting the decision on when to harvest include availability of labour and winery capacity. For instance, if it has been a warm December and January, the harvest window later in winter may become very narrow, placing considerable stress on the operation to pick and process the entire crop while the temperature profile in the vineyard is right.

There are also some general differences in the Icewines from grapes grown and harvested in different areas of Canada. In the study of Nurgel et al., Ontario Icewines had more apricot, raisin, honey, and oak aromas, while those from

British Columbia had more pineapple and oxidized aromas.[11] Further research will look into the effect of region of origin on Icewine, particularly in light of the creation of new sub-appellations in 2006 for the Niagara region.

During the harvest of grapes destined for Icewine, the fruit is both hand and machine harvested. The traditional method is hand harvesting, where the bird netting is opened from the bottom and the frozen bunches picked into boxes. Machine harvesting the frozen berries through the Icewine netting is now the most common way to harvest Icewine grapes in Ontario, and has the considerable advantages of being quicker and able to make use of much smaller harvest windows (Figure 13.3). The frozen grapes are stored in insulated totes until they are ready to be pressed (Figure 13.4).

In contrast with grapes intended for table wine, the sugar levels in autumn of grapes destined for Icewine play a minor role in the final juice sugar concentration of the must. Temperatures at harvest and during pressing are the main determinants. The higher the sugar content in the grape, the lower the temperature must be in order to freeze out the water. The sugar concentration itself is influenced by the physiological ripeness of the fruit, but also by prior dehydration from hang time, weather, and Botrytis. Harvest temperatures at one Ontario winery have varied between minus 17° Celsius and minus 10° Celsius over one three-year period, producing musts of 55 and 38 °Brix, respectively.[12] When temperatures fall much below minus 17° Celsius, it is exceedingly difficult

Figure 13.3 Machine harvesting of Vidal Icewine grapes
Photo: Debbie Inglis.

Figure 13.4 Icewine grapes stored in insulated totes prior to pressing
Photos: Debbie Inglis.

to extract the juice, as most of the water component is frozen. Late Harvest, Select Late Harvest, and Special Select Late Harvest wines are also produced, and are sweet wine styles that have a lower minimum Brix requirement than is needed for Icewine. These may be produced from second (and subsequent) pressings of Icewine grapes or from fruit harvested earlier in winter at (comparatively) warmer temperatures.

Pressing

For some of the larger producers, the harvested fruit is pressed at the vineyard itself in order to process the grapes quickly, while still frozen. Smaller producers transport the fruit to their wineries. In either case, no crushing occurs. Under VQA regulations, pressing must be a continuous process, and a low temperature needs to be retained throughout the operation to obtain a juice with a minimum of 35 °Brix. A single pressing may never be lower than 32 °Brix. During pressing, much of the water is retained with the grape skins as ice, while a juice highly concentrated in sugars, acids, nutrients for fermentation, and aroma compounds is extracted. Work from our group in evaluating 298 Vidal Icewine juice samples (Table 13.1) produced by processors across the Niagara Peninsula shows the range of sugar concentrations from the minimum of 32 up to 46 °Brix, with an average of 39.3 °Brix. The juice is quite acidic, with an average titratable acidity of 10.5 grams per litre of tartaric acid and an upper concentration of 14.6 grams per litre. In contrast to juice used for table wine production, a large proportion of juice acidity is from malic acid, with average juice concentrations at 7.8 grams per litre. Much of the tartaric acid likely precipitates out in the frozen berry while on the vine, presumably as potassium bitartrate.

A range of presses can be and are used, including basket (Figure 13.5), bladder, and membrane-based designs; the major criterion is that sufficient pressure can be obtained to extract juice from the frozen berries. The ideal press will

Table 13.1. Composition of Icewine juices across the Niagara Peninsula[1]

Variety	Soluble Solids (°Brix)	Amino Acid Nitrogen (mgN/L)	Ammonia Nitrogen (mgN/L)	Total Yeast Assimilable Nitrogen (mgN/L)	pH	Titratable Acidity (g/L tartaric acid)	MalicAcid (g/L)	% TA as MalicAcid
Riesling	38.0±2.1	381.7 ±132.6	79.7±38.6	461.3±166.5	3.22±0.18	9.0±1.7	5.39±1.17	67.0±7.5
Vidal	39.3±1.7	497.7 ±105.2	57.3±18.7	555.0±120.2	3.38±0.16	10.5±1.5	7.8±1.3	84.0±10.5
Cabernet Franc	39.4±2.0	292.9 ±116.7	22.3±8.5	315.3±123.0	3.79±0.24	5.7±1.2	4.68±0.83	93.1±12.4

Source: From G. Pickering, K. Ker, D. Inglis, R. Hay, and H. Seltred, "Ice Wine Production Practices," in Proceedings of the Sixth International Cool Climate Symposium for Viticulture and Oenology, Christchurch, New Zealand, 5–10 February 2006 (Havelock North, NZ: New Zealand Society for Viticulture & Oenology, 2006). Reproduced with thanks to Dr. Debbie Inglis.

Data are mean values ± SD from 298 Vidal samples, 24 Riesling samples, and 23 Cabernet Franc samples across two vintages.

Figures 13.5a and 13.5b Hydraulic basket press for pressing Icewine grapes
Photo 13.5a: Debbie Inglis.
Photo 13.5b: B. Grant.

provide high pressure to reduce pressing time, and yield as clean a juice as possible. Pressing can be an exceedingly lengthy process, taking many hours, particularly so at very low temperatures, when the yield is also reduced, and when using low-pressure presses. These considerations can place some challenges on the winemaking team with respect to organization and timing of operations.

For example, during pressing, the juice can be fractionated according to its sugar concentration to enable more than one product to be produced. For instance, juice extracted late in the pressing process that has a soluble solids concentration lower than the 32 °Brix minimum for an individual Icewine pressing could be used for one of the Late Harvest dessert wine styles mentioned above.

Settling and clarification of the juice is typically slow. Depending on juice composition and winemaker preference, fining or suphiting (or both) of the juice are sometimes practised. Generally, we have found that Icewine musts have adequate levels of assimilable nitrogen, and therefore do not require supplementation with Diammonium Phosphate or other sources of nitrogen (Table 13.1).

Vinification

After settling, the juice is inoculated with yeast, and fermentation is typically carried out at 15 to 17° Celsius. The high sugar and acid—and, as fermentation progresses, the increasing ethanol concentration—place significant stress on the yeast. As juice-soluble solids increase, a negative linear correlation is found with yeast growth and sugar consumption.[13] When juice above 42 °Brix is fermented, yeast are hard pressed to achieve a target of 10 percent alcohol by volume—the industry norm for Icewine—before they stop fermenting. There is a strong negative linear correlation between ethanol in the final wine and initial Icewine juice sugar concentration (i.e., the higher the starting Brix, the lower the alcohol in the finished wine). As an example, using Riesling Icewine juice at 46 °Brix fermented over a one-month period, we found only 6.5 percent alcohol by volume was reached before the yeast stopped fermenting. We also found that, in excess of 52.5 °Brix, Icewine juice is theoretically non-fermentable by yeast.[14] Due to the stress imposed by concentrated Icewine juice, yeast multiply no more than three times when fermenting Icewine.[15] If an inadequate amount of yeast is used, the yeast population will not amass sufficiently to reach an adequate ethanol concentration in a practical time frame (i.e., four weeks). We recommend adding yeast at two and a half times the table wine rate (e.g., 50 grams per hectolitre) to allow the population to grow and ferment to a minimum of 10 percent alcohol by volume in a one-month time frame.[16] Favoured strains include Lallemand's K1-V1116 and EC1118.

The indigenous yeast strains present in Icewine juice may impact Icewine quality as they do in table wines—perhaps more so, given the more challenging conditions of Icewine juice. Dr. Debbie Inglis has pioneered research into yeast strain and inoculation procedures for Canadian Icewine. Her team has identified several species indigenous to Vidal and Riesling Icewine juices that may lead to the identification of optimal strains or combinations of yeast strains for Icewine fermentation.[17]

It is the osmotic stress created by the extremely high sugar content that most challenges yeast viability and performance.[18] One specific consequence of the "hyperosmotic stress response" of the yeast during Icewine fermentation is increased production of acetic acid[19]—the major constituent of volatile acidity (VA) in wine. Acetic acid in Icewine is often desirable to help balance the sweetness;[20] however, at elevated levels, VA may detract from overall quality. Its concentration is regulated—the maximum limit set by the International Organisation of Vine and Wine (OIV) and VQA is 2.1 grams per litre (expressed as acetic acid). Hence, the mechanisms underlying VA production and its management in Icewine are a priority to Icewine makers and researchers. As reported by the Inglis research team, several yeast genes are expressed as part of the yeast stress response, which lead to higher amounts of acetic acid and glycerol.[21] Although glycerol is present at significant concentrations in commercial Icewine at 10 grams per litre,[22] there is no evidence that glycerol contributes to the mouthfeel of the wine,[23] contrary to some anecdotal comment from industry. However, the glycerol is important to the yeast during fermentation to assist them to survive the difficult fermentation conditions.[24] To minimize VA production, two general methods have been shown to be effective: the use of yeast nutrients (like GO-FERM)[25] and step-wise acclimatizing the yeast to these fermentation conditions.[26]

Fermentation of Icewine typically takes a few weeks longer to complete than for table wines—"complete" in the sense of achieving an ethanol concentration of approximately 10 percent alcohol by volume, where there is still considerable residual sugar present (approximately 200 grams per litre), imparting the sweetness the style is renowned for. Acidulation of the juice or wine with up to 4 grams per litre of acid is permitted in Canada, and while not usually needed for Riesling, other Icewine varieties do sometimes require small additions to improve the sweetness to sourness balance and, to a lesser extent, to maintain an acceptable pH.

Bentonite fining to ensure protein stability is common, and Icewines often require higher levels than their table wine equivalents. Although a significant portion of tartaric acid has precipitated out in the berry, some remains after fermentation, and cold stabilization is practised. It is, however, more difficult to achieve than with table wine, possibly related to the higher density of Icewine. Once the wines are stable, sulphur dioxide and often sorbate are added to provide sufficient antioxidant and anti-microbial protection. Filtration, followed by early bottling, is encouraged; early bottling is believed to preserve varietal intensity. Most Icewines are filled into 375-millilitre—and sometimes smaller—bottles. Decisions on closures, label design, and overall packaging are often made to reinforce the message that Icewine is a prestigious and "pure" product, and these components are frequently tailored to the specific marketing needs of export countries.

Cultivars and Icewine Styles

Winter hardiness is the prime prerequisite for a variety destined for Icewine. High natural acidity, late ripening, tough skins and stems, and reasonable resistance to disease have been specifically noted as the most desirable characteristics.[27] A wide range of cultivars are used, including Vidal, Riesling, Cabernet Franc, Cabernet Sauvignon, and Gewürztraminer in Ontario, and Riesling, Gewürztraminer, Chardonnay, Pinot noir, Erhenfelser, Kerner, Muscat, and Syrah in British Columbia, and, less commonly, blends of these. However, single varietal wines from Vidal or Riesling account for over 90 percent of Icewine produced. Table 13.2 provides the breakdown, by variety, of Icewine produced in Ontario.

Table 13.2 Varietal breakdown of Icewine produced in Ontario in 2009/2010 vintage

Wine Category	Number of Products	Volume (litres)
Vidal	46	675,864
Cabernet Franc	22	118,836
Riesling	21	49,498
Cabernet Sauvignon	13	30,753
Gewurztraminer	8	7,713
Other	8	15,597
Total	118	898,261

Source: Reproduced with permission from VQA-O 2010 Annual Report.

The tough skins of Riesling and Vidal offer some protection from disease and the physical and mechanical stress of the freeze/thaw cycle, and they impart desirable fruit intensity and—particularly with Riesling—acid balance to the finished wines. Vidal is a French–American hybrid (Ugni Blanc x Seibel 4986) common in many areas of eastern North America where cultivation of *Vitis vinifera* is difficult. The Vidal berries have thick skins and loose bunches, affording them superior resistance to rot; table wines are also made from the variety. It is much more prevalent in Ontario, while the white varieties most used for Icewine in British Columbia are Riesling, Pinot blanc, Ehrenfelser, and Kerner.

The complexity of Icewine flavour is evident in the unique quality of its components. It is distinct among the world's wine styles with respect to its high sugar concentration; the only comparable wine is Tokaji. Fortunately, the titratable acidity (TA) is also elevated during berry dehydration, with final levels in the wine of around 11 grams per litre, allowing for a pleasing sweetness to sourness balance. The other major constituents of Icewine are acetic acid and glycerol.[28] These components, together with other aroma-active and aroma precursor compounds, elicit the complex array of sensory sensations for which the wine is famous (Table 13.3).

A number of producers are also making oak-aged Vidal Icewines, which add further complexity to the style, including an accentuation of caramel, oak, and walnut notes and an increase in colour intensity.[29] Icewine from red-skinned varieties has been a relatively recent trend, with Cabernet Franc and Cabernet Sauvignon in particular producing some outstanding wines. Only low levels of pigments are extracted compared to red table wine, as the grapes are not crushed

Table 13.3 The complexity of Icewine: select descriptive terms used to profile appearance, aroma, and flavour

Appearance	Orthonasal aroma		Retronasal aroma	Oral sensations
Pale yellow	Apple	Floral	Peach	Sweetness
Golden yellow	Apricot	Raisin	Pear	Acidity
Golden copper	Citrus	Walnut	Dried fruit	Bitterness
Clarity	Pineapple	Oxidation	Grapefruit	Heat
	Peach	Honey	Orange peel	Viscosity
	Pear	Butterscotch	Raisin	Density
	Dried fruit	Caramel	Honey	
	Tropical fruit	Oak	Butterscotch	
	Grapefruit	Earthy/vegetal	Earthy/vegetal	
	Banana	Petroleum	Oxidation	
	Orange peel	Ethyl acetate		

Source: Nurgel et al., "Sensory and Chemical Characteristics of Canadian Ice Wines"; G.J. Pickering, C. Nurgel, D. Inglis, A. Reynolds, and I. Brindle, "Deconstructing God's Nectar: Characterisation of Ice Wine Using Descriptive Analysis Techniques," in *Proceedings of the 6th Pangborn Sensory Science Symposium*, 7–11 August 2005 (Yorkshire, UK: Harrogate International Centre, 2005).

and the skins are not present during fermentation. Colour may sometimes be supplemented by the addition of small amounts of teinturier (red-juiced) varieties, such as Dornfelder.

Further extending the range of styles is a new innovation—sparkling Icewine. While production is currently limited to a small number of producers, the quality can be quite extraordinary. By regulation, the "sparkle" cannot be derived from direct carbonation, and instead is usually achieved using a variation on the *Méthode Cuve Close* (closed tank method). The Icewine juice is fermented in a Charmat tank with an open lid for most of the fermentation. Toward the end, the lid is closed, and the evolving carbon dioxide is trapped in the wine. A short period of lees aging precedes final processing and bottling. The carbon dioxide enhances the acidity, producing a more refreshing and perhaps better-balanced wine. Extra complexity is also achieved from the mouth feel sensations elicited from the fine mousse.

Regulation

"Icewine" is a proprietary name owned by VQA, and its use is strictly regulated by the institution. VQA was designated as Ontario's wine authority under the Vintners Quality Alliance Act, 1999. The act sets the framework by which standards for the production of VQA wine—including Icewine—and appellations for wine-growing regions are established. VQA Ontario establishes, monitors, and enforces an appellation of origin system in accordance with the act and controls the use of specified terms, descriptions, and designations associated with the VQA appellation system.

The specific regulations concerning the viticultural and oenological requirements for Icewine are covered by Ontario Regulation 406/00 (and later amendments). The more salient points are summarized below:

- The wine must be a late-harvested wine.
- The wine must be produced entirely from one or more grape varieties naturally frozen on the vine, picked while the air temperature is minus 8° Celsius or lower, and immediately pressed after picking in a continuous process.
- The list of grape varieties permitted for Icewine is regulated. All varieties are *Vitis vinifera*, with the exception of the hybrid Vidal.
- The final alcohol content of the wine must be between 7.0 percent and 14.9 percent by volume.
- The wine must be produced as a varietal wine in accordance with the requirements for varietal wines given elsewhere in the regulations.
- One hundred percent of the grapes must be grown in a designated viticultural area, of which at least 85 percent must be grown in the named viticultural area shown on the label. The pressing must also take place within the viticultural area where the grapes were grown.

- Grapes, juice, or grape must intended for the production of Icewine may be artificially refrigerated to a temperature of not less than minus 4° Celsius.
- No freeze concentration of juice, grape must, or wine is permitted.
- The Brix level of the juice after each pressing shall be at least 32 °Brix when measured after transfer to the fermentation vessel.
- The finished wine must be produced from a must that achieves a computed average of not less than 35 °Brix.
- The residual sugar at bottling must result exclusively from the natural sugar of the grapes, and must not be less than 125 grams per litre.
- The actual alcohol must come exclusively from the natural sugar of the grapes.

Many of these vQA regulations set the *minimum* standards for safeguarding Icewine quality. Typically, in regard to soluble solids content at harvest, for instance, actual practice far exceeds these requirements (see Table 13.1). A guiding principle when these rules were established was to set standards higher than those in place in Europe, and this generally remains true today. An added feature of the approval process before a wine may be granted vQA approval is that an independent tasting panel must first evaluate and approve it. This provides added credibility to the process, and the vQA overall has been one of the most important factors in elevating consumer perception of Canadian wines, including Icewine. These regulations, if coupled with rigorous enforcement, will continue to protect the consumer and guarantee the high quality of the product that has come to be expected.

The Economics of Production

The average retail price in Canada for a 375-millilitre bottle of Icewine is around $50 (CAD), based on products currently available for purchase at the Liquor Control Board of Ontario. A question that is frequently asked is "Why so much?" Part of the answer may lie in consideration of some of the costs and other economic factors that pertain to producing and marketing Icewine:

- *Yield loss.* Yields are typically 15 to 20 percent of that expected for table wine, largely because of loss of volume from freezing and dehydration. This becomes significantly lower if pests or poor weather further reduce the crop—a risk that increases with hang time. Furthermore, there is a realistic risk of no Icewine whatsoever in some years, should the required weather conditions during later autumn and winter not occur.
- *Growing and harvesting costs.* The crop and vineyard must be monitored and maintained for three to five months longer than for table wines. The purchase of bird netting is mandatory, and the operation of other bird-protection devices for an extended period incurs additional costs. If the Icewine grapes are hand harvested, the labour rates are significantly higher than for table wine grapes, and the sometimes torturous conditions in winter can extend the time needed to bring in the crop.

- *Winemaking costs.* General winemaking costs per litre are higher because of the relatively small quantities produced (for most wineries) compared with table wines and the long periods required for pressing and fermentation. In addition, damage or wear on some equipment is greater, particularly press components such as membranes and bladders.
- *Marketing costs.* Domestic consumption of Canadian wines is low and relatively static. An increasing proportion of the rising production of Icewine has to be exported, with the concurrent costs associated with establishing new international markets as well as servicing and growing existing ones. Coupled with this, increased educational and promotional activity is now required to counter the rise in fraudulent products that are successfully competing with Icewine in Asia, and supporting initiatives aimed at enforcing international trade and copyright laws.

In this context, some have argued that Icewine may be significantly *under*priced.

Faux vs. Authentic Icewine

Because of the significant success of Icewine on the international stage, it perhaps shouldn't be surprising that there has been a rapid growth in producers who seek to either simulate the style through artificial means or fraudulently misrepresent their wine as genuine Icewine. In the case of the former, techniques other than natural freezing of the grape on the vine are used to concentrate the sugars and flavours. This most commonly involves cryoextraction, where freezing and partial thawing of the fruit or juice occurs artificially after harvest, mainly through use of commercial freezers. Most international wine conventions and countries do not allow the labelling of these products as ice wine. Because cryoextraction assumes almost none of the risks associated with Icewine production (e.g., the grapes do not need to remain on the vine for months after the normal table wine harvest), the pricing can and normally does seriously undercut that of the authentic product. However, an even greater threat is posed by counterfeiting.

Counterfeiting in this context means the deliberate imitation of Canadian Icewine, and Taiwan and China are the main culprits.[30] Labels, bottles, and other components of the Icewine package are dutifully reproduced to imitate those of specific Canadian producers or to convey the impression that the wine is from Canada through use of iconic imagery on the bottle or label (maple leaf, Canadian goose, beaver, etc.). The content of these counterfeit bottles is, at the very best, sweetened table wine, but is sold as genuine Icewine at a fraction of the cost.

Both of these issues have been identified as a serious threat to the Icewine industry, and Brock University's Cool Climate Oenology and Viticulture Institute is in the midst of a major multidisciplinary research initiative to identify chemical and sensory markers that will help regulators and enforcement agencies to differentiate the real from the faux.

Summary and Future Challenges

Icewine has been largely responsible for driving the—albeit embryonic—success of the Canadian wine industry in the international market, somewhat analogous to what Sauvignon Blanc did for general recognition of New Zealand wines. In the space of a decade, production has increased from moderate levels to a volume that is now much larger than any other ice-wine-producing nation. Concurrent with this development is a widely held belief that the quality is second to none.[31]

Perhaps five factors can be identified as most responsible for this success:

- Reliably cold winters.
- The predominance of the Vidal grape in Ontario, the physiological properties of which are close to perfection for making quality Icewine.
- The combination of technical competence and entrepreneurial flair among the early Icewine producers.
- Sound and innovative business and marketing strategies.
- The establishment of regulations under VQA aimed at maintaining high standards of quality.

While the Icewine industry has undergone significant growth, a number of challenges must be addressed for the sustainability of this sector. The threat posed by counterfeit Icewines is one significant concern; as mentioned, research initiatives aimed at detecting fraudulent wines are in progress. However, in contrast with much of the winemaking world, Canada does not yet have a federal national standard for wine. Such legislated standards exist only at the provincial level. The absence of enforceable national standards makes it more difficult to protect against fraudulent Icewines, and addressing this situation should be a priority. How much Icewine the global market can absorb is unknown, as is the effect on pricing and industry profitability as supply increases. Current supply is outpacing demand, with the recent global recession playing an important role in this unsustainable situation. More research and innovation is needed to address these fundamental economic and marketing challenges.

At the more technical level, novel yeast strains or combinations of yeast species are being investigated specifically for Canadian Icewine production to further define our unique style. Further experimentation is encouraged into the use of teinturier grape varieties for red Icewines, particularly given the high value placed on this colour in some Asian markets and the possible advantages in introducing further novelty into the style. Finally, the impact of climate change, and global warming in particular, on the future suitability of Canada's Icewine-growing regions is largely unknown and warrants serious and perhaps urgent consideration.

From the challenges of bird and disease pressures to harvesting in the middle of bone-chilling nights followed by the yeast's struggle to ferment in such a hostile environment, our treasure comes forward from the extremes: Icewine.

Notes

1 Dedication: We would like to dedicate this chapter to Karl J. Kaiser, LLD. Dr. Kaiser has been a true pioneer of Canadian Icewine in every sense of the word. On behalf of oenophiles the world over, thank you, Karl.

2 J. Schreiner, *Icewine: The Complete Story* (Toronto: Warwick Publishing, 2001).

3 D. Cyr and M. Kusy, "Canadian Ice Wine Production: A Case for the Use of Weather Derivatives," *Journal of Wine Economics* 2 (2007): 145–67.

4 VQA Ontario. "Annual Report 2010," http://www.vqaontario.com/AboutVQA/AnnualReports/2010/Statistics/Overview.

5 J. Schreiner, *Icewine: The Complete Story*; J. Schreiner, *Icewine: The Wine of Winter* (London: The International Wine & Food Society and The International Wine & Food Foundation, 2004).

6 G. Pickering, K. Ker, D. Inglis, R. Hay, and H. Seifred, "Ice Wine Production Practices," in *Proceedings of the Sixth International Cool Climate Symposium for Viticulture and Oenology*, Christchurch, New Zealand, 5–10 February 2006 (Havelock North, NZ: New Zealand Society for Viticulture & Oenology, 2006), 170; O.C. Phillips, Jr., "The Influence of Ovid on Lucan's *Bellum civile*" (Ph.D. dissertation, University of Chicago, 1962), 14.

7 K. Kaiser, "Icewine Concepts and Facts: A Personal View," in *Anatomy of a Winery: The Art of Wine at Inniskillin*, 2nd ed., ed. D.J.P. Ziraldo (Toronto: Key Porter Books, 2006).

8 A. Bowen, A. Reynolds, and G. Pickering, "Effect of Harvest Date and Crop Load on Sensory Profiles of Icewines from Niagara Peninsula," *American Journal of Enology and Viticulture* 57, no. 4 (2006): 526A; A. Bowen, "Elucidation of Odour-Potent Compounds and Sensory Profiles of Vidal Blanc and Riesling Icewines from the Niagara Peninsula: Effect of Harvest Date and Crop Level" (Ph.D. dissertation, Brock University, 2011).

9 Kaiser, "Icewine Concepts and Facts."

10 Pickering et al., "Ice Wine Production Practices"; A. Bowen, A. Reynolds, and I. Leschaeve, "Influence of Harvest Date on the Sensory Profiles of Icewines from the Niagara Peninsula," *American Journal of Enology and Viticulture* 59, no. 3 (2008): 300A; Bowen, "Elucidation of Odour-Potent Compounds and Sensory Profiles of Vidal Blanc and Riesling Icewines from the Niagara Peninsula."

11 C. Nurgel, G.J. Pickering, and D. Inglis, "Sensory and Chemical Characteristics of Canadian Ice Wines," *Journal of the Science of Food and Agriculture* 84 (2004): 1675–84.

12 J. Schreiner, *Icewine: The Complete Story.*

13 C. Pitkin, D. Kontkanen, and D. Inglis, "The Effects of Varying Soluble Solids Concentration in Icewine Juice on Metabolite Production by *Saccharomyces cerevisiae* K1-V1116 during Fermentation," in *Proceedings of the International Bacchus to the Future Conference*, St. Catharines, Ontario, 23–25 May 2002, ed. C.W. Cullen, G.J. Pickering, and R. Phillips (St. Catharines, ON: Brock University Press), 179.

14 G. Pigeau, E. Bozza, K. Kaiser, and D. Inglis, "Concentration Effect of Riesling Icewine Juice on Yeast Performance and Wine Acidity," *Journal of Applied Microbiology* 103, no. 5 (2007): 1691–98.

15 D. Kontkanen, D.L. Inglis, G.J. Pickering, and A. Reynolds, "Effect of Yeast Inoculation Rate, Acclimatization, and Nutrient Addition on Icewine Fermentation," *American Journal of Enology and Viticulture* 55, no. 4 (2004): 363–70.

16 D.L. Inglis, "Make Icewine Easier, at Least for Yeast," *Vineyard and Winery Management*, March/April (2008): 71–75.

17 C. Nurgel, D.L. Inglis, G.J. Pickering, A. Reynolds, and I. Brindle, "Dynamics of Indigenous and Inoculated Yeast Populations in Vidal and Riesling Icewine Fermentations," *American Journal of Enology and Viticulture* 55 (2004): 435A.

18 Kontkanen et al., "Effect of Yeast Inoculation Rate, Acclimatization, and Nutrient Addition on Icewine Fermentation."

19 G. Pigeau and D.L. Inglis, "Upregulation of *ALD3* and *GPD1* in *Saccharomyces cerevisiae* during Icewine Fermentation," *Journal of Applied Microbiology* 99 (2005): 112–25.

20 M. Cliff and G. Pickering, "Determination of Detection Thresholds for Acetic Acid and Ethyl Acetate in Icewine," *Journal of Wine Research* 17, no. 1 (2006): 45–52.

21 Pigeau and Inglis, "Upregulation of *ALD3* and *GPD1* in *Saccharomyces cerevisiae* during Icewine Fermentation."

22 Nurgel et al., "Sensory and Chemical Characteristics of Canadian Ice Wines."

23 C. Nurgel and G. Pickering, "Contribution of Glycerol, Ethanol and Sugar to the Perception of Viscosity and Density Elicited by Model White Wines," *Journal of Texture Studies* 36 (2005): 303–23.

24 Pigeau and Inglis, "Upregulation of *ALD3* and *GPD1* in *Saccharomyces cerevisiae* during Icewine Fermentation."

25 D.L. Inglis, D. Kontkanen, G.J. Pickering, and A. Reynolds, "Impact of Yeast Inoculation Methods for Icewine Fermentation on Yeast Growth, Fermentation Time, Yeast Metabolite Production and Sensory Profiles of Vidal Icewine," *American Journal of Enology and Viticulture* 55 (2004): 433A.

26 D. Kontkanen, G.J. Pickering, A. Reynolds and D.L. Inglis, "Impact of Yeast Conditioning on the Sensory Profile of Vidal Icewine," *American Journal of Enology and Viticulture* 56 (2005): 298A.

27 Karl Kaiser, cited in Schneider, *Icewine: The Complete Story.*

28 Nurgel et al., "Sensory and Chemical Characteristics of Canadian Ice Wines."

29 Nurgel et al., "Sensory and Chemical Characteristics of Canadian Ice Wines."

30 J. Schreiner, *Icewine: The Wine of Winter.*

31 O. Jones, "Ice Winemaking: It's All in the Detail," *Australian and New Zealand Wine Industry Journal* 19, no. 2 (2004): 22–23.

A Cultural Perspective on Niagara Wines

Fourteen

New Wine in Old Wineskins: Marketing Wine as Agricultural Heritage

Michael Ripmeester and Russell Johnston

Introduction

Niagarans know their history. Ask anyone on the street to name events that shaped the region and they will tell you about frontier battles between empires, the building of the Welland Canal, or the Rebellion of 1837. Some take pride in the region's role in the Underground Railroad. Others may tell you about the Loyalist migration during the American War of Independence and the pioneer spirit that carved farms from forests. These events tell two related but distinct narratives. The first is an epic tale of international intrigues and combat, while the second is an intimate history of families, family farms, and proud agricultural traditions.

In 2005, we surveyed Niagarans to explore popular memory of the region's history. Niagara's role in international events is marked everywhere, most prominently by a series of monuments, forts, and battlegrounds that line the American border. Our survey participants surprised us, however, when the vast majority preferred contemporary people, events, and places, rather than historical ones, as symbols of the region. Among their answers, grapes and wine figured prominently.[1] The viticulture we know today was nurtured in the 1970s, and its reputation took years to establish. Still, it became clear that Niagarans associated the new wine industry with the region's agricultural heritage. How did something so recent become linked so prominently in the popular mind with the region's heritage?

The answer, we believe, lies in wine marketing. Our survey suggested that there is a weak relationship between the national history marked by memorials and the ways Niagarans viewed the region. By contrast, wine marketing draws inspiration from the intimate history of family farms: the wineries have positioned themselves as the caretakers of long-held local traditions.

New Wine, Old Narrative

Life in Niagara lends itself to the inclusion of grapes and wine in popular memory. Farms and small towns dominate the regional landscape. Its climatic and soil conditions made possible extensive tender-fruit farming that once branded the region "Canada's Fruit Belt." This is still an important facet of regional identity, even though many farmers have replaced tender-fruit orchards with grape vines and greenhouses. A.S. Hill reminds us that farmhouses and outbuildings surrounded by acres of open fields mean that farming is highly visible and, whether they are conscious of it or not, shapes residents' perceptions of the landscape and local culture.[2] Even city folk can take pride in this conception of the region. There are twenty annual events celebrating agriculture in Niagara, including festivals for strawberries and pumpkins. Grimsby's hockey team is proudly known as the Peach Kings.

Where tender-fruit farming has declined, wine seeks to fill the void. The industry and its allies have reimagined Niagara as "wine country," an expression proclaimed on regional welcome signs. A "wine route" rambles through vineyards and small towns, with signs directing traffic to wineries and restaurants. Among the region's annual festivals, three celebrate wine: the Grape & Wine Festival, New Vintage Festival, and Icewine Festival. The Grape & Wine Festival deserves special mention. Founded by the City of St. Catharines and the Grape Growers Marketing Board in 1952, it now draws a half million participants each year.[3] As wine tourism develops, marketers have increasingly drawn upon local agricultural heritage to promote local fine dining, and chefs distinguish themselves through their use of local produce, wines, and traditions. A culinary tour complements the wine route. It seeks to raise the everyday visibility of both fine dining and wine, to secure the place of the culinary arts in the regional economy.

These efforts promote general economic development, but there is no question that they appeal to specific sets of beliefs and values. Through their on-site architecture, landscaping, advertising, and web presence, the wineries and their allies speak in a singular voice. We examined all of the pamphlets available to tourists at a local information kiosk—140 pieces in all. Like all good advertising, a pamphlet must tell a coherent, compelling story about the brand, and it must distill that message into a single sheet of imagery and text. Presumably, the final presentation is designed in a way that the company's target market will find compelling; the message will play sympathetically to their values and beliefs. In our sample, the largest group of pamphlets touted tourist attractions in Niagara Falls itself (27.1 percent), but wine and wine-related activities were a close second (22.9 percent). If those for food and dining were added (5.0 percent), then wine, food, and dining occupied more shelf space than any other category of pamphlets. By contrast, surprisingly few pamphlets described historic sites and museums (5.7 percent).

The imagery and language of the wine industry pamphlets evoke many cultural touchstones for an urban middle class. They conjure a release from frenetic city confines into soothing natural landscapes with sweeping views and pastoral charms. Target markets in Toronto and upstate New York are constantly reminded that Niagara is nearby and ideal for a weekend escape. EastDell Estates bids readers to experience not just its wines but "The nature of Niagara": "[it] rules in every aspect of our integrated operation—from the vineyards and wine-making practices, to the restaurant's focus on seasonal cuisine, to the preservation of our property's spectacular Carolinian woodlot and nature trails."[4] The pamphlets also suggest a particular sense of refined taste. Illustrations portray smart, well-heeled couples in lush surroundings. They appear charmed in equal measure by the warmth of winery stores conceived in exposed brick and pine planks, and refined restaurants settings sparkling with polished crystal glasses and silver set on crisp linens. Pamphlets bid readers to indulge their passion for good food and drink.

Each pamphlet invites readers to discover more by visiting a website. In print or on the Internet, Niagara wineries stake claims to rustic authenticity. We visited sixty-one winery websites, starting with those mentioned in our pamphlets and then linking to others. A majority claimed family ownership and evoked the romance of local, small-scale production—the two concepts were deployed in equal measure. A significant number emphasized their history as farms and their part in local farming traditions. Ravine Vineyard told us, "[In] 1869, David Jackson Lowrey planted one of the earliest commercial vineyards with 500 vines. Five farming generations of Lowreys would eventually reap the benefits of his decision, growing all kinds of tree fruits and grapes. ... [In] 2003 ... Lowrey descendant Norma Jane (Lowrey) Harber and her husband Blair Harber prepared to return the upper farm to grapes."[5]

The pamphlets' claims reflect a genuine aspect of Niagara. A study commissioned by the regional government found that the average size of Niagara farms is smaller than those of other regions in Ontario, that almost half have been worked by multiple generations of the same family, and that three-quarters employ family members.[6]

By contrast, references to the dramatic history of conflict and industry in the region were rare. A notable exception is Henry of Pelham, which links its owners' family to the region's founding: "Owned and operated by the Speck family since 1988, the land was first deeded to our great, great, great grandfather, Nicholas Smith, in 1794. Nicholas fought with Butler's Rangers in the American Revolutionary war. His youngest son Henry of Pelham built the building that houses our wine store and hospitality rooms. Henry built the former carriage house in 1842."[7]

Still, the Speck story seems to emphasize the family's local pedigree rather than its role in moments of national significance. More tellingly, it is rare to find labels at any winery that commemorate the region's military past like that of Hillebrand Estates' Cuvée 1812, now out of production.

Many wineries invoked cosmopolitan associations. Beyond their obvious pride of place in Niagara, some wineries claimed pedigrees rooted in ancestral homelands. This point is particularly common among growers and vintners with roots, either through kinship or training, in the wine regions of Europe. Those with kinship ties often provide touching family histories that focus on the immigrant experience.[8] These intimate histories are clearly useful on at least two levels. Consumers may appreciate stories that put a face to the wines they drink, that deepen a sense of personal connection with the winemakers. References to European traditions also serve the wineries' efforts to strengthen their legitimacy in a market that lionizes European wines.

Tensions in the Narrative?

Tensions exist within the local wine industry. Despite the success of local wineries, grape growers face two significant challenges. The first stems from provincial farm marketing regulations. Under provincial laws, wineries may sell wines that contain up to 70 percent imported grapes under the designation International–Canadian Blends—a term crafted to draw Ontario consumers to seemingly Canadian wines. The original policy was instituted in 1972 to help wineries while grape growers shifted their fields from *Vitis labrusca* grapes to *Vitis vinifera* grapes. That goal was accomplished, but the wineries still lobby for similar policies because local growers cannot match the international market price for grapes. The result has been catastrophic for growers. In 2009, an estimated 10,000 tons of quality grape juice were unsold; this number was expected to rise to 16,000 tons in 2010.[9]

The second challenge facing grape growers is the rising cost of land. Owning a vineyard and producing wine has become fashionable, a fact underscored by the local investments of celebrities such as Wayne Gretzky, Dan Aykroyd, and Mike Weir. The success and glamour of the wineries has brought new investors to the region, and this trend has served to increase land prices. For established farmers, it has become more profitable to sell their lands than to develop their own operations.[10] One local realtor explained, "Your local farmer can't afford to expand. The costs are too great. The majority of all land purchases are from out-of-town-buyers that have made money elsewhere, not by farming.... Most have a dream, a passion for wine, and also realize Niagara is still in its infancy stage when it comes to that industry."[11]

An associated issue grew out of a reconception of the Grape & Wine Festival after 2000. Originally, the festival was the city's version of a rural fall fair. It celebrated a unique aspect of the fruit belt by staging parades, licensed concerts, a midway, and a beauty contest crowning the "grape queen." More recently, tourism officials and wine marketers sought to use the festival to lure tourists to the region through event marketing. The marketers were especially interested in affluent tourists seeking fine wines. To make the event more attractive to this target market, festival organizers dropped the word "grape" from its name and tweaked the parade to focus strictly on the wine theme. The midway, staged in a downtown park, became dominated by restaurants selling examples of their fare paired with local wines; meanwhile, local arts and crafts vendors and children's games were pushed to the fringes of the park. And an annual rock concert known as the Event in the Tent—generally a raucous evening where beer outsold wine—was killed.

Taken together, the pressures faced by grape growers and the changes made to the Grape & Wine Festival suggest thinking that runs counter to the vision of Niagara articulated by the wineries' marketing efforts. The wineries are keen to portray themselves as family farms that are firmly rooted in local culture and traditions, a portrayal that melds well with the presumed escapist desires of their target market. At the same time, the wineries' business decisions seem to undermine the sustainability of grape growers, and have altered some of the most prominent local traditions associated with grapes and wine. This contrast leads us back to a consideration of local residents and their perceptions of the industry.

A Captive Market

It is one thing to promote a particular aspect of local heritage, and quite another to have it embraced by local residents. After our 2005 survey, we sought to discover, through a second survey in 2009, the extent to which Niagarans have adopted the wineries' vision for the region. Our interviews began with one simple question: If you had to choose one thing that identifies Niagara, what would it be? The most common answer surprised us. Of 219 participants, 29.7 percent responded with reference to the wine industry and vineyards, while another 9.6 percent referred to farming, tender fruit, and other agricultural produce. By contrast, the world-renowned attraction that is Niagara Falls garnered only 19.6 percent of their responses, and national historical attractions such as the 1812 sites drew 1.4 percent. One person surprised herself, stating, "Wine—that's not what I thought I would have said. I'm not even a wine lover." Another asserted, "Our agriculture is the best thing to share with outsiders. We identify with the fruit and wines. It is the image and substance of local existence—it's all around." Though many participants volunteered that early Niagara wines were poor in quality, the vast majority agreed that grapes and wine were part of local heritage generally (80.3 percent) and local agricultural heritage in particular (87.5 percent).

We asked our participants where they get information about the local wine industry. The mass media were their most important source of information: 52.9 percent of all answers. Newspapers were the most commonly mentioned source of information (20.5 percent of respondents), followed by pamphlets and brochures (9.4 percent). Our participants also indicated that personal contact and experiences shaped their impressions of the wine industry, registering in 44.2 percent of all responses. In fact, stories and recommendations from friends and family were cited slightly more often than newspapers (20.8 percent). This was followed closely by visits to wineries and festivals, and participation in tours and tastings (16.6 percent). Taken together, these responses suggest that Niagarans do hear the wineries' messages. Also, whatever curiosity is twigged by the wineries' messages is legitimated by the credibility of family and friends.

Media coverage of the wine industry tends to be boosterish, emphasizing both its economic and cultural benefits. The opening of a winery is treated as positive economic news, connoisseurs review local labels for weekly columns, and the three wine festivals are celebrated as major cultural events. Indeed, articles on the Grape & Wine Festival almost always revel in its tradition of community involvement. The St. Catharines *Standard* provides a ready example with the following passage, quoting a local resident: "When my daughter was two, we took her to the parade and to my delight she was so excited that I thought that she would burst.... That would have been enough to have been my best memory of the festival, but no, it was only second best. My favourite memory was years later when we took her son to his first parade."[12] Positive stories like this, told by the wineries through their marketing and their boosters in the media, place the industry in both intimate family narratives and local heritage narratives. At the same time, the local press does not ignore controversial issues, such as the problems faced by grape growers and disagreements among players within the wine industry itself. Such articles are spread evenly between local news sections and the business sections of local papers. Still, the overall tone of these articles is that this is a thriving and glamorous industry that is important to the region's economic future.

When we talked with our survey participants, we heard similarly boosterish comments about the wine industry and its role in local culture. The Grape & Wine Festival was prominent in many responses. Most were proud that the city hosted the event. Moreover, most of the participants in this research echoed sentiments expressed in the St. Catharines *Standard*, that family participation is an important facet of the festival's place in local heritage. For many, it is an annual highlight event and the main link between themselves and the wine industry. One person told us, "Grape and Wine has been going on forever. Most of us grew up with it. It's just a part of our lives." Another said, "Grape and Wine is something we grew up with, it provides an outlet for cultural events. It identifies the region and evolves with the community."

That said, not all Niagarans who spoke with us were enthusiastic about a Niagara reconceived by the wineries. Many linked the region's history of tender-fruit farming to today's grapes, but there were some who distinguished between the old orchards and the new vines. Many reminded us that orchards were torn out to plant the new varieties of grapes that made the wineries possible. Visually, there is a striking difference between acres of trees laden with blossoms or fruit, and rows of short, closely cropped vines. One person told us, "Grapes and wine are important in an economic sense, but otherwise they are negative. The expansion of the grape and wine industry has destroyed fruit farms all over the region." A few participants went a step further and asserted that wine is not food. They linked the demise of local agricultural diversity with growth of a simple luxury. As one person noted, "We used to have lots of different crops, but the fields have all been given over to grapes. It is not for food, but for rich people to have fun." Similarly, another opined, "Time was when tender fruits came from Niagara. They were part of a way of life when I was a kid. Grapes have become important for the wrong reasons; they are going to wine instead of to food. Hope time will change this. Orchards have been lost to vineyards because of the growing wine industry." More commonly, however, participants explained the demise of tender fruits in relation to global shifts in food production. "So many orchards are being plowed under," one person commented, because "there is no real market for local soft fruits. People are buying American produce."

The popular memory of soft fruit farms also challenged the wineries' claims to a local agricultural pedigree. In this respect, the wineries' marketing rang hollow with many of our participants. One insisted, "Wine is a part of our local heritage, but not in the historical meaning of the term.... It's a recent phenomenon." Others questioned the link between any farming, conceived of the land, and winemaking, an industrial practice. One observed, "Grape growing is a part of our agricultural heritage but wineries now are different. Farmers have been here longer. The wineries have only been here a short time—they are the outsiders." Another put it more bluntly, stating that grapes and wine "are taking over everything. I'm really glad my parents are not around, they wouldn't support this. They were real farmers." Hence, residents may rely on marketing and the mass media to get information about local wineries, but popular memory of the region's past persists. For this group of residents, perhaps from an older generation, a new emphasis on the heritage of rural Niagara is welcome, but the wineries' place in that heritage is questioned.

The separation of grapes from wine was another common thread in participants' responses. The grape growers have been engaged in local agriculture since the nineteenth century, and created the Grape & Wine Festival almost sixty years ago. If respondents felt that grapes and wine could be associated with the region's heritage, their feeling seemed to be rooted in the authenticity of the grape growers rather than the vintners. Our participants noted this separation in relation to the debate over blended wines. Those who raised the issue all supported the grape growers, who are believed to be authentic farmers. By contrast, some view the wineries almost as bullies in their dealings with the grape growers. One participant commented, "Wine gets all the advantages over fruit. Nobody cares much about the grapes."

When participants criticized the wineries for their role in local culture, their criticism always led back to the Grape & Wine Festival. The fact that so many people cherished the festival as a civic and family institution meant that they also resented the tone of the changes that had been made over the last ten years. They told us in no uncertain terms that they felt excluded after the changes were made, and that this exclusion was defined along class lines. Although organizers claimed to have heard these complaints, and revised their approach in 2009, our participants were still skeptical. One of our participants described her family's experience of the new wine environment, and one gala event known as "Cuvée":

> We used to go to wine events. But we went to one Cuvée, it cost us hundreds of dollars. The same thing happened at the Grape and Wine Festival. A bottle of $10 wine cost us $27! The same is true of the Jazz Fest they hold at one of the wineries. These events are not of the people; the average person doesn't have the disposable income to pay that much. If they want to continue to include locals, they need more things that are not so expensive.

While we were conducting our survey, several letters to the editor of the *Standard* echoed these sentiments in surprisingly similar terms. One letter writer commended the organizers for their plan to rethink their approach: "By that, I assume the organizers plan to tone down the 'hoity-toity,' upscale, cosmopolitan flair that has dominated the festival over the past few years now, to the point of eliminating parade participants that didn't fit that image and giving us a parade that was only 55 minutes long."[13]

Niagara's wineries have conceived a new image for the region. In alliance with fine dining, upscale hotels, the Shaw Festival, and golf courses, they have invested the landscape with a set of values and practices designed to appeal to well-heeled consumers seeking cultured pleasures. The wineries have done so by linking their own often brief histories to a much older local narrative of pioneer endeavour, tender-fruit farming, and rural life. In short, they have written themselves into the agricultural heritage of the region, implicitly declaring that grapes and wine provide a new yet tangible link to the region's past. Local food movements, "the simple life" trend, and eco-tourism have only deepened the commercial possibilities of this strategy. The appeal to non-residents, then, is relatively clear: for those who seek such things, the wineries provide an experience that is thick with local tradition, pastoral charm, and natural settings.

For locals, the wineries serve as an economic and cultural resource. Niagara has experienced first-hand the erosion of heavy industry. The recession of 2009 shuttered plants for major employers such as General Motors and John Deere. It is not surprising, then, that Niagarans seek a new identity rooted in the past, one that speaks to contemporary aspirations for cultural cachet as well as economic and environmental sustainability. Popular memory recalls the golden age of Canada's Fruit Belt; today's residents see the ads and media support for the wine industry, hear the stories and recommendations of family and friends, and experience the vineyards, wineries, and festivals for themselves. The wineries, though selling themselves, nonetheless offer a coherent and desirable vision that is reinforced through their labels and their landscapes, their ads and their websites. The intangible heritage of tender-fruit farming and rural life has inspired the wineries, and the wineries have in turn reanimated this heritage for locals. That this process has been embraced by residents speaks to life in Niagara—or at least a life that many aspire to live. The reinvention of Niagara as "wine country" allows current events to be understood as the latest episodes in a cherished history.

Despite the pride our participants had in the local wine industry, however, this sentiment is by no means guaranteed. The sustainability of any heritage narrative requires that it remain relevant in the imaginations of those who foster it.[14] We suggest this relevance could slip away if the wine industry's actions betray its words. By claiming a local heritage narrative as their own, by invoking the values of small-scale family production, the wineries should carefully nurture those values in ways that ring true with local residents. If their use of these values and narratives are solely geared toward marketing their products, then they will alienate local residents—particularly those who perceive that social class is a barrier preventing them from participating. An emphasis on drawing wealthy patrons may unintentionally remove grapes and wine from local relevance, regardless of wineries' claims to a place in local lore.

Notes

1 Russell Johnston and Michael Ripmeester, "'That Big Statue of Whoever': Material Commemoration and Narrative in the Niagara Region," in *Placing Memory and Remembering Place in Canada,* ed. James Opp and John C. Walsh (Vancouver: UBC Press, 2010), 130–56.

2 A.S. Hill, "A Serpent in the Garden: Implications of Highway Development in Canada's Niagara Fruit Belt," *Journal of Historical Sociology* 15 (2002): 495–515.

3   H. Clark, "Residents' Attitudes and Perceptions toward the Niagara Grape and Wine
    Festival" (unpublished B.A. Thesis, Brock University, Recreation and Leisure Studies
    program).
4   EastDell Estates, "About Us," company website, http://www.eastdell.com/about_us.
5   Ravine Vineyard, "Past," company website, http://www.ravinevineyard.com/about
    -ravine/past.
6   Regional Municipality of Niagara, *Regional Agricultural Economic Impact Study* v. 1–2
    (Thorold, ON: Niagara Region, 2003).
7   Henry of Pelham Winery, "About Us," company website, http://www.henryofpelham
    .com/aboutus.php.
8   See, for example, Pilliteri Estates Winery, "About Us," company website http://www
    .pillitteri.com/pages/about_us.cfm.
9   Ontario Viniculture Association, "Information Release #20: Is Disaster Fermenting in
    Ontario's Greenbelt?" association website, http://www.realontariowine.ca.
10 Regional Municipality of Niagara, *Regional Agricultural Economic Impact Study* v. 1–2.
11 Larry Bilkszto, quoted by Tiffany Mayer, "Family Farms Being Plowed Under,"
    St. Catharines *Standard* website, http://www.scstandard.com/ArticleDisplay.aspx?
    archive=true&e=927490.
12 Susan Cooper, quoted in "Favourite Wine Festival Moments," St. Catharines *Standard*
    website, http://www.scstandard.com/ArticleDisplay.aspx?archive=true&e=1771173.
13 Jim Smelle, letter to the editor, St. Catharines *Standard*, 8 (27 August 2009).
14 John Bodnar, *Remaking America: Public Memory, Commemoration, and Patriotism
    in the Twentieth Century* (Princeton, NJ: Princeton University Press, 1992); Sanford
    Levinson, *Written in Stone* (Durham, NC: Duke University Press, 1998); Roy
    Rosenzweig and David Thelen, *The Presence of the Past: Popular Uses of History in
    American Life* (New York: Columbia University Press, 1998).

Fifteen

Constructing Authenticity: Architecture and Landscape in Niagara's Wineries

Nick Baxter-Moore and Caroline Charest

Introduction

N iagara is a relatively new wine region in comparison to the old wine-producing clusters of Western Europe—Bordeaux, Burgundy, Champagne,
Mosel, Rhineland-Palatinate, Emilia-Romagna, or Tuscany. So how does a relatively new wine-producing region convince the world—wine critics, general
consumers, journalists, oenophiles, tourist authorities, and tourists—that it
deserves to be taken seriously, or, to put it another way, how does the Niagara
wine industry authenticate itself?

The concept of "authenticity" is a contested term; that is, there is little general
agreement on its meaning. If there is a relatively accepted, everyday definition
of authenticity, it is "realness," a sense that something is genuine as opposed to
constructed, natural as opposed to artificial, or principally motivated by artistic
rather than commercial imperatives.[1] Authenticity, therefore, is what we might
call a relational concept, in that it is often easier to understand in terms of what
it is not, rather than what it is. At the same time, Michael Beverland argues in
his study of brand authenticity among luxury wines that authenticity "is often
more contrived than real"—in other words, it, too, is something that is created.[2]

According to Michael Lundeen, "The histories of fine wine and of fine winery architecture have intertwined over the centuries. The design of the winery
building can be central to the quality of the wine produced as well as to the
winemakers' marketing image."[3] In this chapter, we consider one of the means
by which individual wineries and the Niagara wine industry collectively have
attempted to authenticate themselves—that is, to establish their credentials as
part of a winemaking region that can compete on the world stage with other
wine regions, old and new. Our principal focus is on the literal "construction" of
authenticity in the architecture and landscaping of Niagara wineries as part of

what our colleague Christopher Fullerton has described elsewhere in this volume as Niagara's "winescape."

We have borrowed the concept of "constructing authenticity" from the late Richard Peterson, whose theorization of the "fabrication" of authenticity was originally applied to the study of country music.[4] Peterson's approach was, in turn, more broadly based on the so-called production of culture perspective in media and cultural studies. Briefly stated, this approach argues that culture does not emerge spontaneously, nor does it enjoy absolute autonomy from economic, social, and political influences; rather, the production of culture is powerfully influenced by economic and political institutions, markets and legal arrangements, interests and power.[5] Tastes, for example—in wines, popular music, or anything else—are not formed in a vacuum; they are the product of carefully orchestrated campaigns waged by those who have the power and influence to impose their preferences, or interests, on others. For Peterson, authenticity likewise is not a natural phenomenon; it is also the outcome of strategically constructed campaigns.

One strategy of authentication, according to Peterson, is naturalization. Because, as we have noted, authenticity is often defined in terms of what it is not (i.e., the "inauthentic"), Peterson argues that "one of the most effective ways to assert authenticity is to claim that an action, object, or person is 'natural' and without 'artifice.'"[6] By way of illustration, he departs from his central focus on country music to the world of wine and, in particular, the invention of the French winemaking tradition.[7] Robert Ulin demonstrates, for example, that the superiority of Bordeaux wines was not always accepted elsewhere; rather, it was the product of a series of political campaigns that first established these wines' reputation and later allowed them to withstand competition from other winemaking regions.[8]

Moreover, by defining wine as a "natural product," rather than an industrial one, French winemaking elites were not only able to "naturalize" claims of authenticity based on soil, climate, topography, and region—those factors that contribute to the increasingly fashionable concept of "terroir"—but also to "differentiate" themselves and their product from their competition.[9] The importance of differentiation as an authentication strategy in the Niagara region may be seen in the adoption in 1988 of the Vintners Quality Alliance (vQA—see Chapter 4), the designation of Niagara as a separate appellation, and the more recent identification of ten sub-appellations within the Niagara Peninsula.[10] These moves served both to confer a certain status on Niagara as a wine-producing region and to distinguish it from other such regions throughout the world.

Constructing Authenticity: Architecture and Landscape

Niagara wineries compete on a number of levels. On the global scale, the Niagara wine region competes with other wine-producing regions of the world, old and new, for a stake in international markets; more locally, it competes for the loyalties of Ontario wine consumers and for position in Liquor Control Board of Ontario (LCBO) outlets.[11] Nationally, Niagara wines face competition from British Columbia and other Ontario wine-producing regions, most notably the north shore of Lake Erie and, more recently, Prince Edward County. Within the Niagara region, both large conglomerates and independent wineries compete among themselves for public sales, through the LCBO and other retail outlets, for exposure through restaurant wine lists, and for shares in the tourism trade. Winery tours, on-site restaurants, and direct sales in the winery gift shop are important sources of income for many local wineries, especially those with production runs too small to gain access to the LCBO. Every winery, therefore, both large and small, presents a consciously constructed face to the public that is linked to its particular authentication strategy, and this is particularly manifested in its choice of architectural and landscape design.

At first glance, some wineries seem to care little about their external appearance. The architectural style of the Malivoire Winery on the Beamsville Bench, for example, seems to be inspired primarily by the Quonset hut—as indeed it was (Figure 15.1). In 1995, having acquired a property large enough for both vineyard and winery operations, co-owners Martin Malivoire and Moira Saganski decided to locate their winery on a northeast-facing slope that would be less suitable for farming. Rather than tear down an existing farm building, they opted to incorporate it into the winery, which, according to Martin Malivoire, combines utilitarian "Bauhaus form with Newtonian function,"[12] the latter apparent in the use of the naturally sloping land to create the first "gravity-flow" winery in Ontario. (Allowing the fruit to be moved by nature, rather than by machines, is supposed to bring out the more subtle flavours of terroir.) Whether viewed as a multi-level Quonset hut, with rounded, ridged roofs, or, more imaginatively, as a series of waterfalls tumbling down the escarpment, Malivoire's architectural design is generally consistent with its branding image: that of a craft winery, more concerned with the process and the product of its winemaking than its architecture.[13]

The design of the 20 Bees Winery in Niagara-on-the-Lake is purely utilitarian. Originally established in 2004 as a co-operative venture by nineteen grape growers and one winemaker, 20 Bees went into receivership in February 2008.[14] It was subsequently purchased by Diamond Estates Wines and Spirits Ltd., which already owned other wineries and brands in the region, including EastDell Estates in Beamsville, Lakeview Cellars, and the celebrity-endorsed Dan Aykroyd Wines. The 20 Bees entry on the Diamond Estates website announces that

Figure 15.1 Malivoire Winery
Photo: Nick Baxter-Moore.

"Our goal is simple, to produce great-tasting, unpretentious wine at an unpretentious price,"[15] which might account for the "unpretentious" appearance of the winery and its sales outlet, a large steel-sided industrial barn fronted by what looks like a double-wide trailer housing the retail store (Figure 15.2). Although artist's renditions are displayed in the store of a new state-of-the-art retail facility for Diamond Estates, designed by local architects Allen Chui Inc. and to be built on the 20 Bees site,[16] at time of writing, construction had not yet begun.

For most wineries in Niagara, architectural design and landscaping are intended to convey a particular image to the public and, often, to particular taste makers. In the following analysis, we identify three principal approaches in constructing authenticity in the physical appearance of Niagara wineries. The first two of these are varieties of what Peterson identified as the strategy of naturalization. The first is labelled "Transplanting Tradition." Here, we see the relocation of architectural styles associated with established wine-producing regions elsewhere in the world—most notably, Western Europe—into the southern Ontario countryside. Second is the approach we have called "(Re)Inventing Local Tradition," whereby wineries have blended themselves into the local agricultural environment and Ontario's rural heritage, real or imagined. The third broad strategy we label "Differentiation: Envisioning the Future." This

Figure 15.2 20 Bees Winery (now Diamond Estates)
Photo: Nick Baxter-Moore.

encompasses those "futuristic" or "synergistic" architectural styles that combine features designed to enhance the winemaking process and a commitment to environmental sustainability.

Naturalization: Transplanting Tradition

In Chapter 6 in this volume, Maxim Voronov, Dirk De Clercq, and Narongsak Thongpapanl identify a paradox of the Niagara wine cluster, in that it is "an Old World wine region located in the New World." This tension, we argue, is evident in much of the winery architecture in the region. One obvious example is the transplantation of architectural styles from the "Old World" to the new. The wineries of Château des Charmes, Peller Estates, and Konzelmann, all in Niagara-on-the-Lake, are often described as examples of the "faux château" style, although in the same spirit, perhaps, Konzelmann would be more accurately labelled an "ersatz Schloss."

The oldest of the three is Château des Charmes. Founded by Paul Bosc and family, who left behind their wineries in Algeria following the uprising against French colonial rule there in the 1960s, the winery was opened in 1994, at which time Bosc said, "I hope it will be here in two hundred years, producing the best quality wines."[17] Built at a cost of $6 million, the winery's architectural style fits

its name (see Chapter 16, Figure 16.9). While acknowledging that "(t)he château-like architecture of some Niagara wineries might seem a bit imitative," Kenneth Bagnall, writing in *The Globe and Mail* newspaper in 1999, extolled Château des Charmes as "the most stunning of all buildings in the region…. It rises in an ocean of vineyards, almost 100 hectares, its walls and roof the work of European stonemasons, mostly from Italy but living in Niagara. The strength of its lines, the burnished lustre of its stone, its presence beneath a canopy of brilliant blue create a stage-like, utterly memorable impression."[18] Château des Charmes is situated on York Road, just across from the family homestead, modelled after a French manor house, and is conveniently close to a major intersection on the Queen Elizabeth Way (QEW), the highway linking Toronto and Buffalo, which helps it attract somewhere close to 150,000 visitors a year.[19]

A second château winery, Peller Estates, was opened close to the Niagara River Parkway in 2001. It is the public face of Andrew Peller Ltd., formerly Andrés, which also owns Hillebrand Estates wineries and Thirty Bench wineries in Niagara, and others in British Columbia. [20] Although it is surrounded by vineyards, no wine is actually produced on-site at Peller Estates Winery. It may have a wine cellar, containing wines made elsewhere maturing in oak casks, which is an important component of the wine tour experience, but the Peller Estates château is principally a tourist attraction, housing a store purveying wines and all

Figure 15.3 Peller Estates Winery
Photo: Nick Baxter-Moore.

manner of related items, a well-reviewed and, frankly, expensive restaurant, and conference facilities rentable by the day. Surrounded by vineyards, and offering splendid views of the escarpment and Isaac Brock's monument on the southern horizon, it is a picturesque spot (Figure 15.3) in which to spend an summer afternoon or evening—in the heart of Niagara, but not quite a part of it.

Konzelmann Estate Winery's new retail and hospitality building (see Chapter 16, Figure 16.3) opened in 2008. According to Jansin Ozkur, the director of marketing, "I think everybody has an opinion of what a winery should look like. I think everybody's going for a different look. Some will go for modern. Some will go for the castle look. Herbert [Konzelmann] wanted to be different."[21] The new stone building has been described as "cathedral-looking" with its central tower, arched doorway, and huge wooden doors, although the many smaller windows set in the red-tiled roof are more suggestive of a religious retreat—an abbey or a monastery—which of course, in a European setting in particular, has powerful associations with the production of fine wines or other alcoholic beverages. Whether the architectural inspiration is religious or secular, both the building and the site are clearly Germanic in style, reflecting the Konzelmann family's long tradition of winemaking, extending back to the original winery established by Friedrich Konzelmann in Uhlbach, near Stuttgart, in 1893. More evidence of its historical and geographical links may be found in a faux street sign marking the entrance to the winery from Lakeshore Road reading "Uhlbacher Str." (Uhlbacher Strasse).

In the western part of the region, Angels Gate Winery, which opened in 2002 on the Beamsville Bench, offers a further example of transplanting tradition. The land, once occupied by the Congregation of Missionary Sisters of Christian Charity, and which also has "failed mink farm" as part of its resumé, was converted into a vineyard in 1995. To reflect its Christian heritage, Hamilton-based architect Lorne Haverty designed a mission-style building.[22] It is unlikely, however, that such a mission ever graced the Niagara Escarpment; indeed, Haverty's design (Figure 15.4) seems more inspired by a French nunnery, or possibly the Spanish colonial style of missions in the American Southwest. Whether an attempt to incorporate an element of local heritage or to evoke associations with other wine regions (France, Spain, California's Napa Valley), the winery is an architectural anomaly in the Niagara countryside. For one critic, Mark Criden writing in *Buffalo Spree* magazine, Angels Gate is "another example of the Disneyfication of the local wine business. [It] is Falcon Crest on the lake: big, opulent, showy, and obviously developed by a bunch of rich guys, busy turning big fortunes into small ones. You know the formula: More is less."[23]

The faux château, ersatz Schloss, ambivalent abbey, or misplaced mission does not naturally belong in the Niagara Peninsula. Nor do such structures always receive good press: they are frequently criticized, if not directly, then at

Figure 15.4 Angels Gate Winery
Photo: Nick Baxter-Moore.

least by unfavourable comparison to other wineries and architectural styles; for example, according to Lloyd Alter, "Jackson-Triggs is an Ontario winery that we have praised before for breaking the mold in architectural design, not building a tacky faux château winery but hiring a decent architect."[24] In a similar vein, when the design for Jackson-Triggs won a *Canadian Architect* Award of Excellence, one of the jurors commended it for eschewing "the homespun romanticism or ersatz historicism of wineries that have been created in recent years."[25] At the same time, tourists often have preconceived notions of what a winery should look like, based on romantic associations with architectural styles from other places. Given the origins of the owners of some of these wineries, and given the continuing influence of European winemaking traditions, it is not surprising that such transplanted "foreign" architectural designs are among the authentication strategies found in the Niagara wine region.

Naturalization: (Re)Inventing Local Tradition

The great majority of Niagara wineries are situated in rural contexts;[26] many have sought consciously to emphasize that fact, even the "faux chateaux." But, rather than looking elsewhere for their architectural inspiration, many wineries following a strategy of naturalization have reimagined and reinvented local rural and agricultural heritage. In this approach to authenticity, the winery and the process of winemaking are portrayed as "natural" parts of the rural tradition in the Niagara region. In some cases, this means simply adapting an existing farm, and its buildings, to a new industry; in others, new structures have been designed to blend more or less seamlessly into the rural environment.

A paradigmatic example of this strategy is Vineland Estates Winery. Established on the site of a former Mennonite homestead dating back to the 1840s, Vineland Estates has taken advantage of existing buildings, or their foot-prints, and the natural rolling topography of the Beamsville Bench part of the escarpment to create for the visitor the twin illusion of being simultaneously in both rural Ontario and a historic European winemaking region. The tasting and retail centre and some of the winery operations are housed in a Century Barn, somewhat adapted, but retaining some sense of rural authenticity (see Chapter 16, Figure 16.6). On-site also is a nineteenth-century carriage house, now used for weddings, other functions, and small-scale conferences; it burned down in 1992, leaving only the side walls more or less intact, but—thanks to the construction background of the deGasperis family, which owns the estate—has been expertly restored to the point where only those knowing the history of the site are likely to see the join.[27] The restaurant, in a restored 1845 farmhouse, and its outdoor patio look out over vineyards and rolling countryside reminiscent of the Mosel region, home to the original planter of the vineyards, Hermann Weis.

Inniskillin Winery, in Niagara-on-the-Lake, is one of the founding institu-tions of the modern Niagara wine industry, and, at the same time, an exemplar of the industry's inherent contradictions. Its history is the stuff of legend: in 1975, its co-founders, Karl J. Kaiser and Donald J.P. Ziraldo, were granted the first winery licence in Ontario since the Prohibition era.[28] Having begun business on the Ziraldo family farm for the first couple of years, Inniskillin moved in 1978 to the Brae Burn Estate, where its Ontario operations have remained. Since 1993, Inniskillin has been a subsidiary of Vincor International, which, in turn, was taken over by Constellation Brands, one of the world's largest wine companies, in 2006. Co-founders Kaiser and Ziraldo are no longer directly associated with the company and, hence, cannot presumably be blamed for its recent architec-tural decisions.

The dominant feature of the Brae Burn Estate is the barn (see Chapter 16, Figure 16.1), dating from 1926, its original design often attributed to the influ-ence of the great American architect Frank Lloyd Wright.[29] For many years, the barn housed both tasting rooms (on the upper floor) and a retail store, while more or less retaining its external appearance as a functional, yet traditional, part of the rural landscape. In 2008, however, Inniskillin was given a major facelift. The interior of the barn was thoroughly modernized, while preserving elements of tradition such as the post-and-beam support structures; outside, a brick-paved piazza now leads visitors to a brand-new entrance to the visitor centre/tasting rooms—a modernistic steel and glass appendage incongruously inserted into the side of the old wooden barn.[30]

Many other Niagara wineries have been built in and around farm buildings that lend a sense of agricultural tradition and help to naturalize this new industry in its rural setting. The Calamus Winery (Figure 15.5—its slogan is "Very rustic, very scenic, and very good wine!"), on the Vinemount Ridge near Ball's Falls Conservation Area in Jordan, and the Daniel Lenko Vineyard and Estate Winery in Beamsville are but two examples in Niagara of the basic "roadside farm turned winery" without visible evidence of much money spent on architects or builders. At the other extreme, Peninsula Ridge Estates Winery, on the western end of the Beamsville Ridge sub-appellation, almost in the town of Grimsby, is another expensive heritage renovation project (Figure 15.6). The winery opened in 2000 after a reported $6 million had been spent on renovating the property. An old post-and-beam barn was rebuilt using locally sourced barnboard, after raising the floor and installing a modern interior, to house the winery and a second-floor wine boutique. The most prominent feature, however, is the William D. Kitchen House, an 1885 red brick Queen Anne Revival Victorian manor house, where the restaurant is situated.[31]

Figure 15.5 The Calamus Winery
Photo: Nick Baxter-Moore.

But the (re)invention of rural traditions sometimes also involves bringing other parts of rural Ontario to Niagara. For example, publicity material for the Hernder Estate Winery in the Short Hills Bench, west of St. Catharines, creates a certain set of expectations for the future visitor: "A quiet country road, the gentle swell of vineyards beneath the brow of the Niagara Escarpment; turn down the lane, through a unique wooden covered bridge up to an immense, perfectly restored 1867 Victorian barn."[32] The barn has, indeed, been well restored and is central to both the winery's operations and the architectural style of the other buildings around it—it also appears on the winery's labels. What is truly "unique," however, is the covered bridge (see Chapter 16, Figure 16.4) through which visitors must pass en route from the road to the winery. While there are covered bridges elsewhere in southern Ontario and other parts of Canada, there are none in the Niagara Peninsula, other than the entrance to Hernder Estate.

Further west, at Fielding Estate Winery on the Beamsville Bench, Toronto architects Superkül Inc designed "a long shed under an open gabled roof, with the retail component overlooking the production areas on the ground floor visually and spatially integrating it with the winemaking process. The shed form speaks to the family's agricultural background and their interest in southern Ontario vernacular architecture."[33] According to the winery website, the location

Figure 15.6 William D. Kitchen House and Peninsula Ridge Winery
Photo: Nick Baxter-Moore.

reminded the Fieldings of their family home in the Haliburton Highlands; rather than "a shed," therefore, they portray the building as a "lodge" that "pays tribute to Northern Ontario's cottage country."[34] Surrounded by trees, with Muskoka chairs on the deck and at the entrance to the store/visitor centre (Figure 15.7), Fielding represents, for its owners at least, a piece of Ontario's rural heritage transplanted from cottage country to Niagara.

Finally, in this section, Cave Spring Cellars in Jordan Village offers a more comprehensive example of the reinvention of agricultural (and industrial) heritage in a rural setting. Cave Spring Cellars are nowhere near the Cave Spring Vineyards, the latter located on the Beamsville Bench. The Cellars, the principal tourist site, where visitors are invited to discover and buy their wines, is located in the heart of the village in one of the oldest wine cellars in Ontario, dating back to 1871. For much of the twentieth century, it was occupied by Jordan Wines, one of the "Big Four" industrial wine producers (along with Andrés, Brights, and Château-Gai) back in the "bad old days."

Since its purchase by the Pennachetti family in 1990, not only the wine cellar but the whole village has been fundamentally transformed. What was once, in the words of its owner, "a neglected, derelict company town, almost like an old mining town," is now a tourist site where visitors may enjoy gourmet food, relax at the spa, or rest in a luxurious hotel room.[35] Today, the façade of an old industrial building, now transformed into the winery, is embellished with elaborate

Figure 15.7 Fielding Estates Winery: Entrance and Muskoka chairs
Photo: Nick Baxter-Moore.

ornamental wood mouldings around the doors, windows, and roof cornices (Figure 15.8). Across the street, an old Jordan warehouse is now an upscale historic twenty-four-room hotel, Inn on the Twenty, whose suites are named after local pioneer families and furnished with a mix of modern pieces and crafted antiques rescued from heritage buildings throughout Niagara and western New York. Other warehouses have been converted into antique shops and restaurants, while, more recently, Jordan's 1842 roadhouse, once a biker hangout, was transformed into the fine fourteen-room Jordan House and Tavern.[36] The winery, in other words, has been integrated into a carefully reconstructed village streetscape that emphasizes fine living in a rural setting.

Figure 15.8 Cave Spring Cellars Wine Shop, Main Street, Jordan Village
Photo: Nick Baxter-Moore.

Differentiation: Envisioning the Future

In stark contrast to those wineries that evoke history, rurality, and agricultural connections to nature in their authentication and branding strategies are such enterprises as Southbrook, Stratus, and Jackson-Triggs—wineries that differentiate themselves by seeming to reject the past and embrace the future through renewed synergy with nature, both in their functionally modern or postmodern architectural styles and in their conscious embrace of modern approaches to environmental sustainability.

Southbrook Vineyards winery was one of four Canadian projects to win a prestigious International Architecture Award in 2009.[37] It is also the first Canadian winery to receive "gold" certification under the LEED (Leadership in Energy and Environmental Design) Green Building Rating System; it is also recognized by Demeter International for its biodynamic agricultural methods.[38] Its landscaping features a wetland filtration system to treat waste water produced on site, a flock of sheep is used both to "mow" the grass and to fertilize the vines, and the visitor centre takes advantage of a wide array of energy-saving technologies.

The dominant visual image of Southbrook is a three-metre-high wall, extending back from the road. Once blank and subject to graffiti art, it is now painted in a colour variously described as blue, purple, periwinkle, lavender, even "grape," punctuated some two-thirds along its 200-metre length by a set of glass doors leading to the tasting rooms and retail store (Figure 15.9). For architect Jack Diamond, the wall invites attention, anticipation; visitors want to walk through it to see what lies behind.[39] It leads the tourist's gaze to the visitor centre, where that curiosity might be assuaged. Once inside, brushed steel, blond butcher-block wood, Scandinavian ultra-modern décor (think upscale IKEA) prevail, a wall of glass providing natural light within, and without, a view over the vineyards.[40]

In some respects, the wall, relatively low in comparison to its length, might be seen as a metaphor for Southbrook's eco-friendly brand image—to make an impression (through its wines) without leaving one (on the environment). But it is also a symbol of hubris. Since it is the first winery encountered along Highway 55 after visitors leave the QEW on their way to Niagara-on-the-Lake, the physical symbol of the wall and the visitor centre are portrayed, respectively, as the boundary of and the gateway to the Niagara wine region.[41]

A few miles down the road, having passed by or stopped to visit 20 Bees (a major disappointment after Southbrook), Hillebrand (a long-established major producer, owned by Andrew Peller Ltd.) in the village of Virgil, Pillitteri's "fruit stand" winery (another variation on the naturalization of rural heritage), and Joseph's Estate cottage winery, the wine tourist will encounter, just before entering the historic town of Niagara-on-the-Lake, first Stratus and then Jackson-Triggs, two of the earlier modernist, even postmodern, wineries in the region.

Figure 15.9 The Wall at Southbrook Vineyards
Photo: Nick Baxter-Moore.

Viewed from the outside, Stratus might be an office building, even an insurance company, its exterior a minimalist high-tech box in black and white and glass and western red cedar siding, which will turn silver with age (see Chapter 16, Figure 16.11). While owner David Feldberg had flirted with ideas of a faux château or something like the cross between a spaceship and a Greek temple found in the Opus One winery in Napa, California, architect Les Andrew recalls that "The client basically fired everyone from the project, then rehired me because I had told him the truth.... I told him the best thing would be to create a simple, elegant box with a kind of minimalist theme, nothing showoffy."[42]

If the exterior is not "showoffy," the interior—the tasting room and retail space—was created by Burdifilek, the company that designed the flagship Holt Renfrew store on Toronto's Bloor Street. In other words, here art and commerce collide, but the idea that wine appreciation is part of elite culture prevails. Its leading brands are *assemblage* wines, Stratus Red and Stratus White, which combine the best varietals in each year, engineered by the skill and art of the winemaker, and the impression of wine-as-art is reinforced to some extent by the high-end retail design of the public spaces.[43]

Notwithstanding the experiential impressions given to the visitor, Stratus also trades on its environment-friendly reputation. Stratus was the first Ontario winery to receive LEED recognition (albeit silver, versus Southbrook's gold certification). It is another gravity-flow winery, following the principle that pumping the grape juice upwards could damage it. It was also constructed on the footprint of, and reusing materials from, existing buildings. Geothermal wells assist with heating and cooling, and other energy-efficient and water-saving measures were instituted, right down to the use of a Toyota Prius hybrid vehicle for all local wine deliveries.[44]

Next door to Stratus lies Jackson-Triggs Niagara Estate, which is, architecturally speaking, the paradigmatic postmodern winery in Niagara. Variously called "a vast implement shed" or "a Japanese aircraft hangar,"[45] and completed in 2001 at a time when other new wineries were embracing rural and traditional styles, Jackson-Triggs was then, and to some extent remains, an outrageously bold innovation (Figure 15.10) and, at the same time, the yardstick against which other designs are, usually unfavourably, measured. Designed by Marianne McKenna of Toronto-based KPMB Architects, the building is a rectangular box, situated on an east–west axis to parallel the escarpment and make maximum use of natural light. It consists of two functional sections—the working winery and the hospitality facilities—unified by a floating metal roof and linked by an atrium, the Great Hall, which may be converted into a covered open space by the simple expedient of opening huge barn-like sliding doors. That space takes the visitor from the road and the parking lot, the kitchen garden and demonstration vines at the front of the building to the real vineyards at the back: on a bigger scale, it connects the lakeshore to the Four Mile Creek plain and, beyond that, to the Niagara Escarpment.

The oblong box, however, features a number of innovative design elements. On the rear of the building, an external fieldstone ramp leads tour visitors up to the "business" end of the winery; according to tour guides, it is intended to reflect the slope and composition of the escarpment, taking visitors up to a higher level from which they can observe the vineyards below and be introduced to some of the preliminary operations of the winemaking process before they re-enter the winery. Reverse trusses and support beams for the roof are visible from the outside, and all construction materials, many of them recycled, have been left in their natural state, reflecting a commitment both to terroir and to sustainable practices.[46] The winery uses a gravity-flow system; the underground barrel and storage cellars not only reduce the winery's footprint but also its energy requirements—indeed, had the LEED system been instituted by 2001, Jackson-Triggs would likely have been the first winery to win that coveted certification.

The alluvial plain that is the distinguishing topographical feature of the Four Mile Creek sub-appellation does not permit much variety in the way of landscaping, at least in the absence of major intervention that might fly in the

Figure 15.10 Jackson-Triggs Niagara Estate
Photo: Nick Baxter-Moore.

face of the ecological commitments of Southbrook, Stratus, and Jackson-Triggs; hence, each of these wineries sits four-square on the land, surrounded by and providing vistas of rolling, relatively flat vineyards. Elsewhere in the region, other future-oriented wineries have made use of their surroundings to combine architecture, landscaping, and winemaking process. Both Flat Rock Cellars and the Tawse Winery have exploited the rolling hillsides of the escarpment and the Twenty Mile Bench to create multi-storey gravity-flow wineries.

At Flat Rock, the glass-surrounded tasting room, located in a hexagonal pavilion raised 10 metres off the hillside, allows visitors to enjoy a glass of wine and 360-degree views of vineyards and, to the north, the Toronto skyline across Lake Ontario (Figure 15.11). If Flat Rock's design—a gazebo on stilts, or perhaps a land-bound lighthouse—is quirky and fun, the Tawse Winery, nestled in a hollow on the Jordan Double Bench and partly buried in the escarpment, looks from the outside like an expensive ski resort (Figure 15.12). Its winemaking operations progress down through six levels to the barrel-aging cellars dug deep in the earth to maintain temperature and humidity. Both Flat Rock and Tawse make use of ponds to enhance their landscapes and to facilitate geothermal cooling and heating systems to regulate temperatures throughout the respective facilities, along with wetlands to manage waste water.[47] Like their three counterparts to the

Figure 15.11 Flat Rock Cellars
Photo: Nick Baxter-Moore.

east, these wineries look to the future rather than the past, in both their architectural designs and their winemaking practices, suggesting that, in a relatively new winemaking region, as Christopher Hume puts it, "the best is yet to come."[48]

Conclusion

Since approximately ninety wineries now operate in the Niagara wine region, our survey has been necessarily selective.[49] Moreover, it must be acknowledged that architecture and landscape design constitute only one element in the authentication strategies of wineries. Nonetheless, external appearance is often linked to other strategies as part of the overall image or brand that each winery endeavours to construct. Wine labels often bear visual representations of the winery buildings or their décor; local examples include Peninsula Ridge (the Kitchen House), Vineland Estates (the tower), Inniskillin (the original unadulterated barn), Château des Charmes (a pencil-sketch outline) and Fielding (a Muskoka chair). Most wineries' websites contain pictures or virtual tours of the winery and its surroundings. In many cases, physical attributes of the winery are linked, directly or indirectly, to the quality of the wine produced; this is particularly true of the gravity-flow wineries, but it is also generally the intention, we argue, that architecture becomes associated in the minds of tourists and wine consumers

Figure 15.12 Tawse Winery
Photo: Nick Baxter-Moore.

with particular styles of winemaking or with a particular attitude toward wine as art or commerce. Through our "reading" of visual images and the words of architects, winery owners, and others, we have identified three principal strategies of authentication through architecture and landscape design: transplanting tradition, (re)inventing local tradition, and differentiation. Further research remains to be undertaken on the extent to which such strategies actually influence the perceptions of authenticity and the behaviour of tourists and wine consumers.

Notes

1 On this last point, see the discussion of artistic and commercial rationalities in the wine industry in Maxim Voronov, Dirk De Clercq, and Narongsak Thongpapanl, "The Ontario Wine Industry: Moving Forward," Chapter 6 in this volume.
2 Michael B. Beverland, "Crafting Brand Authenticity: The Case of Luxury Wines," *Journal of Management Studies* 42, no. 5 (2005): 1007.
3 Michael Lundeen, "LEED Winery," *Architecture Week*, September 6, 2006, E.1.1.
4 Richard A. Peterson, *Creating Country Music: Fabricating Authenticity* (Chicago: University of Chicago Press, 1997).
5 See Richard A. Peterson and N. Anand, "The Production of Culture Perspective," *Annual Review of Sociology* 30 (2004): 311–34; and *Cultural Industries and the Production of Culture*, ed. Dominic Power and Allen J. Scott (New York: Routledge, 2004).

6 Peterson, *Creating Country Music*, 211.

7 For various approaches to the invention of tradition and their connection to authenticity, see *The Invention of Tradition*, ed. Eric Hobsbawm and Terence Ranger (Cambridge: Cambridge University Press, 1983).

8 Robert Ulin, "Invention and Representation as Cultural Capital: Southwest French Winegrowing History," *American Anthropologist* 97, no. 3 (1995): 519–27.

9 Ulin, "Invention and Representation as Cultural Capital," 523; Beverland, "Crafting Brand Authenticity," 1008.

10 On the appellations and sub-appellations of the wine industry in Ontario, and especially in the Niagara region, see http://www.vqaontario.com/Appellations/NiagaraPeninsula.

11 The Liquor Control Board of Ontario (LCBO) has a virtual monopoly on the sale of wines and spirits in Ontario. Access to the wider public through LCBO outlets is crucial for expansion of production and sales for many of the smaller wineries (see Voronov, De Clerq, and Thongpapani, Chapter 6 in this volume).

12 Malivoire Wine, "Combining the Elements," http://www.malivoire.com/index.php?page=viniculture.

13 The waterfall interpretation was proposed by a staff member in the tasting room/store. She also suggested that Martin Malivoire admits that, if he had slept on the proposed design for the winery, he might have settled on something different (personal conversation, 2011). At the same time, it must be recognized that Malivoire often uses stories about his self-acknowledged "mistakes" or "happy accidents" to promote his wines; see, for example, the labels of Malivoire's "2009 Guilty Men" or "2010 Melon" wines.

14 Tony Aspler, "Canada's First Co-op Ends in Disaster," *Decanter*, www.decanter.com/news/wine-news/485955/canada-s-first-co-op-ends-in-disaster.

15 Diamond Estates website, "20 Bees," http://www.diamondestates.ca/canadian-wines/alphabetical-listing/20-bees.

16 See the designs for the new retail outlet at "Projects – Commercial – Diamond Estates Wines" at the Allen Chui Architects website, http://www.ac-architects.com.

17 Cited in Wade Rowland, "Tasting Niagara, Part 2 – Château des Charmes," at http://www.worldtravelguide.com/travel-gourmet/canada/ontario/niagara/niagara-ontario-wine-tour2.html.

18 Kenneth Bagnell, "The Vineyards of Niagara," *Globe and Mail*, June 5, 1999, F1.

19 Whatever the architectural and touristic merits of the Château des Charmes winery, the Bosc family was also responsible for a major blight on the Niagara winescape, since they led the way in importing and subsequently securing Canadian marketing rights for the US-made wind machines that both disperse cold air in nighttime frosts and disturb the sleep of rural residents living close to vineyards; see Rowland, "Tasting Niagara."

20 Andreas Peller immigrated to Canada from Hungary in 1927 and, on entering the wine business in British Columbia in the early 1960s, decided that a French-sounding name would appear more authentic—hence "Andrés" (see http://www.andrewpeller.com/History.php). Later, seeking an entrée into the now predominantly anglophone Canadian wine business, and, perhaps, to encourage wine consumers to ignore Andrés' best-known contribution to Canadian wine history, the confection known as "Baby Duck," in 2006 Andrés Wines Ltd became Andrew Peller Ltd.

21 Cited in Monique Beech, "Sleek Look at Konzelmann Winery," St. Catharines, *Standard*, November 16, 2007, www.stcatharinesstandard.ca/2007/11/16/sleek-look-at-konzelmann-winery.

22 Angels Gate Winery website, http://www.angelsgatewinery.com/history.html.
23 Mark Criden, "Wine: Aspiring to the Bench," *Buffalo Spree Magazine*, Sept./Oct. 2005. Criden is much less critical of the quality of the wines produced by this "Disneyfied" winery.
24 Lloyd Alter, "Which Is Greener, Wine Bottle or Box?" Treehugger website, http://www .treehugger.com/files/2008/08/green-wine-depends-on-box.php.
25 Barry Sampson, cited in "Award of Excellence: Jackson Triggs Winery," *Canadian Architect*, 44, issue 12 (1999): 2224.
26 There are obvious exceptions. Both Hillebrand Estates Winery in Virgil and Cave Spring Cellars in Jordan (see below) have blended into their respective small-town or "village" environments. Other, more obviously large-scale manufacturing operations, such as Kittling Ridge in Grimsby and the the long-established Vincor facility in Niagara Falls, neither of which has obvious connections to any local vineyard, have melded into the landscape of the industrial estate.
27 Peter C. Newman, "His Skin May Be Canadian, 'but My Blood Is Italian,'" *Maclean's*, September 26, 2005, 47–53.
28 Inniskillin website, "History," http://www.inniskillin.com/en/about/history.asp.
29 There is actually little evidence for the claim that the Brae Burn Barn was designed by Frank Lloyd Wright. The link has been much promoted by Inniskillin co-founder Donald Ziraldo, a self-confessed fan of both art deco design and Frank Lloyd Wright's architecture, but, according to Rick Haldenby, Director of the School of Architecture at the University of Waterloo (in correspondence with one of the authors), "the claim seems very far fetched given the design" and is at best "speculative."
30 According to architect Michael Allen, "The uniqueness of the Inniskillin Brae Burn part of the renovations was designing the program within the confines of the existing barn structure. This was a great opportunity to display the grandeur of the wood structure and focus the accents and details around the mystique of the Frank Lloyd Wright inspired barn," http://www.docstoc.com/docs/49621039/Unique-Winery -Architecture-and-Stunning-Scenery.
31 In 2009, the restaurant (in the William D. Kitchen House) and the nearby Coach house were purchased from the Beale family (which owns Peninsula Ridge) by Chef Ross Midgley and his wife, Wendy, and are now run as a separate operation, "The Kitchen House at Peninsula Ridge." Nonetheless, the house still appears on Peninsula Ridge wine labels. The Kitchen House is designated a "Historic House" by the Town of Lincoln under the Ontario Heritage Act. See http://www.niagaragreenbelt.com/ listings/54-historic-houses/192-william-d-kitchen-house.html.
32 Hernder Estate Wines home page, http://www.hernder.com.
33 Superkül Inc, "Projects," http://superkul.ca/projects/fielding-estate-winery#.
34 Fielding Estate Winery, "About Fielding," http://www.fieldingwines.com/page/about. Note that the architects and their clients have somewhat differing perspectives on both the design of the winery and Ontario's geography—at least with respect to concepts of "northern" and "southern" Ontario.
35 Len Pennachetti, cited in Susan Pigg, "The Valley of Grape Expectations," Toronto *Star*, September 15, 2007, T1.
36 Pigg, "The Valley of Grape Expectations."
37 Don Fraser, "Winery Garners Top Architectural Award," St. Catharines *Standard*, September 14, 2009, http://www.stcatharinesstandard.ca/2009/09/11/winery-garners -top-architectural-award. See also Monique Beech, "Winery's Wall an 'Architectural Statement," St. Catharines *Standard* (n.d.), http://www.stcatharinesstandard.ca/ ArticleDisplay.aspx?archive=true&e=1065927.

38   Irene Seiberling, "Southbrook Makes a Case for Biodynamic Wine," *Regina Leader-Post*, August 1, 2009, http://www2.canada.com/reginaleaderpost/columnists/story .html?id=be1ce7f7-63e0-4ed5-b1d1-6b606d8da692.

39   Fraser, "Winery Garners Top Architectural Award."

40   Interestingly, even an architecturally modern winery such as Southbrook emphasizes the rural, agricultural roots of the Redelmeier family, farmers for three generations spanning back to 1941, on its website. See Southbrook Vineyards, "Our History: From Bovines to Grape Vines," http://www.southbrook.com/our_history.

41   See Southbrook Vineyards, "The Gateway to Wine Country," http://www.southbrook .com/hospitality_pavilion.

42   Cited in Christopher Hume, "Wine in a Box," *Azure* Magazine, September 2006, 13.

43   The synergy created by "assemblage" reflects the age-old theory that the sum is greater than the parts.

44   Enermodal Engineering, "Case Study – Stratus Winery," http://www.enermodal.com/ Canadian/pdf/Stratus_1page.pdf,

45   Wade Rowland, "Tasting Niagara – Part 5, Jackson Triggs Winery, Niagara-on-the -Lake," http://www.worldtravelguide.com/travel-gourmet/canada/ontario/niagara/ niagara-ontario-wine-tour5.html.

46   Marco Polo, "In Vino Veritas," *Canadian Architect*, 46, issue 10 (2001): 19–23.

47   See Flat Rock Cellars, "The Winery – Environmental Sustainability," http://www .flatrockcellars.com/pages/winery/environment; and Tawse Winery, "Biodynamics," http://www.tawsewinery.ca/index.cfm?fuseaction=page.display&page_id=52.

48   See Christopher Hume, "A Modern Spin on Vintage Art," Toronto *Star*, April 2, 2009: U2.

49   In addition to omitting many wineries already constructed, lack of space here precluded discussion of what might have been the most spectacular winery in Niagara, had it ever been built. Frank Gehry's design for a new winery for Le Clos Jordanne was unveiled in 2002: see ArchNewsNow.com, "Unveiled Vintage," http:// www.archnewsnow.com/features/Feature46.htm. The winery was never begun and the takeover of Vincor (partners in Le Clos Jordanne with Groupe Boisset of Burgundy) by Constellation Brands, combined with the economic downturn of fall 2008, effectively killed the project; see John Schreiner, "Le Clos Jordanne motors on without Frank Gehry," http://johnschreiner.blogspot.com/2009/03/le-clos-jordanne-motors-on -without.html.

Sixteen

Wine and Culinary Tourism in Niagara

David J. Telfer and Atsuko Hashimoto

Introduction

Niagara is no longer just about seeing the Falls, but is home to a burgeoning wine and culinary tourism industry. Niagara wine and regional food are rebranding the area as a culinary tourism destination with wineries, festivals, restaurants, farmers' markets, and cooking schools all celebrating the bounty of Niagara. Wine and culinary tourism have received attention from every level of government in Canada, with Niagara being at the forefront of many of these plans. Clusters of wineries along the Niagara Escarpment and in Niagara-on-the-Lake form the heart of the Niagara Wine Route, which includes approximately seventy wineries. Many wineries are incorporating a larger hospitality component through a range of tourist services, such as guided tours, elaborate tasting venues, wine-related souvenirs, second-language services, barrel clubs, home delivery, and, in some cases, restaurants. Celebrity chefs and restaurants are focusing on Niagara wine and sourcing local food for their menus. Annual wine and food festivals have become important tourism events to celebrate the culinary resources of Niagara. The trends in Niagara are reflective of the growing interest in local wine and food in many countries around the world.[1] This chapter provides an overview of some of the dynamic changes occurring in Niagara and highlights some of the challenges the region faces in further developing the wine and culinary tourism product.

The Niagara Region and Niagara Wine

The Niagara Peninsula is home to a favourable microclimate, as it is surrounded by water on three sides by Lake Erie, the Niagara River, and Lake Ontario. Although known as a New World cool climate wine region, the latitude in Niagara, between 41 and 43 degrees, is actually similar to some of the Old World wine regions in Europe. The Niagara Escarpment, a UNESCO Biosphere Reserve, also enhances the climate. The area to the north of the Escarpment toward Lake Ontario has between ten and twenty extra frost-free days compared to the area to the south of the Escarpment.[2] Together with the climate and fertile soils, Niagara is one of the few places in Canada that can successfully produce soft fruit such as peaches, nectarines, plums, and grapes. The importance of this prime agricultural land in Niagara can be seen in the inclusion of parts of it in the Ontario Greenbelt, designed to protect environmentally sensitive areas and farmland from urban development.

While there has been a long tradition of soft fruit orchards in Niagara, it is only over the last thirty years that Niagara has really become known for wine. Wine was originally produced from native *Vitis labrusca* and *Vitis riparia* vines and was not typically known for its high quality.[3] Over time, however, a number of key developments in the evolution of Niagara wines have driven wine and culinary tourism, some of which are outlined below. In 1975, the Ontario government granted the first new winery licence in fifty years to Inniskillin, which had its own vineyard. Previously, it was rare for Canadian wine producers to have their own vineyards or have much control over the grapes being grown by independent growers.[4] With pioneers like Inniskillin, the shift soon began toward premium wine production using *Vitis vinifera* grape varieties. The move toward these varieties was further enhanced through the Grape Acreage Reduction Program (GARP) negotiated by the federal and provincial governments. This program was developed in association with the loss of protection for Ontario wines under free trade with the United States in 1989. The GARP led to the removal up to 3,300 hectares of grapes.[5] The grapes were replaced by new plantings of the *Vitis vinifera* varieties,[6] and now the vines of France (Bordeaux, Alsace, the Rhône) and Germany are all grown in Canada.[7] The introduction of the Vintners Quality Alliance (VQA) in 1988 was also very important in improving the quality of wines; the VQA Act became law in 1999. VQA Ontario is the regulatory agency that sets up the wine appellations in the province as well as enforcing winemaking and labelling standards.[8] All VQA wines must be made from 100 percent Ontario grapes, which are taken from the appellation presented on the label. There are four wine appellations in Ontario, including Lake Erie North Shore, Niagara Peninsula, Pelee Island, and Prince Edward County. Within the Niagara Peninsula there are two regional appellations: the Niagara Escarpment and Niagara-on-the-Lake. Table 16.1 includes a list of the further ten sub-appellations in Niagara.

Canadian wines arrived on the international stage in 1991, when Inniskillin (Figure 16.1) won the Grand Prix d'Honneur for its 1989 Icewine at Vinexpo in Bordeaux, France, making Icewine a signature Canadian product. Icewine is produced by harvesting the grapes in winter when they are frozen on the vines,

Table 16.1. Niagara Appellations

Niagara Peninsula Appellation	
Niagara Escarpment sub-appellations	**Niagara-on-the-Lake sub-appellations**
Beamsville Bench	Niagara Lakeshore
Creek Shores	Niagara River
Lincoln Lakeshore	St. Davids Bench
Short Hills Bench	Four Mile Creek
Twenty Mile Bench	
Vinemount Ridge	

Source: VQA Ontario 2010.

resulting in a very sweet dessert wine. This high-end product has become a sought-after souvenir by many tourists. To meet this demand, Inniskillin, for example, developed sampler-sized bottles and targeted the Japanese tourist market with Icewine.[9] Niagara wines have continued to improve in quality. Moving away from Concord grapes, the vineyards in Niagara have now had success with varieties including Chardonnay, Riesling, Syrah, Pinot Noir, Gamay, and Bordeaux reds such as Cabernet Sauvignon.[10]

Government assistance and industry associations have also helped in the branding and marketing of Niagara as a wine and culinary destination. The Wine Council of Ontario is the non-profit trade association representing eighty-two Ontario wineries. It markets the Wine of Ontario brand through a number of initiatives, including a guidebook for tourists.[11] It also manages Ontario's 2009 Wine Strategy, which focuses on supporting VQA wine, as well as marketing tourism and the grape sector.[12] The strategy calls for investment in the promotion of tourism-related events, such as Cuvée and the Niagara Wine Festival.

Figure 16.1 Inniskillin Wines (East)
Photo: D. Telfer & A. Hashimoto.

Niagara Wineries

With about seventy wineries making up the Niagara Wine Route, the wineries have become known not only for their wines, but also for their hospitality and architecture. The size and scale of the wineries varies from small, historic farm buildings to large-scale, elaborate tasting venues. Due to government restrictions on the sale of alcohol, the cellar door has become vital for wine sales. To attract visitors and to educate tourists about wines, the wineries offer a wide range of products and services, including tastings, tours, concerts, barrel clubs, weddings, interactive websites, food and wine pairings, or even dinners with the winemakers, as seen in Table 16.2.

Figure 16.2 contains a continuum of wineries and the products and services offered, ranging from those wineries focusing only on wine production to those wineries that are fully integrated into wine tourism. Some of the wineries have progressed through the continuum, adding tourism services, facilities, and products over time, while others are purpose-built tourism attractions offering a wide range of wine products and services in very high-end, landscaped venues. Two examples of elaborate architecture include Konzelmann Estate Winery (Figure 16.3) and Hernder Estate Winery (Figure 16.4). As one moves

Table 16.2. Selected Niagara wine tour companies

Name of Tour Company	Example of Tour
Crush on Niagara Wine Tours	Sip & Savour: four winery visits and a winery luncheon
Elite Wine Tours	Custom Winery Tour package: four-hour tour and tasting at three wineries
Grape and Wine Tours	Afternoon Delight: extensive tour of one winery and tasting at three wineries
Lincoln Limo & Cab	Hidden Gems Winery Tours: three- or five-hour rental of limousine or winery van and tour of up to five wineries
Niagara Adventures	Wine Adventures
Niagara Classic Transport	The Vine & Dine: four-winery tour and lunch at Angel Inn
Niagara Wine Tours International	Luxurious Niagara-on-the-Lake Charles Inn and Wine Gourmet Lunch Tour: two nights' accommodation, continental breakfast, visits to four wineries, lunch at winery restaurant
Niagara Winery Tours	Wine Country Bus Tour: four-hour tour to four wineries and a visit to Niagara-on-the-Lake
Steve Bauer Bike Tours Inc.	Thirty-kilometre bike ride, including visit to winery with picnic lunch
Wine Country Tours	The Happy Spirits Tour: tour of winery, distillery, micro-brewery and picnic lunch

Source: Tour operator websites.

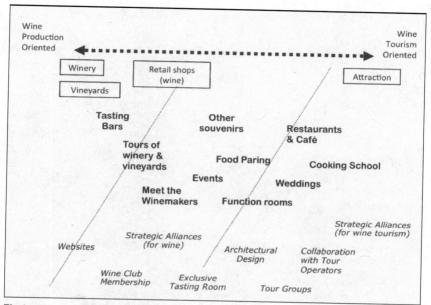

Figure 16.2 Continuum from wine production to wine tourism

toward the inclusion of tourism, the services offered expand to include tours, special events, and links with food, such as restaurants and cooking schools. The networks of strategic alliances also expand and there is a trend toward adding elaborate architectural designs to increase the attractiveness of the facilities, as seen in many of the images in this chapter. Wineries that offer food have made the commitment to using local area produce, which is identified on their menus. An important element in developing the Niagara Wine Route has been the networking and collaboration that has occurred among many of the wineries, as well as with government and the tourism industry.[13] Under the banner of "Wineries of Niagara-on-the-Lake: Twenty-Two Distinctive Wineries, One Unforgettable Place," the wineries collaborate on a number of joint events and marketing campaigns.[14]

The wineries have experimented with a range of unique branding, labelling, and marketing strategies, such as the Organized Crime Winery or the Sibling Rivalry brand from Henry of Pelham Family Estate Winery. Celebrities including Wayne Gretzky (hockey player), Mike Weir (golfer), and Dan Aykroyd (actor, musician, comedian) have linked their names to wineries. While many of the wineries are family owned, there have also been acquisitions of wineries by larger companies that in fact now own several Niagara brands. For example, Vincor brands in Niagara include Inniskillin, Jackson-Triggs Estate Winery, and Le Clos

Figure 16.3 Konzelmann Estate Winery (East)
Photo: D. Telfer & A. Hashimoto.

Figure 16.4 Hernder Estate Wines (West)
Photo: D. Telfer & A. Hashimoto.

Jordanne.[15] Vincor itself is a division of Constellation Brands, which is the world's largest wine company.[16] Diamond Estates Wines and Spirits Ltd. own a number of Niagara brands, including 20 Bees, Birchwood Estate Wines, Dan Aykroyd Wines, De Sousa Wine Cellars, Eastdell Estates Winery, and Lakeview Cellars. (Diamond Estates Wines and Spirits Ltd. is a Canadian-based wine and spirits distribution and marketing organization.)[17] A third example is Andrew Peller Limited, from Andrés Wines; its premier Niagara brands include Peller Estates, Hillebrand Estates, Trius, and Thirty Bench.[18]

As we saw in Table 16.2, Niagara-based tour operators offer prepackaged and custom tours of area wineries. The tours often involve multiple winery visits, as well as stops at restaurants featuring local foods. In addition, tour operators are bringing in buses with international tourists.[19] While it is impossible to provide a detailed account of every winery here, a few wineries from both the Niagara Escarpment and Niagara-on-the-Lake appellations will be highlighted to illustrate the wide range of options available for visitors to the region. The inclusion here does not imply that the noted winery, and its respective wines, is better than others; certainly, other authors writing this chapter would likely choose other wineries.

Niagara Escarpment Appellation Wineries

The Niagara Escarpment wineries are located to the west of the city of St. Catharines. Featherstone Estate Winery (Figure 16.5) is located on 23 acres and has a traditional 1830s farmhouse where guests can have lunch on the veranda. This winery employs eco-friendly farming methods, producing wine only from grapes grown in their vineyard. Avoiding pesticides, having sheep graze on the leaves of the vineyard, and using falcons to help control the bird population add to the uniqueness of the winery.[20] On a larger scale, Henry of Pelham Family Estate Winery has 170 acres and was opened in 1988. The carriage house on the property was built in 1842, and the Coach House Café and Cheese Shop is open for lunch. One of the interesting summer events at Henry of Pelham is "Shakespeare in the Vineyard"; in July 2012, *As You Like It* was performed.[21] Other wineries, such as Vineland Estates Winery (Figure 16.6), Peninsula Ridge Estates Winery, and Eastdell, are well known for their restaurants, which feature local ingredients.[22] Dinner menus for Vineland Estates are stamped with the logo "Local Food Plus," which is a non-profit organization committed to creating local sustainable food systems.[23] Ridgepoint Wines (Figure 16.7) also offers casual dining. Rockway Glen Estate Winery features an eighteen-hole champion golf course, along with a restaurant and banquet facilities.[24] One of the important tourism destinations along the Niagara Escarpment is the village of Jordan, which can be described as a wine tourism village centred on a winery.[25] Along the

Figure 16.5 Featherstone Estate Winery & Vineyard (West)
Photo: D. Telfer & A. Hashimoto.

Figure 16.6 Vineland Estates Winery (West)
Photo: D. Telfer & A. Hashimoto.

main tourist street in Jordan are found Cave Spring Cellars, restaurants including "On the Twenty," antique shops, "Inn on the Twenty," and a variety of other shops, as seen as in the sign post in Figure 16.8. The village itself has become a culinary destination in its own right.[26]

Figure 16.7 Ridgepoint Wines (West)
Photo: D. Telfer & A. Hashimoto.

Figure 16.8 Jordan Village Sign Post (West)
Photo: D. Telfer & A. Hashimoto.

Niagara-on-the-Lake Appellation Wineries

The wineries on Niagara-on-the-Lake are geographically clustered in and around historic Niagara-on-the-Lake, which is also home to the Shaw Festival Theatre. The diversity of the wineries continues here in the east end of the Peninsula. Château des Charmes (Figure 16.9) takes the form of a large French château,[27] while Strewn Winery (Figure 16.10) hosts a cooking school and restaurant. The cooking school offers culinary vacations and the restaurant, Terroir La Cachette, serves French Provençal-style cooking with local Niagara ingredients.[28] Niagara College Teaching Winery is home to a restaurant, culinary institute, winery, and vineyard.[29] Hillebrand Estates Winery Restaurant has a forager who sources local food from area farmers for the restaurant. It offers special Harvest Table lunches and dinners where local products and growers are highlighted.[30] Peller Estates Winery is also a grand French-style château set on a ten-acre vineyard that includes a restaurant focused on local food.[31] A very different winery in terms of architecture is the very modern Stratus Winery (Figure 16.11). With a focus on sustainability, it was the first building in Canada to receive LEED (Leadership in Energy and Environmental Design) certification.[32] Reif Estate Winery (Figure 16.12) has the first wine sensory garden, which includes plantings of the different grape varieties grown in Niagara.[33] Finally, on a much smaller scale is Frogpond Farm (Figure 16.13), the only certified organic winery in Ontario.[34]

Figure 16.9 Château des Charmes (East)
Photo: D. Telfer & A. Hashimoto.

Figure 16.10 Strewn Winery (East)
Photo: D. Telfer & A. Hashimoto.

Figure 16.11 Stratus Winery (East)
Photo: D. Telfer & A. Hashimoto.

Figure 16.12 Reif Estate Winery (East)
Photo: D. Telfer & A. Hashimoto.

Figure 16.13 Frogpond Farm Organic Winery (East)
Photo: D. Telfer & A. Hashimoto.

The Niagara Wine Tourist

An important element for the wineries is to understand who their customers are. Wine tourism has been defined as "visitation to vineyards, wineries, wine festivals and wine shows for which grape and wine tasting and/or experiencing the attributes of a grape wine region are the prime motivating factors for visitors."[35] Based on surveys in Niagara, Carmichael found that overall, visitors to Niagara wineries are "mainly middle aged, have high incomes, have high levels of education and primarily drive short distances to travel to wineries," and a high percentage come from the urban area of Toronto.[36] A study of visitors to Niagara wineries by Hashimoto and Telfer revealed that all respondents evaluated equally the purchase of wine, tasting, socialization, a day out, relaxation, and a learning experience as important factors in visiting a winery.[37] In addition,

Figure 16.14 Tourists at Inniskillin Wines (East)
Photo: D. Telfer & A. Hashimoto.

they found that the environment, the winery's cleanliness, the display of goods, the smell inside the building, and the architecture were important attributes to visiting a winery. An interesting finding from the study was that there were differences between visitors to the wineries at the east and west ends of the wine route. For example, the visitors to the west-end wineries tend to buy more wine from the winery and place less importance on the tour, while those visiting the east end tended to buy more wine from the Liquor Control Board of Ontario (LCBO) or wine shops and placed higher importance on the on-site tour (Figure 16.14).[38] As Stewart, Bramble, and Ziraldo suggest, additional research is needed on visitors in order to build a growing target customer base.[39]

Wine and Food Festivals

Food festivals and events are becoming increasingly important, not only for their value in food retailing, marketing, and promotion, but also because they coincide with other social, economic, and political concerns that relate to the nature of contemporary agricultural systems, the conservation of rural land-scapes, concern over food quality, and the maintenance of rural lifestyles and communities.[40] In Niagara, wine and food are linked to many local festivals held throughout the year. Table 16.3 presents a small selection of festivals and events that celebrate local food and wine; however, many other events hosted by individual wineries to highlight their products are not included in the table (e.g., Figure 16.15, Hillebrand Estate Winery). There are three main wine festivals held throughout the year: they focus not only on wine, but also on local food, and are attracting many tourists to Niagara.[41] In January is the three-week Niagara Icewine Festival. Celebrating Icewine, the festival begins with the Xerox Icewine Gala, which features Icewine, local cuisine, entertainment, and other Niagara wines. Throughout the festival, wineries offer a variety of events, such as Icewine bars made of ice, ice sculptures, and pairings of food and wine. The Niagara New Vintage Festival takes place in June, and focuses on the new vintages. As part of that festival, the New Vintage Tailgate Party is held in the Grower of the Year's vineyard and features wine from over thirty wineries as well as celebrity Niagara chefs using local ingredients.[42] Discovery Packages are sold to visitors to experience food and wine pairings at VQA wineries. The third major annual festival is the Niagara Wine Festival, held in September to celebrate the harvest. It is advertised as having over one hundred events across the region and is well known for its festival parade. The festival, which began in 1952, attracts a large number of tourists.[43] In 2003, organizers dropped the word "Grape" from the fes-tival name. The move was highly controversial among residents and also raised concerns among grape growers.

Table 16.3. Selected wine and culinary tourism festivals/events

January
- Niagara Icewine Festival
- Twenty Valley Winter WineFest
- Niagara-on-the-Lake Icewine Bar
- White Meadows Sugar Bush Adventure (January–April)

February
- Days of Wine and Roses
- Cuvée

March
- Empty Bowls

April
- Shaw Festival (April–October)

May
- Niagara Folk Arts Festival
- Niagara-on-the-Lake Wine & Herb Experience
- A Spring Evening in the Wine Sensory Garden (Reif Estate Winery)

June
- Niagara New Vintage Festival
- Welland Rose Festival
- Springlicious Beer and Wine Festival

July
- Hillebrand Estates Winery Jazz and Blues
- Flavours of Niagara International Food, Wine and Jazz Festival

August
- The Extravaganza of Niagara Passion

September
- Niagara Wine Festival
- Niagara Wine and Food Classic

October
- Niagara Food Festival
- Ball's Falls Thanksgiving Festival

November
- Wrapped Up in the Valley
- Niagara Winter Festival of Lights

December
- Winterglow
- New Year's Eve

Source: www.tourismniagara.com, www.onlineniagara.com, www.insideniagara.com.

A food festival to note for linking food and wine is the Niagara Food Festival, held in Welland in early October. The festival showcases growers, producers, restaurants, chefs, and cooking demonstrations, along with entertainment. An interesting part of the festival is the "Experience Niagara Market," where local growers can highlight their produce.[44] The market is hosted by "Buy Local Niagara," an organization dedicated to getting Niagara residents to purchase local products. Finally, Savour Niagara is an annual September event, sponsored by the Small Business Club Niagara, where restaurants, wineries, and chefs work together to pair food and wine for the evening event.[45]

Farm-to-Table Cuisine and Niagara Culinary Innovations

Terms such as "slow food" (as opposed to fast food), "farm-to-table cuisine," "local food," "the hundred-mile diet," and "organic food" have all emerged as important culinary issues across Canada, including the recognition of culinary tourism as a destination branding tool.[46] In the Ontario Culinary Tourism Strategy and Action Plan (2005–2015), Niagara has been identified as one of the key regions in Ontario to drive culinary tourism.[47] The Ontario Culinary Tourism Alliance

Figure 16.15 Hillebrand – Taste the Season (East)
Photo: D. Telfer & A. Hashimoto.

is an industry-driven organization, focusing on making Ontario an international culinary tourism destination;[48] the Savour Ontario Dining program allows tourists to go online and find restaurants in the Niagara region that feature local food.[49] In Niagara, the Niagara Culinary Trail has ninety-four members under the following categories: accommodations, bakeries, cafés, farms, markets, restaurants, shops, and wineries. The Trail map allows tourists to wander the back roads of Niagara, visiting farms and sampling local foods and wine. In addition to the map, the Niagara Culinary Trail website offers a variety of information on local food, including a description of the five culinary regions of Niagara.[50] Taste Niagara is held in the spring, summer, and fall; during the promotion, participating restaurants offer a three-course meal featuring local food and wine, while shops offer local foods to take away. A few of the participating restaurants include AG Inspired Cuisine, Escabeche at the Prince of Wales Hotel (Figure 16.16), Riverbend Inn & Vineyard, and Zest Restaurant.[51]

A second culinary trail is the Niagara AgriTourism Circuit, created by the Niagara Agri-Tourism Centre. This bilingual resource centre promotes rural economic development in Niagara. The tourist map has fifty-six entries, including farms, restaurants, greenhouses, artists' studios, ranches, stables, and vineyards.[52] A number of cooking schools also operate in Niagara. As mentioned

Figure 16.16 Prince of Wales Hotel, Niagara-on-the-Lake
Photo: D. Telfer & A. Hashimoto.

above, the Wine Country Cooking School is located in Strewn Winery. The Good Earth Food and Wine Co. is also a known leader in the farm-to-table trend, offering cooking classes set against a backdrop of orchards and vineyards.[53] The Niagara Culinary Institute, located at the Niagara-on-the-Lake campus of Niagara College, offers a dining room that is open to the public, as well as culinary courses.[54] Sourcing local food continues to be opened up to the consumer through area farmers' markets, the Greenbelt website, and the Niagara Local Food Co-op. Brock University operates a farmers' market on Fridays during the warmer months and, along with its Cool Climate Oenology & Viticulture Institute and a new cafeteria highlighting local food, is trying to promote education about local food and wine.

Challenges in Wine and Culinary Tourism in Niagara

The challenges facing wine and culinary tourism come from a variety of sources. External to the region, a number of issues have reduced the overall numbers of visitors to Niagara. For example, in recent years, the rising Canadian dollar, the requirement for Americans to have a passport when returning to the United States, and the H1N1 virus have reduced the number of Americans—the primary international market—who visit Niagara. Within Niagara, conflicts over grape prices and, most recently, the International–Canadian Blends have created some tension. (These blended wines contain imported grape juice and retail at lower prices than VQA wines. Some of the larger wineries producing

International–Canadian Blends have left the Wine Council of Ontario, the promoter of the Niagara Wine Route.)[55] Growers unable to find a buyer for their grapes have removed some local vineyards. The loss of local fruit-processing companies has also resulted in peach orchards being pulled out. The challenge of improving the local processing and distribution of Niagara products remains. Although there is an abundant variety of local food, local residents continue to buy imported products at supermarkets. A recent report on local food by the Niagara Community Observatory at Brock University stated that locals are just not buying local food.[56] Regardless of these challenges, there are many opportunities for wine and culinary tourism in Niagara. Partnerships have been instrumental in developing the product and promoting it to tourists as the Niagara Region rebrands itself around wine and culinary tourism.[57] The Niagara Wine Route, which runs through the Peninsula, is leading tourists to innovative wineries, restaurants, cafés, and theatres in an area abundant with orchards, vineyards and farmlands, all of which contributes to what Peters calls a "winescape."[58] The Niagara winescape is taking tourists off the beaten path to Niagara Falls and promoting rural development.

Notes

1 Michael C. Hall and Liz Sharples, "Food Events, Festivals and Farmers' Markets: An Introduction," in *Food Tourism Around the World: Development, Management and Markets,* ed. C. Michael Hall and Liz Sharples, 3–22 (London: Butterworth-Heinemann, 2008); Atsuko Hashimoto and David J. Telfer, "Selling Canadian Culinary Tourism: Branding the Global and Regional Product," *Tourism Geographies* 8, no. 1 (2006): 31–55.

2 Paul Chapman, "Agriculture in Niagara: An Overview," in *Niagara's Changing Landscape,* ed. Hugh Gayler, 279–99 (Ottawa: Carleton University Press, 1994).

3 David J. Telfer, "The Northeast Wine Route: Wine Tourism in Ontario, Canada and New York State," in *Wine Tourism around the World: Development, Management and Markets,* ed. C. Michael Hall, Liz Sharples, Brock Cambourne, and Niki Macionis (Oxford: Butterworth Heinemann, 2000), 253–71.

4 John Schreiner, *The Wineries of Canada* (London: Mitchell Beazley, 2005).

5 Telfer, "The Northeast Wine Route."

6 Chapman, "Agriculture in Niagara."

7 Schreiner, *The Wineries of Canada.*

8 vqa Ontario, http://www.vqaontario.ca/Wineries.

9 David J. Telfer and Atsuko Hashimoto, "Niagara Icewine Tourism: Japanese Souvenir Purchases at Inniskillin Winery," *Tourism and Hospitality Research: The Surrey Quarterly Review* 2, no. 4 (2000): 343–56.

10 Schreiner, *The Wineries of Canada.*

11 Wine Council of Ontario, http://winesofontario.org/content/pdfs/MediaKit.pdf 2010.

12 Ministry of Consumer Services, "Ontario's Wine Strategy," *Newsroom,* http://news.ontario.ca/mcs/en/2010/04/ontarios-wine-strategy.html.

13 David J. Telfer, "From a Wine Tourism Village to a Regional Wine Route: An Investigation of the Competitive Advantage of Embedded Clusters in Niagara,

Canada," *Tourism Recreation Research* 26, no. 2 (2001): 23–33; David Telfer, "Strategic Alliances along the Niagara Wine Route," *Tourism Management* 22, no. 1 (2001): 21–30.

14　Wineries of Niagara-on-the-Lake, http://wineriesofniagaraonthelake.com/#.

15　Vincor Canada, http://www.vincorinternational.com.

16　Constellation Brands, http://www.cbrands.com.

17　Diamond Estate Wines & Spirits, http://www.diamondestates.ca.

18　Andrew Peller Ltd., http://www.andrewpeller.com.

19　Telfer and Hashimoto, "Niagara Icewine Tourism."

20　Featherstone Estate Winery, http://www.featherstonewinery.ca.

21　Henry of Pelham Family Estate Winery, http://www.henryofpelham.com.

22　Vineland Estates Winery, http://vineland.com; Peninsula Ridge Estates Winery, http://www.peninsularidge.com; EastDell Estate Winery, http://www.eastdell.com.

23　Local Food Plus, http://www.localfoodplus.ca; Vineland Estates Winery, http://www.vineland.com/index/page/name/dine.

24　Rockway Glen Golf Course & Estate Winery, http://www.rockwayglen.com.

25　Telfer, "From a Wine Tourism Village to a Regional Wine Route."

26　Cave Spring Cellars, http://www.cavespringcellars.com; Twenty Valley Tourism Association, http://www.20valley.ca.

27　Château des Charmes, http://www.chateaudescharmes.com.

28　Strewn Winery, http://www.strewnwinery.com.

29　Niagara College Teaching Winery, http://www.nctwinery.ca.

30　Hillebrand Estates Winery, http://www.hillebrand.com.

31　Peller Estates, http://www.peller.com.

32　Stratus, http://www.stratuswines.com.

33　Reif Estate Winery, http://www.reifwinery.com.

34　Frogpondfarm, http://www.frogpondfarm.ca.

35　C. Michael Hall, "Wine Tourism in New Zealand," in *Tourism Down Under II, Conference Proceedings*, ed. G. Kearsley (Dunedin, New Zealand: Centre for Tourism, University of Otago, 1996), 109–19.

36　Barbara Carmichael, "Understanding the Wine Tourism Experience for Winery Visitors in the Niagara Region, Ontario, Canada," *Tourism Geographies* 7, no. 2 (2005): 185–204.

37　David Telfer and Atsuko Hashimoto, "Food Tourism in the Niagara Region: The Development of *Nouvelle Cuisine*," in *Food and Tourism from around the World*, ed. C. Michael Hall and Liz Sharples (London: Butterworth Heinemann, 2003), 158–77.

38　Atsuko Hashimoto and David J. Telfer, "Positioning an Emerging Wine Route in the Niagara Region: Understanding the Wine Tourism Market and Its Implications for Marketing," in *Wine, Food and Tourism Marketing*, ed. C. Michael Hall, 61–76 (London: Haworth Hospitality Press, 2003).

39　Jeffrey Stewart, Linda Bramble, and Donald Ziraldo, "Key Challenges in Wine and Culinary Tourism with Practical Recommendations," *International Journal of Contemporary Hospitality Management* 20 (2008): 302–12.

40　Hall and Sharples, "Food Events, Festivals and Farmers' Markets."

41　Niagara Wine Festival, http://www.niagarawinefestival.com/.

42　Niagara Wine Festival, http://www.niagarawinefestival.com/.

43　Niagara Wine Festival, http://www.niagarawinefestival.com/.

44　Niagara Food Festival, http://www.niagarafoodfestival.com.

45　Small Business Club Niagara, http://www.sbcn.ca/events/savourniagara.html.

46　Telfer and Hashimoto, "Food Tourism in the Niagara Region."

47 Ministry of Tourism, "Culinary Tourism in Ontario Strategy and Action Plan 2005–2015," http://ontarioculinary.com/wp-content/uploads/images/documents/research/ontario_2005_culinary_tourism_strategy.pdf; Ontario Culinary Tourism Alliance, http://ontarioculinary.com/.

48 Ministry of Tourism, Ontario Culinary Tourism Alliance, http://ontarioculinary.com/.

49 Savour Ontario, http://www.savourontario.ca/en/index.html.

50 Niagara Culinary Trail home page, http://niagaraculinarytrail.com/.

51 Niagara Culinary Trail, http://niagaraculinarytrail.com/.

52 Niagara Agritourism Centre, http://www.agrotourismeniagara.com.

53 The Good Earth, http://www.goodearthfoodandwine.com/.

54 Niagara College, Teaching Winery, http://www.netwinery.ca.

55 Julie Gedeon, "Largest Producers Break off from the Ontario Wine Council," http://www.winebusiness.com/news/?go=getArticle&dataId=70025.

56 Brock University, "Brock research institute releases report on local food in Niagara," News around Campus, http://www.brocku.ca/news/10343.

57 Telfer, "Strategic Alliances along the Niagara Wine Route."

58 Gary Peters, American Winescapes: The Cultural Landscape of America's Wine Country (Boulder, CO: Westview Press, 1).

Seventeen

Conflict in the Niagara Countryside: Securing the Land Base for the Wine Industry

Hugh J. Gayler

Introduction

Tne Niagara wine region, the area between Lake Ontario and the Niagara Escarpment from the Grimsby–Hamilton border to the Niagara River (Figure 17.1), presents an idyllic landscape of vineyards and approximately ninety wineries in their midst, attracting hundreds of thousands of tourists each year and contributing over half a billion dollars to the regional economy. This success story is relatively recent. For the first hundred years, until the 1970s, the Niagara wine industry was largely a non-event, and its recent development could easily not have happened. Since the Second World War, approximately 40 percent of this wine region has succumbed to suburbanization and urban sprawl; much of the remainder would have disappeared had it not been for the determined efforts of a few individuals, a new and conservation-minded planning process, and a local economy where urban development has slowed to a crawl. This chapter will explore how the loss of valuable agricultural land led to demands for better governance and a regional plan, and later how the growth of a new wine industry forced the Ontario government to rethink the local and regional responsibilities for land-use planning and secure the agricultural land base in perpetuity.[1]

The Demise of the Agricultural Land Base

Until the 1970s, it was comparatively easy in Ontario to convert agricultural land for any number of urban-related uses.[2] Being contiguous to an existing urban area meant the land could be more easily serviced and offered the seller the best price, but unserviced rural subdivisions and individual lots were permitted almost anywhere, providing there was access to a public right-of-way and no undue health problems were presented by wells and septic beds. Niagara's prosperous industrial economy, the mass appeal of the automobile,

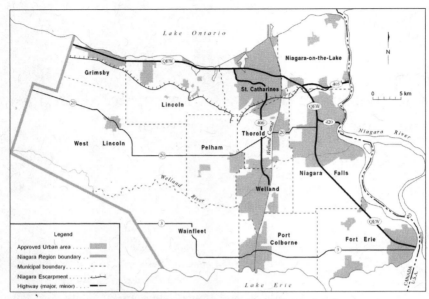

Figure 17.1 Niagara urban areas (Regional Municipality of Niagara 2007)

the promotion of home ownership through more accessible mortgage financing, and the unprecedented population increases of the 1950s and 1960s fuelled a land-hungry and haphazard expansion of suburbia through east Hamilton, Stoney Creek, and north St. Catharines, as well as growth in small towns such as Grimsby and sprawl in the rural areas in between. This was the era when formal land-use planning applied only to urban areas; land in the rural areas beyond was invariably not classified or was regarded as "holding" or "vacant" until some higher and better (i.e., more profitable) urban use could be found. The vagaries of tender-fruit and grape growing, the major agricultural land uses in this area, were numerous, including issues of surpluses or shortages because of weather conditions, small and inefficient land holdings, cost-price squeezes, and competition in the marketplace. Farmers were often only too glad to supplement their incomes through urban land conversions. Indeed, it was considered an owner's right to do what he or she wanted with the land, a belief that is still mistakenly held even in today's highly regulated planning environment.

This loss of agricultural land soon attracted attention in the 1950s, because it occurred on some of Canada's most valuable and productive lands, and in one of the very few areas in Canada that would support viable commercial tender-fruit and grape operations.[3] For the late Dr. Ralph Krueger, a professor of geography at the University of Waterloo, the saving of these lands became a *cause célèbre* from the time he was a graduate student in the 1950s.[4] Other geographers joined the fray with academic and government reports that similarly called for urban development to be reined in and rural resources protected.[5] The reports were directed to the Ontario government because local government lacked the professional expertise, financial capability, and political will to put good planning principles into practice. In fact, the 1960s saw increasing calls from Niagara for local government to be reorganized to better fit the needs of the second half of the twentieth century rather than the nineteenth.[6] The Ontario government heeded the calls and a commission was appointed to review possible local government structures. It later recommended that the two counties (Lincoln and Welland) and twenty-six municipalities be reduced to a two-tier system of the Niagara Region and twelve predominantly urban-centred municipalities.[7] Better planning and a better use of urban and rural resources were among the eventual outcomes of how a more professional regional and local government system would operate. The Ontario government largely endorsed the commission's report and the Niagara Region was born on 1 January 1970.

Toward a Regional Plan

The Regional Municipality of Niagara was required to have a regional plan approved by the Ontario government by December 1973, but the inability to think and act regionally, as opposed to twelve times locally, held up final approval until 1981. The sticking points were the setting up of realistic urban-area boundaries for future growth and planning in rural areas on the basis of land quality.[8] The first involved land being serviced, or not, and thus whether a farmer would overnight become a multi-millionaire through land being converted to urban uses; the second could relinquish all hope of the farmer selling out for any urban-related use if that land was deemed to be of high agricultural quality.

The views of professional planners were consistently overridden as local and regional politicians—imbued by the economic outcomes of growth (were it to happen) in their municipalities, and backed by farmers and developers—went for a huge potential urban-land grab. The result can only be described as a vulgar overestimate of the amount of urban land needed in Niagara; the proposed plan was conservatively estimated to accommodate a 1991 population of 640,000 (a figure that did not take into account that residential densities were likely to rise in the future). Meanwhile, the figure projected by Statistics Canada was closer to 500,000 (and falling by the year); twenty years later, a 1977 estimate of 415,000 by 1996 proved to be not far wrong!

Opposition to these proposals rested on the speculative nature of the development and the valuable agricultural lands that would be placed under threat, specifically in the Niagara wine region, the most accessible and fastest-growing part of Niagara. Moreover, excessive urban land would raise false hopes and needless competition between local municipalities and, more importantly, lead to costly and wasteful overservicing. The Ontario government's attempts to lend some reality to the situation were constantly thwarted at the local level. In the end, the government took the politically easy way out and referred the finalizing of the regional plan to the quasi-judicial Ontario Municipal Board (OMB). The ensuing deliberations over a two-year period were the longest and most bitter on record, and a truly David and Goliath situation as the might of local and regional government, the development industry, and landowners, with their easy access to high-priced legal and professional expertise, was pitted against a hopelessly underfinanced public-interest group, the Preservation of Agricultural Lands Society (PALS), which was dependent on pro bono work, fundraising from raffles and bake sales, and, over Goliath's objections, legal aid.[9]

Reason was to prevail. The OMB, in its 1981 decision, acknowledged the work of PALS as the sole spokesperson in the public interest; and while the OMB set generous urban-area boundaries, this was to allow municipalities in the critical wine region time to adjust to the new reality of permanent urban-area boundaries and to plan for the redirection of development to poorer areas south of the Niagara Escarpment.[10] (It is interesting to note that thirty years later, these boundaries are only just being reached in some northern Niagara municipalities.) Meanwhile, outside these boundaries, the newly designated Good Grape and Good Tender Fruit areas were restricted to agriculturally related development only.

The Regional Plan in Action: Death by a Thousand Cuts

In no time at all after the 1981 decision, and continuing for the next twenty years, it could be seen how this bottom-up approach to land-use planning was not going to work for Niagara's valuable agricultural lands. The OMB's ban on farm retirement severances was reversed on appeal to the Cabinet; and the rule of having to be a full-time farmer for twenty years before 1973 to be eligible for such a severance was later struck down in favour of any twenty-year period. Thus, the resulting problem of sprawl, which would have eventually gone away as the number of eligible farmers diminished, would therefore remain with us for all time. The notion of permanent boundaries had to be removed because the Ontario Planning Act allows for official plan amendments, which could extend an urban-area boundary. Indeed, it was probably the 500-acre extension to the urban-area boundary of Fonthill in the Town of Pelham in 2000, on some of Canada's best agricultural land, that confirmed the weakness inherent in the Planning Act.[11] While there was more than enough potential developable land in adjacent municipalities, Pelham argued that it did not have sufficient land to meet future local residential needs as required under the Planning Act. The Town and the Niagara Region approved the expansion, against the advice of planning staffs; an appeal by PALS to the OMB was turned down on the basis of the developer's evidence (a consultant arguing that this was not valuable agricultural land) somehow being more convincing than the PALS consultant and the evidence he presented.

In spite of restrictive policies in Niagara's rural areas, studies have shown the ease by which non-agricultural developments could take place. During the 1990s, Niagara was the second-worst area in Ontario for lot severances.[12] Aside from the impact of retirement severances, especially given Niagara's low average farm sizes, severances were approved for a host of reasons, many bearing little relationship to agriculture. Many were clearly economic severances, farmers looking for a way out of agriculture who were lured by development proposals that were arguably "sympathetic" to agriculture (but usually not in any way agriculturally related) and that hopefully would gain planning approval. Niagara has seen an array of rural churches with large parking lots that were clearly intended for nearby urbanites, golf courses and driving ranges, recreational and social clubs, and farmers' markets (as opposed to farmgate sales). Such was the value of these rural lands that even allowable and long-recognized agricultural uses came under scrutiny. These included the expansion of greenhouses and industrial operations associated with wineries, where the actual soil was not used and the land uses could conceivably have gone to an urban industrial park or less valuable lands elsewhere.

Aside from the expansion of Fonthill, Niagara's position as the slowest-growing area in the Greater Golden Horseshoe of southern Ontario has resulted in only small adjustments to its urban-area boundaries over a thirty-year period. Moreover, there has been a concerted effort to oppose or redirect a number of major development proposals in rural areas, especially those that would require the extension of urban services, and in so doing invite a plethora of other urban land uses. However, early in 2000, in response to a ten-year review of the Regional Policy Plan, stakeholders in Niagara's wine industry were becoming dismayed at the effects of urban sprawl on their business—the so-called death by a thousand cuts—and called for more restrictive planning legislation. Niagara's adoption of smart growth principles, one of which is preservation of agricultural land, brought the matter to the fore.[13] Meanwhile, Jim Bradley, provincial Member of Parliament for St. Catharines, introduced a private member's bill to establish an agricultural preserve in the Niagara wine region. While this type of bill from an opposition backbencher had little chance of success, it did raise awareness levels. The Niagara Region's Smarter Niagara Steering Committee, at its Smarter Niagara Summit in 2002, followed this up with a debate between Mr. Bradley and a winery owner on the one hand, and two grape growers, who wanted no such land-use restrictions, on the other.

To defuse what this author witnessed as an ugly, name-calling scene between saving the farmer versus saving the land, the chair of Niagara's Regional Council agreed to form a task force on the long-term viability of agriculture. Its report represented an important meeting of minds, in that all four government levels, agricultural organizations and the public at large agreed to the preservation of valuable agricultural land in return for support for the economic viability of agriculture.[14] This optimistic, "made in Niagara" solution was always in doubt. Preserving agricultural land use only required the casting in stone of existing planning policies; however, securing economic viability for farmers in the long term relied on a moving target of government policies and market forces in an increasingly globalized environment. Also, before any actions could be taken, Niagara's task force found its report overtaken by the wider provincial debate on land-use planning.

Greenbelt and Places to Grow Acts, 2005

The newly elected Liberal government in 2003 honoured its election promises to strengthen the Planning Act, the OMB, and the Provincial Policy Statement, and to introduce more restrictive planning policies by introducing a greenbelt for the Greater Golden Horseshoe area (Figure 17.2 and Figure 17.3).[15] The immediate freeze on rural, non-agricultural development while a task force studied the greenbelt issue, and the resulting recommendations to the Ontario government, drew contrasting reactions in Niagara.[16] The farming community felt betrayed. The very people the Greenbelt Act was supposedly designed to help, by saving their land from speculation and being paved over, turned it down because it interfered with their age-old right, whether real or imaginary, to make more money, if and when they chose, by selling all or part of their land for urban-related activities. This right had been seriously curtailed by the 1981 Regional Policy Plan, but under the existing Ontario Planning Act, there was always the hope that an official plan amendment would be approved and not successfully appealed. Under the Greenbelt Act, however, the right would be lost forever, or at least until the act was repealed. Furthermore, given the specialty cropland status for Niagara, all urban-area boundaries within the greenbelt were frozen. Also, there was no attempt by the Ontario government to tie land-use planning to agricultural viability; while they remain related, and farmers certainly need the one to enhance the other, they were considered separate jurisdictional areas.

In Niagara, unlike elsewhere, the public hearing process became a one-issue campaign dominated by angry farmers, many sadly using scare tactics to whip up support and help their cause: for example, invoking the word "expropriation," as if the government were compulsorily purchasing their land, or claiming that the greenbelt was preventing them from selling their land—which was only

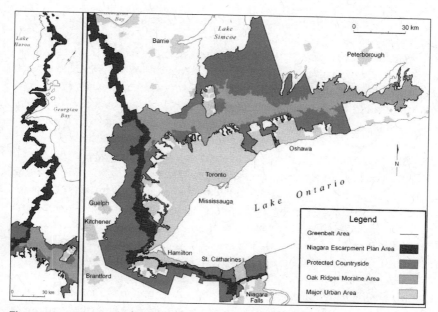

Figure 17.2 Greenbelt Plan area (Government of Ontario 2005)

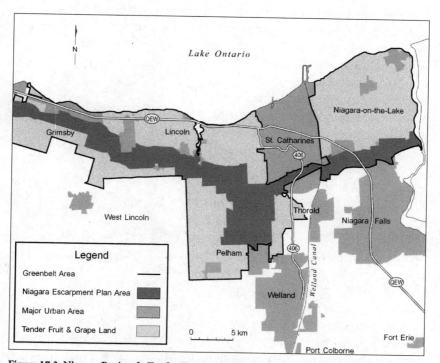

Figure 17.3 Niagara Peninsula Tender Fruit and Grape Area (Government of Ontario 2005)

true in that certain types of buyer (i.e., urban developers) would no longer be interested in buying. Placing blame on the Greenbelt Act for adversely affecting farm incomes hardly rang true. In the Niagara case, the act was in effect endorsing what already existed under the Regional Policy Plan, providing exceptions to the rule were not allowed.[17] The abolition of retirement severances was now law, and could be regarded as vindication for the original OMB decision being successfully appealed. In fact, Niagara was simply catching up with most other jurisdictions where such an invitation to urban sprawl had not been allowed.

The Greenbelt Act was viewed by most as a logical next step to protect an irreplaceable resource in Niagara for all time. It was casting in stone what the 1981 Regional Policy Plan had attempted to do. While there were cries of anguish from towns such as Grimsby about stifling traditional suburban development, the City of St. Catharines had long since embraced the freezing of its urban-area boundaries and sought redirection, intensification, and brownfield redevelopment in order to facilitate growth yet protect its valuable agricultural land.

Meanwhile, the Places to Grow Act, 2005, aimed to take urban pressures off the greenbelt by focusing development on twenty-five urban centres in the Greater Golden Horseshoe, one of which is downtown St. Catharines, and intensifying growth in both existing and new greenfield developments.[18] Seven of Niagara's twelve local municipalities, in fact, were welcoming of the act, since redirection policies already in effect would now set them up as "places to grow."

Overnight, it seems, opposition to the Greenbelt Plan has largely ameliorated in Niagara. However, the very institution of a greenbelt remains a litmus test at the present time for a disgruntled part of the wine industry—namely, many grape growers who find themselves without contracts with wineries or a surplus of grapes to sell (Figure 17.4 and Figure 17.5). The greenbelt itself can hardly be blamed for these outcomes, but it can be held up for preventing that tried, but not necessarily true, fallback position of selling a lot here, a lot there, for whatever short-term, money-making pur-

Figure 17.4 Promoting Ontario's Greenbelt
Photo courtesy of the author.

pose. While the agricultural industry has never considered this urban sprawl outcome a long-term solution to the issue of agricultural viability, it was often politically expedient in the short term. Now government fiat forbids such a practice, forcing an agricultural solution to what is an agricultural problem. With the land base protected, government agricultural, tax, and liquor policies are

Figure 17.5 Protesting Ontario's Greenbelt in Niagara
Photo courtesy of the author.

being amended in various ways to improve the economic condition of grape growers and to try to settle the disputes between growers and local wineries.

It has not been an easy transition. The phenomenal increase in the number of estate and boutique wineries in the last twenty-five years and the resulting expansion of new vineyards have not relieved the uncertainties faced by the grape grower.[19] A combination of good weather conditions and insufficient contracts with wineries has resulted in unwanted surpluses. However, the wine industry has argued that surpluses relate to too great an emphasis on quantity, whereas it is felt that better pruning and the improving of grape quality are demanded by winemakers for today's ever more sophisticated consumers.

Grape growers continue to be frustrated by low price agreements that are negotiated with wineries; surpluses could be reduced if the two largest wineries, producing over 80 percent of Ontario's wine, were not permitted to import up to 70 percent foreign content and blend it with local wine (see Chapter 4). Blended wines are vigorously defended by the industry's largest players on the grounds of lower consumer prices; furthermore, a large proportion of grape growers who contract with these wineries recognize that they could be out of business were it not for blended wines. Meanwhile, it has to be admitted that many consumers, who do not read the labels carefully enough, can be hoodwinked into thinking they are purchasing a truly Canadian wine. This is something that the smaller wineries, which can produce wine only from 100 percent Ontario grapes and invariably at higher prices, resent as they strive to improve overall quality and raise the stature of the industry. In a move to help these smaller wineries, the Ontario government has increased the tax on blended wines; at the same time, it plans by 2014 to remove the 70 percent foreign content limit altogether, as a challenge to growers to improve domestic grape quality and lessen the need for imported wine.

Whatever the internal conflicts in the grape and wine industry in Niagara, the expansion of that industry and the growing stature of Canadian wines on the national and international scene demand first and foremost the protection of the land base where the grape is grown and the wine is produced. While grapes are grown in other parts of eastern Canada, the vast majority of the vineyards and wineries, and the multi-million-dollar investment associated with them, are situated in the Niagara wine region. They cannot be easily moved; thus, it is a small step in logic to say that the greenbelt in some form is here to stay.

Conclusions

A greenbelt plan for Niagara comes as no great surprise. Debates to save its valuable agricultural lands have been ongoing for half a century, and a Regional Policy Plan has tried to do so since the early 1980s. Sadly, a bottom-up planning process under the Ontario Planning Act was too readily abused by local and regional politicians, listening to a narrow constituency rather than considering the wider public good. Together with pressures from other threatened environments in the Greater Golden Horseshoe, the Ontario government put in place in 2005 a more effective top-down process (familiar for decades in many European countries) to stem, indeed halt, urban sprawl and the loss of valuable rural resources. For Niagara's wine industry, a greenbelt affords that protection; with that in place, it is hoped that the many problems in the industry associated with the long-term viability of the farmer can also be tackled so that this relatively new and evolving wine industry can grow and prosper.

Notes

1 The author would like to thank the many professional people who have assisted in this research by freely offering opinions and information. These include local and regional planners in Niagara, farmers, winery owners, and winemakers. The usual disclaimer applies.

2 H.J. Gayler, "The Niagara Fruit Belt: Planning Conflicts in the Preservation of a National Resource," in *Big Places Big Plans*, ed. M. Lapping and O. Furuseth (Aldershot, UK: Ashgate, 2004), 55–82.

3 P. Chapman, "Agriculture in Niagara: an Overview," in *Niagara's Changing Landscapes*, ed. H.J. Gayler (Ottawa: Carleton University Press, 1994), 279–99.

4 R.R. Krueger, "Changing Land Use Patterns in the Niagara Fruit Belt," *Transactions of the Royal Canadian Institute* 32, no. 5 (1959): 39–140; R.R. Krueger, "Urbanization of the Niagara Fruit Belt," *The Canadian Geographer* 22 (1978): 179–94; R.R. Krueger, "The Struggle to Preserve Specialty Crop Land in the Rural–Urban Fringe of the Niagara Peninsula," *Environments* 14 (1982): 1–10.

5 L.O. Gertler, *Niagara Escarpment Study Fruit Belt Report* (Toronto: Ontario Department of Treasury and Economics, 1968); *Factors Affecting Land Use in a Selected Area in Southern Ontario*, ed. R.M. Irving (Toronto: Ontario Department of Agriculture, 1957); L. Reeds, *Niagara Region Agricultural Research Report* (Toronto: Ontario Department of Treasury and Economics, 1969).

6 J.N. Jackson, *Land-Use Planning in the Niagara Region* (Thorold, ON: Niagara Region Study Review Commission, 1976).

7 Niagara Region Local Government Review Commission, *Report of the Commission* (Toronto: Ontario Department of Municipal Affairs, 1966).

8 H.J. Gayler, "The Problems of Adjusting to Slow Growth in the Niagara Region of Ontario," *The Canadian Geographer* 26 (1982): 165–72; H.J. Gayler, "Conservation Versus Development in Urban Growth: Conflict on the Rural–Urban Fringe in Ontario," *Town Planning Review* 53 (1982): 321–41.

9 G. Janes, "Preservation of Agricultural Lands Society (P.A.L.S.): The Fight for the Fruitlands," in *Environmental Stewardship: Studies in Active Earthkeeping*, ed. S. Learner, 249–73 (Waterloo, ON: University of Waterloo Department of Geography, 1993).

10 J.N. Jackson, "The Niagara Fruit Belt: The Ontario Municipal Board Decision of 1981," *The Canadian Geographer* 26 (1982): 172–76.

11 Ontario Municipal Board, *Decision Relating to Official Plan Amendments to Allow Urban Expansion in Fonthill*. File no. PL980963 (Toronto: Ontario Municipal Board, 2000); Regional Municipality of Niagara, *Proposed Expansion of the Fonthill Urban Area, Town of Pelham*, DPD 136-99 (Thorold, ON: Regional Municipality of Niagara, 1999).

12 W. Caldwell and C. Weir, *A Review of Severance Activity in Ontario's Agricultural Land during the 1990s* (Guelph, ON: University of Guelph School of Rural Planning and Development, 2002).

13 Regional Municipality of Niagara, *Smart Growth for Niagara: A New Approach to Development*, DPD 118-2000 (Thorold, ON: Regional Municipality of Niagara, 2000).

14 Regional Municipality of Niagara, *Securing a Legacy for Niagara's Agricultural Land: A Vision from One Voice*, Report of the Agricultural Task Force CAO 13-2004 (Thorold, ON: Regional Municipality of Niagara, 2004).

15 Government of Ontario, *Greenbelt Plan* (Toronto: Ontario Ministry of Municipal Affairs and Housing, 2005); Greenbelt Task Force, *Toward a Golden Horseshoe Greenbelt* (Toronto: Ontario Ministry of Municipal Affairs and Housing, 2004).

16 H.J. Gayler, "Ontario's Greenbelt and Places to Grow Legislation and the Future for the Countryside and the Rural Economy," in *The Next Rural Economies: Constructing Rural Place in Global Economies*, ed. G. Halseth, S. Markey, and D. Bruce (Wallingford, UK: CABI Publishing, 2010), 75–88.

17 Regional Municipality of Niagara, *Regional Niagara Policy Plan: Office Consolidation* (Thorold, ON: Regional Municipality of Niagara, 2007).

18 Government of Ontario, *Places to Grow: Growth Plan for the Greater Golden Horseshoe* (Toronto: Ontario Ministry of Public Infrastructure Renewal, 2006).

19 H.J. Gayler, "Niagara's Emerging Wine Culture," in *Covering Niagara: Studies in Local Popular Culture*, ed. J. Nicks and B. Grant (Waterloo, ON: Wilfrid Laurier University Press, 2010), 195–212.

Eighteen

The Niagara Wine Festival's Grande Parade: The Public Geography of "Grape and Wine" Controversy

Phillip Gordon Mackintosh

E arly on the first night of the 58th annual Niagara Wine Festival, held in St. Catharines' Montebello Park in 2009, an already unsteady festivalgoer lurched along Duke Street away from the park as my party headed toward it. As the man got near to us he blurted, as much in hurt as anger, "They're making ya pay to get into the park … you gotta pay to get into the park!" On that auspicious evening, RBC Opening Night (sponsored by Royal Bank of Canada), ex–Bare Naked Ladies singer Steven Page opened the festival at a premium. Ticket sellers at a temporary gate, blocked by security guards, informed us the cover charge was $15 a person (not unreasonable for a concert, but price was not the point). We stood agape, holding the empty wineglasses we had purchased at the previous year's festival. "You're charging us to go into a public park?" I asked incredulously. "It's a concert with Steven Page," the ticket seller responded. "Well, we're not paying $60 for admission to Montebello Park," I answered. "But it's going to be a great concert," she added as we turned away, as if this would ameliorate our discontent. "But it's a public park," I grumbled.

I relate this personal anecdote to introduce a chapter commenting on a rarely mentioned element of the Niagara wine issue: the repercussions of marketing Niagara wine on the community use of public space. I am interested in the growing disparity between the motivations of Niagara Wine Festival organizers (hereafter, "the Festival") and the festival expectations of Niagarans. In the last decade, the "grape and wine" festival and parade have undergone specific, industry-focused changes, although it has been noted sarcastically that Niagara wineries themselves boycott the parade.[1] These intentional modifications have not only annoyed long-time parade- and festival-goers, but also have affected interpretations of public space in St. Catharines at festival time. Wine industry boosters tend to construe the public as a legitimate venue to market its product. The locals seem to like public space that represents their perceived (and doubtless ironic) sense of collective identity, which includes wine, but invariably transcends it.

Economistic Visions of Public Space

In its blurring of the public space of St. Catharines with grander economic aspirations, especially at "grape and wine" time, the Festival hardly stands alone. Cities throughout Ontario struggle with diminishing revenues as a result of trade liberalization, tax cuts, and tax moratoriums offered as incentives to retain increasingly footloose businesses and manufacturers. Higher than usual unemployment rates, especially since the global financial collapse of 2007–8, abet the continuing problems of municipal amalgamation enacted by the Harris government in the 1990s, including the downloading of increased responsibility for funding and operating municipal social services.[2] This, too, is an uneven process. Some municipalities are more burdened than others, as Ontario municipalities scramble to exploit all assets within their city's precincts.[3] Chief among an Ontario city's marketable resources at the turn of the twenty-first century, it seems, is its public space.

The problem with such a well-intentioned but democratically indelicate interpretation of the city's public spaces is its logical consequence: the reduction of all public actions in public space to the level of economic transaction.[4] The historical significance of public space as a receptacle for a wide array of activities, behaviours, and practices atrophies to fit a predominating but narrow view of the public as a yet to be fully exploited site for consumption. Citizen-pedestrians become clients, shoppers, and tourists, increasingly members of a "creative class" of bourgeois urban professionals whose role is to save the city by having it designed to meet their consumption requirements.[5] New "creative economy" development strategies target "creative class" consumers through a consistent appeal to an ever-growing list of "commonsense" urban design rules of thumb: infrastructural beautification and bourgeois aestheticization, or, as one critic has cleverly called it, "domestication by Cappuccino," through the use of cafés and restaurants, boutique shops, galleries, theatres, and the like (what St. Catharines hopes to effect with the construction of the Niagara Centre for the Performing Arts and its cappuccino corollaries on St. Paul Street).[6] This also includes hypersensitive policy responses to recidivist street behaviours: graffiti eradication, closed-circuit TV, zero-tolerance policing, and behaviour bylaws, such as St. Catharines' recently enacted public nuisance bylaw, which includes an identification-on-demand ordinance.[7] This fits a general view. Municipal policy makers everywhere in the last thirty years have subscribed to the idea that "a downtown can be designed and developed to make visitors feel that it—or a significant portion of it—is attractive and the type of place that 'respectable

'people' like themselves tend to frequent."[8] Such pandering to bourgeois expectations is certainly indicative of the type of thinking guiding the Niagara Wine
Festival's reconsideration of the public space of St. Catharines as a wine tourist
destination, as I will show.

By way of a brief comment on method, much of this chapter involves
research gleaned from Niagara's daily and weekly newspapers and, largely, the
St. Catharines *Standard* (*Standard* hereafter), which is delivered free to Niagarans
in St. Catharines, Thorold, Niagara-on-the-Lake, St. Davids, Queenston, and
Virgil at least once a week (I also include *st.catharinesstandard.ca* and the
Standard Facebook page). The *Standard* is owned by Quebecor's neoconserva-
tive Sun Media division, which spent much of 2010 developing an alleged "Fox
News" cable channel in Canada, using old Reform-Alliance party stalwarts Kory
Teneyke and Eric Duhaime to shepherd it.[9] Nevertheless, the *Standard* arguably exhibits an unevenly progressive conservative editorial position, usually
articulated by syndicated editor Michael den Tandt (for example, den Tandt was
one of the myriad voices that castigated the Harper government for neutering
the long-form census in the summer of 2010).[10] Thus, on the one hand, the
Standard's regular wine column propagandizes for the Niagara wine industry,
thinly disguised advertising for the wineries' latest consumables and marketing
efforts. On the other, the newspaper seems sincerely conflicted by the public
distress at changes wrought by the Niagara Wine Festival Board to the Niagara
Wine Festival and Grande Parade, to make both more spectacular for international tourists but alienating the Niagara public. Letters to the editor lament
changes to the festival that make it less "local," though the paper waxes enthusiastic about "improvements." This genuine ambivalence of the *Standard* regarding
the Festival persuades me to allow its letters, columns, editorials, and online
comments to speak for Niagarans, and to represent the circumstance reasonably
accurately in what follows, in part because of the role that the *Standard* "thinks"
it plays in influencing public opinion in St. Catharines.

What Is a Parade?

The Festival's difficulties began in October 2002, when the *Standard* reported "a
huge outcry in the community" after the Festival voted to remove "Grape" from
"Niagara Grape and Wine Festival"; thenceforth it was "Niagara Wine Festival."[11]
This change incensed Niagarans. Yet, in their resulting fury, they perhaps for-
got that the Festival's own promotional film in the 1980s was called "Niagara's
Winefest," or they willfully ignored that the Festival has regarded the festival and
parade as a tourism product since the 1980s.[12] For the Festival, the official name
change simply initiated a process that openly acknowledged that "[the wine]
industry has changed and marketing to a worldwide audience has to reflect

that," as Charlie Pillitteri, of Pillitteri Estates Winery, and Festival director, noted at the time.[13]

The Festival's decision to make the festival more industry friendly eroded Niagarans' patience further when, in 2007, the Festival established rules that both excluded certain Niagarans from participating in what they thought was their own Grande Parade and restricted the behaviour of entrants. This put the Festival on the front page of the *Standard*, accused of "Raining on their Parade."[14] Niagarans, it turns out, may associate themselves affirmatively with the Niagara wine industry, as Ripmeester and Johnson show in Chapter 14. Yet Niagarans also exhibit a rather proprietary ownership over their "grape and wine," the local term for the celebration that began in 1951. And they are convinced, rightly or wrongly, that the new parade and festival are inferior to the parades and festivals of their childhoods.[15] Parade-goers commonly hear other parade-goers say things like, "When I was a kid, the parade was two and half hours long, there were bands from all over the US, and half a dozen steel bands," as I heard one Niagaran explain to a visitor as both walked down Queen St. toward Montebello Park, just after the 2010 parade. Nothing illustrates this local ownership of grape and wine better than the Festival's remaking of the Grande Parade, the unofficial climax of the two-week celebration that, for many, *is* grape and wine.

A community parade is about more than the costumes, bands, floats, and general impedimenta organized for temporary occupation of a city's streets. Such parades not only express shared public cultural attitudes broadly conceived, but also attempt to impose dominant representations of "community," whatever that means to them, as normal, *as public consensus*.[16] Nothing depicts social orderliness better than the faux unanimity of a parade, especially one that vets its entrants as the Grande Parade does, so parades perform another function: they represent hegemonic constructions of social order and identity—local identity, in the case of the Grande Parade. As "street theatre," parades communicate this powerful understanding by saying, visually, "Look, this is how things should be, this is the proper, ideal pattern of social life" in this community.[17] No wonder, for example, that many of Toronto's social conservatives balk at the annual gay pride parade's confrontational and spectacular assertion that queerness is a shared community value.

As we might guess, a parade also expresses power relations, meaning that even a procession as putatively innocuous as the St. Catharines Santa Claus parade (recently reinstated in the city) effects a type of urban passive-aggression aimed at the temporary or ephemeral expression of ownership of the public. Strangely enough, a parade—with its face-painted children, clowns, draft animals, local luminaries, bands, floats, and vehicles adorned with logos of local businesses—is an intentional, if often overlooked, display of power. Thus, "as political acts," as historian Susan Davis writes, "parades and ceremonies take

Figure 18.1 "Contest and confrontation." A moment of not-so-subtle "contest and confrontation" in the 2010 grape and wine parade: the banner of the "Noble Thirteen" Loyal Orange Lodge 998, Niagara Falls, leads its followers not far ahead of the Niagara Region Métis, as the parade moves down Lake Street.
Photo: Phillip G. Mackintosh.

place in a context of contest and confrontation," even given the frequent subtlety or misapprehension of that contest and confrontation (see Figure 18.1).[18]

A parade's preferential display of identity and power becomes more complicated when compounded with the supposed neutrality of public space. The public space of the city is, as Peter Goheen avers, always contested and negotiated, largely because precise definition of public space eludes us.[19] Its very liminality, or its inability to manifest unequivocal publicness or privateness, is signified through the perpetual challenge to neutrality posed by the confusing proximity of public sidewalks and infrastructure to private businesses, entranceways, store windows, and walls. Such a circumstance naturally confuses and polarizes citizens' opinions about the public.[20] On one side, order-mongers, often those promoting the aims of chambers of commerce, align themselves with the commercial ownership and symbolism of the architectural space of the public. They insist that public space, as an overt space of economy, functions best when planned, regulated, and secured according to the needs of orderly consumption and daily business.[21] On the opposite side are the public-space-at-all-cost advocates. They suppose a primordial guarantee of right of access to and use of the public spaces of the city.[22] Deploring business and its increasing colonization of municipalities and policy makers, they assert an inalienable right to the public that trumps even urban capitalist modernity's paradoxical requirement that we must shop to live.[23] It is no coincidence that as twenty-first-century cities succumb to the frequently corrupting influence of political corporatism, social movements emphasizing pedestrianism, anti-consumerism, and free access to public space foment non-transactional uses of what is public—public space used for anything other than spending or earning money—in major cities around the world (the Occupy Wall Street movement is a powerful exemplar). Here, then, is an urban antinomy ripe for ideological battles over "proper" uses of public space.

In the context of the Niagara Wine Festival and the Grande Parade, a few questions, then, need answering: Whose conception of normalcy, of "public consensus," is demonstrated by both festival and parade? In what way has the parade become a political statement, in the wake of forsaking the Niagara public's beloved "grape"? Whose public *is* the public of St. Catharines, as the parade briefly claims the streets of "the Garden City"?

Who Is the Parade?

"This is ... not the Macy's Thanksgiving Day Parade."[24]

"It's great to reach for the stars, but we can't lose sight of what this parade is about—our community."[25]

Who does, or even should, the parade belong to or represent? Niagarans, who want it to express community values? Or the Festival—and the Niagara Region and City of St. Catharines—which hope to transform it into a Macy's-type spectacle to increase tourism revenues (the 2010 parade even included "Macy's Thanksgiving Day Parade–style balloons," though two of the four large helium balloons were pulled from the parade because of the wind, while the WestJet jet balloon "taxied" the entire route in 2012)?[26] In the last decade, the Festival has increasingly understood the role of the Festival and parade in the context of a globalizing wine trade and consumerism and tourism, regarding the parade as its "marquee event." As past Festival president Ken Weir has suggested, "improving the quality of th[is] marquee event has been the No. 1 issue he's heard about from board members, wineries and consumers."[27] Such parade considerations are hardly new: even in the 1970s, the Festival board was concerned to mediate the parade "in the interest of maintaining standards" (though whether insisting on standards in the parade equated with extant tourism concerns is another issue).[28] Walter Sendzik, the festival board chair who oversaw the name change in 2002, and 2012's RBC Business Citizen of the Year, contends that "[i]t was a good move the festival took to focus its marketing to a larger audience beyond Niagara."[29] This mirrors the position of Kimberley Hundertmark, the Festival's executive director. As city tourism manager, she told the *Standard* that "[o]ur tourism product in St. Catharines really does hinge on the festival," which would suggest that a "better" parade equals better success in tourism.[30] St. Catharines city councillor Peter Secord, the city's delegate to the Festival, agrees: "The festival is … a tourist draw and a high-profile event for a local industry that employs many people."[31] Perhaps the most salient of comments belongs to Sendzik, who, in response to the furor over the name change in 2002, candidly admitted to Niagarans, in an open letter to the *Standard*, that "with VQA consumption hovering in Ontario around 30 per cent and surpluses of grape juice on the market year after year, the festival needs to be a conduit in attracting more wine consumers to Niagara."[32] After all, as Pillitteri asks, "What are we presenting? We are presenting wine. And if we really want to be a truly international destination for the world to come and taste our wines, we need a wine festival."[33] For the Festival, the Grande Parade is a business.

If this parade-as-business approach surprises some Niagarans, they need only to peruse the numerous provincial and regional reports declaring how the Niagara wine industry, "Poised for Greatness," has plans for "Energizing Niagara's Wine Country Communities." Importantly, these plans "leverage heritage and cultural tourism for economic development in Niagara Wine Country"—meaning repackaging grape and wine as a wine attraction with "an expectation for quality."[34] A recent comment by Hundertmark could not be more clear: even the provincial "government recognizes that the grande parade ... is a tourism driver for the city and does provide economic impact for the region."[35]

Perhaps Niagarans should have paid closer attention to the 1998 Festival program guide. Here the Festival's three conflicting mission statements portend a showdown:

1. To realize the tourism potential of the Festival and its position as an important component of the tourism market.
2. To promote the grape and wine industries and their products along with the Host City of St. Catharines and the Region of Niagara.
3. To provide a community event and celebration that will contribute to the quality of life and civic pride within the community.[36]

It is easy to imagine how the privileging of statements [1] and [2] could clash with statement [3], especially when considering the following by the *Standard*'s wine reporter, Chris Waters, regarding the name change: "The festival's new name, The Niagara Wine Festival, simply represents what consumers experience at Montebello Park and other festival events."[37] So here is the predicament: the Festival, like Waters, has reimagined the festival and parade as wine products, but Niagarans patently do not think of themselves as "consumers" of their half-century celebration. They believe they are a primary purpose for its existence.

Did the Niagara public realize that the implementation of the Niagara Wine Festival's tripartite mission statement would elevate tourism as the first among equals at grape and wine? No. Do they resent the intimation that they are tourists in their own community? I think they do. Hence the contest over the Grande Parade.

The Festival's need to shape the parade to suit wine tourism imperatives has not been lost on the public, though some may not fully comprehend why their "grape and wine" has changed. We can speculate: in our world of trade liberalization and production outsourcing to China and the developing world, both of which contribute to reduced government revenues and transfers, cash-needy municipalities go where the money is: the private sector. With its twin pillars of profit and efficiency (though the latter is arguable), corporate interest drives cities to adopt regulations and bylaws that encourage order, spectacle, and, most especially, shopping, while discouraging disorder, inefficiency, and non-consumerist spontaneity: the entertaining use of public space that does not involve spending money. This means, of course, that in Niagara's sickly economy, the Festival and Region have hitched their economic aspirations to the "[r]olling vineyards and the fine wines crafted from their bounty," which form "Niagara's lure as an international destination."[38] It is quite plausible that many Niagarans misapprehend the new burdens the parade has to carry as it passes them on the streets of St. Catharines, in the process distorting the image they once held of themselves.

We know this from what Niagarans say. Their answer to the question of whom the parade should represent is unvarying: them. They wax nostalgic about the "old" parade: "I remember watching the parade as a child and looking forward to all the different acts that would follow—the clowns, the cars, the different queens—but it seems as though there are a few snobs in charge today."[39] This "snob" theme recurs in the *Standard* online comments as well. "They're trying to turn it into a wine industry event instead of a community event for people in the city" is a typical reaction.[40] "The GRAPE festival is slowly sliding down the chute," writes "electionwatcher" online. "[T]he entertainment has lost focus, the parade is insulting to those who have supported it during our lifetimes. Then throw on what the wine snobs have done.... The harvest, the Grape, the hometown festival

Figure 18.2 The "old" parade. "The charm of the steel bands and the beautiful costumes of visiting performers add[ing] a touch of the exotic," c. 1980s (Brock University Archives B3F21).
Photo: Phillip G. Mackintosh.

has been stolen by the revenue seeking snobs ... [who should] go back to their board rooms and leave us our Grape festival."[41]

Some feel the Festival has "taken [the] Grande Parade away from the people who support it": "I was so disappointed in this year's Grand Parade.... Maybe we're not sophisticated enough to understand the new rules and regulations involved in being allowed to enter in our parade."[42] For one parade-goer, the new parade is not novel, but a sad reminder that the traditional parade ideal "like everything else connected with downtown St. Catharines is dying."[43] And with the decline of local Caribbean representation in the parade, which began in 1968, when the City of St. Catharines was twinned with Port of Spain, Trinidad, another Niagaran charges the new parade and Festival with racial intolerance: "The entire festival was filled with colour and flavour, the steel bands that played music in the streets and the people who jumped up following them, the fabulous costume creations that brought the city streets to life if only for one day ... [why not] hold the entire event at the St. Catharines Golf and Country Club. Then all of the beer-drinking, grape-stomping, fun-loving, cultural community members could really be excluded."[44] Yet we know that cultural pluralism once mattered to the Festival, as its president Dan Russo explained in his welcoming message in the 1983 festival guide, "the charm of the steel bands and the beautiful costumes of visiting performers add[ing] a touch of the exotic" (Figure 18.2).[45] And while the letter above makes a stinging allegation, it also slights the participation of York Lions Steel Band, "One of the Best Youth Steel Bands North of the Caribbean."[46]

Remembering this noteworthy "Caribbean" parade element, one Niagaran wrote in response to the *Standard*'s Facebook poll "Going to the wine festival parade? Tell us what you think,"

> Not going ... it's a very commercial parade. I can thumb through the phone book to see businesses and advertising. I'm old enough to remember when this festival twinned with Trinidad-Tobago and there were beautiful floats and it was about culture. I do enjoy the pipe and drum bands and love to see the schools getting involved, but I will go the NOTL [Niagara-on-the-Lake] Santa Claus parade to enjoy those! ... this parade is a "sad shadow" of what it was when I was a child.

Another Facebook reply stated curtly, "I'll be sleeping in.... I no longer support this event" after attending the parade "all through [my] childhood and youth."[47]

Still other Niagarans wonder why the overt "gentrification of our parade" has not made it "better for the changes."[48] This gentrification, or deliberate upscaling (which I could argue is overstated, having watched the parade closely since the name change), includes the parade's streamlining: in 2008, the festival organizers "partnered" with the Niagara Artists Centre and the St. Catharines and Area Arts Council to use both organizations as consultants for all parade float entrants.[49]

The accusation of gentrification implicitly targets the parade's new orderliness regimen: the desire of organizers to control the tourists' experience of spectacle by "Rolling out a tighter, focused festival" and parade—and eliminating community spontaneity.[50] This is done by screening parade entries and "not backing down on [new] methods to improve the parade" in 2007, though "organizers hope to avoid [2007's] public outcry by doing a better job of communicating the new standards to the groups and agencies that want to participate."[51] The "rules and regulations" are clear: "All applicants must provide detailed information outlining the entertainment value of their entry. All applicants will be evaluated on: creativity, colour/pageantry, pertinence to the Niagara region, safety and cohesive presentation."[52] Thus, the organizers set costume standards; not meeting them is why Gymnastics Energy, the baton-twirling young gymnasts, was excluded in 2007. So, too, the Rankin Cancer Run: its float of cancer survivors wasn't exciting enough.[53]

The Festival monitors disorder. Young Jedi Niagara, the Star Wars marchers who raise money for the St. Catharines women's shelter, Gillian's Place, were put "on probation and ... giv[en] one more chance to prove [them]selves"; apparently, the entrants' loose formation causes their Star Wars characters to mix with one another (Figure 18.3).[54] The dogs from the Lincoln County Humane Society were denied entry because of a late application. Space limitations and funding changes left the Niagara Memorial Militaires Alumni Drum and Bugle

Figure 18.3 Jedi in Niagara. Children swarm Young Jedi Niagara's Darth Vader character during the 2009 Grande Parade.
Photo: Phillip G. Mackintosh.

Corps out of the 2012 parade.[55] Yet, while the gentrification of the parade worries Niagarans, I suggest the greater fear is that community identity and tradition—authenticity—are being incrementally eradicated from the parade through this winnowing of spontaneity, even if that means accommodating the banal but local. The desire to make the parade a Macy's-type spectacle (certainly the 2012 parade was the closest yet to the ideal, and dependent on imported participants) means that the community element, whatever that means to Niagarans and for whomever Niagarans recognize as Niagaran, is diminished.

Conclusion

The Festival's efforts in the last couple of years have begun to "recognize … the importance of th[e] grape component to the Niagara community," in Hundertmark's words, and are winning favour. Moreover, John Potts, the Festival co-chair, has committed to "listening to the community" and "ensur[ing] the old traditions are going to be there."[56] The 2010 parade, according to Hundertmark, "focused on our audience and what the audience enjoys and what they feel is successful," though this also "meant examining best practices at other festivals and ensuring the event responds to what the community and destination visitor is looking for"—*destination visitor* being an odd euphemism for wine-culture consumer.[57] "Thousands crowded sidewalks and parking garage ledges through downtown St. Catharines" for the 2011 parade, while 2012's undertaking, "the biggest, longest parade in the last five years," virtually conceded that grapes are a chief reason for the celebration, with the grape theme liberally interspersed throughout.[58]

The result? Who can say? Some "loved seeing all the local groups and schools and clubs" in the 2010 parade.[59] Others' feelings were summed up tersely: "The largest presentation was by the company Just Junk … need I say more!!!"[60] The *Standard* championed the 2010 parade only hours after it ended, reporting "Big crowds, soaring spirits at Niagara Wine Festival Grande Parade." One St. Catharines resident said it best, at least from *Standard*'s perspective: "All it can do is get better every year."[61] It had better do so, although from the crowd reactions to the 2012 parade, it might be on its way; one especially attention-grabbing but non-local entrant—"ROTC," or Righteously Outrageous Twirling Corp, a lesbian, gay, bisexual and transgender community group from Toronto—drew uncharacteristic whoops, whistles, and applause. As close to a big deal as 2012's parade and festival may have been—Hundertmark called it "one of the best festivals that we've ever had in Niagara"—a tenacious perception that the "parade and what follow[s] at the park [a]re a mere shadow of what once was!!!" or that the festival in Montebello Park "has become a gigantic open wine store … nothing more!!" persists year upon year—and you can hear patrons make similar observations as they line up to pay (too much, apparently) for their wine and food tokens inside the park.[62]

Has uncritical sentimentality driven the opposition to the new festival and parade, exaggerating the value of the old one, and indulging in nostalgia for a romanticized parade and festival? Probably. The *Standard*'s Doug Herod reminded readers after the 2010 Grande Parade that the parade was not always what Niagarans recalled, or even in the best of taste: the 1974 parade featured a float "which was purporting to show an attack on a white settlement … complete with tomahawk and fake blood" and a mock scalping scene.[63] Notwithstanding, the parade may have lost something; that same 1974 parade ran for three hours, and featured 131 entries, ninety-one floats and forty bands.

That said, do Potts, Hundertmark, the *Standard*, and others, in their enthusiasm for the economic success of the festival, misunderstand something basic? It is not clear that Niagarans particularly care that the 2010 festival burned through 20,000 wineglasses alone on the last weekend (at $3 each, while bottles of wine sold for $21 to $27 each; in 2012, the prices were even higher); or that "the [Niagara Wine Festival] finance guys say there's been record-breaking sales," as Hundertmark told the *Standard*; or even that "a bustling Saturday" in Montebello Park found some vendors running out of wine following 2011's parade.[64]

Niagarans, of course, want their region to thrive: they saw unemployment reach 12.4 percent in October 2009; in September 2012, it was a lower but still worrying 10.9 percent.[65] However, economic success for the wine industry appears not to equate with local approbation for the Festival's fiddling with a local tradition, even if the festival and parade have been identified as a major component of economic development in a declining region.[66] And until political economic circumstances change in Ontario (and Canada), until politicians and voters recognize that the success of regional economies depends on more than market logic and one or two globalizing industries, that trade liberalization and industrial outsourcing not only diminish wages and work in regions and communities but also debase their traditions, then the Festival will continue to operate a public parade serving wine industry purposes for purely economic development reasons. And many Niagarans will continue to line the streets of St. Catharines to watch a parade they grudgingly feel misrepresents them, their past, and perhaps their public.

Notes

1 "If the wineries don't want to be involved (beyond hawking product during the week) and insist on dropping grape from the festival name, then I suggest the board at least be honest and drop 'wine' from the name of the parade." "Where are the winery floats?" St. Catharines *Standard*, 1 November 2002, A7.

2 Enid Slack, "Preliminary Assessment of the New City of Toronto," *Canadian Journal of Regional Science* 23, 1 (2000): 13–29; Roger Keil, "'Common-Sense' Neoliberalism: Progressive Conservative Urbanism in Toronto, Canada," *Antipode* 34 (2002): 578–601.

3 David Siegel and Joseph Kushner, "Effect of Municipal Amalgamations in Ontario on Political Representation and Accessibility," *Canadian Journal of Political Science* 36 (2003): 1035–51.

4 Guy Debord, *The Society of the Spectacle* (New York: Zone Books, 1995); Cindi Katz, "Hiding the Target: Social Reproduction in the Privatized Urban Environment," in *Postmodern Geography: Theory and Praxis*, ed. Claudio Minca, 93–110 (Oxford: Blackwell, 2001); *Postmodern Geography: Theory and Praxis*, ed. Claudio Minca (Oxford: Blackwell, 2001); *Spaces of Neoliberalism: Urban Restructuring in North America and Western Europe*, ed. Neil Brenner and Nick Theodore (Oxford: Blackwell, 2002); Nicholas Blomley, "Un-Real Estate: Proprietary Space and Public Gardening," *Antipode* 36 (2004): 614–41; Jamie Peck, "Struggling with the Creative Class," *International Journal of Urban and Regional Research* 29 (2005): 740–70; *The Politics of Public Space*, ed. Setha Low and Neil Smith (New York: Routledge, 2005); Lynn Staeheli and Don Mitchell, *The People's Property? Power, Politics, and the Public* (New York: Routledge, 2007).

5 Richard Florida, *The Rise of the Creative Class: And How It's Transforming Work, Leisure and Everyday Life* (New York: Basic Books, 2002); Peck, "Struggling with the Creative Class," 740–70.

6 Sharon Zukin, *The Cultures of Cities* (New York: Blackwell, 1995), xiv.

7 Minca, *Postmodern Geography*; Don Mitchell, *The Right to the City: Social Justice and the Fight for Public Space* (New York: Guilford Press, 2003); Steve Herbert and Elizabeth Brown, "Conceptions of Space and Crime in the Punitive Neoliberal City," *Antipode* 38 (2006): 755–77. By-Law No. 2007-295, A By-law to address public nuisances, *By Laws, City of St. Catharines*, http://www.stcatharines.ca/en/documentsBylaws.aspx.

8 David Milder, "Crime and Downtown Revitalization," *Urban Land*, September (1987): 18.

9 "Quebec Throws Its Very Own 'Tea Party,'" *Toronto Star*, 5 September 2010, A8.

10 Michael den Tandt, Editorial: Harper Should Go to Timmy's, St. Catharines *Standard*, 10 September 2010, A14.

11 "Councillors Have Gripe with Lost Grape," St. Catharines *Standard*, 4 October 2002, A3.

12 "Niagara's Winefest," CRTC Canadian Program Recognition Number: C06706, 14:30 mins. 1985/07/17; http://www.crtc.gc.ca/canrec/n1980-89.txt. The festival began marketing itself to New Yorkers in the 1980s: "Grape Fun at the Niagara Grape and Wine Festival," *Western New York Motorist*, September 1989, 83, 9; Box 2 Folder 42, *Niagara Grape and Wine Festival Records, 1955, 1964–1991*, n.d.—RG75-3, Brock University Archives.

13 "Absent Festival Director Favoured Name Change," St. Catharines *Standard*, 19 October 2002, A6.

14 "Raining on Their Parade," St. Catharines *Standard*, 21 September 2007, A1.

15 As someone "from away" who has been watching the parade only since 2004, I have to say that I don't really get the fuss. With or without "grape," the parade rivals every other small-city parade I've ever seen. But if Niagarans see it as a snobby, gentrified imposter, who am I to argue? Their perceptions, and sense, of tradition matter to them. For as Mary Ryan rightly suggests, "a parade is like a text in its susceptibility to multiple interpretations" (Mary Ryan, "The American Parade: Representations of

the Nineteenth-Century Social Order," in *The Making of Urban America*, 2nd ed., ed. Raymond A. Mohl [Oxford: Rowman & Littlefield, 1997], 74).

16 Peter Goheen, "Parading: A Lively Tradition in Early Victorian Toronto," in *Ideology and Landscape in Historical Perspective: Essays on the Meanings of Places in the Past*, ed. Alan Baker and Gideon Biger (New York: Cambridge University Press, 1992), 332.

17 John Skorupski, *Symbol and Theory: A Philosophical Study of Theories of Religion in Social Anthropology* (New York: Cambridge University Press, 1976), 84.

18 Susan Davis, *Parades and Power: Street Theatre in Nineteenth-Century Philadelphia* (Berkeley, CA: University of California Press, 1988).

19 Peter Goheen, "The Public Sphere and the Geography of the Modern City," *Progress in Human Geography* 22, 4 (1998): 479.

20 Sharon Zukin, *Landscapes of Power: From Detroit to Disney World* (Berkeley, CA: University of California Press, 1991), 28–29.

21 Don Mitchell, "The End of Public Space: People's Park, Definitions of the Public, and Democracy," *Annals of The Association of American Geographers* 85, 1 (1995): 115.

22 Mark Kingwell, "Masters of Chancery: The Gift of Public Space." In *Rites of Way: The Politics and Poetics of Public Space*, ed. Mark Kingwell and Patrick Turmel (Waterloo: Wilfrid Laurier University Press, 2009), 3–22.

23 Sharon Zukin, *Point of Purchase: How Shopping Changed American Culture* (New York: Routledge, 2005).

24 "Give the Parade Back to Those Who Support It," St. Catharines *Standard*, 28 September 2007, A14.

25 "Don't Change the Parade at the Expense of Our Kids," St. Catharines *Standard*, 25 September 2007, A6.

26 "The People's Parade," St. Catharines *Standard*, 25 September 2010, A2.

27 "Raining on Their Parade," St. Catharines *Standard*, 21 September 2007, A1.

28 Media release from Niagara Grape and Wine Committee, undated (Box 2 Folder 24, *Niagara Grape and Wine Festival Records, 1955, 1964–1991, n.d.—RG75-3*, Brock University Archives).

29 "Some Still Gripe over Missing Grape," St. Catharines *Standard*, 15 September 2009, A1.

30 "Wine Festival an Economic Boon," St. Catharines *Standard*, 22 September 2007, A6.

31 Marlene Bergsma, "City Bumping Up Its Support," Welland *Tribune*, 23 June 2009, http://www.wellandtribune.ca/ArticleDisplay.aspx?archive=true&e=1669231.

32 "Grape Growers Knew about Festival's Name Change," St. Catharines *Standard*, 14 September 2002, A13.

33 "Absent Festival Director Favoured Name Change," St. Catharines *Standard*, 19 October 2002, A6.

34 Ontario Wine Strategy Steering Committee, *Poised for Greatness: A Strategic Framework for the Ontario Wine Industry* (Toronto: Ontario Ministry of Consumer and Business Services, 2003); Niagara Economic Development Corporation, *Energizing Niagara's Wine Country Communities* (Niagara Economic Development Corporation and Peter J. Smith & Company, 2007), 17.

35 "Grape and Wine Focused on Strong Close," St. Catharines *Standard*, 28 July 2012, A6.

36 From the brochure/guide to the 47th Annual Niagara Grape and Wine Festival, 18–27 September 1998.

37 "Name Change Reflects What Festival-Goers Experience," St. Catharines *Standard*, 14 September 2002, E3.

38 Niagara Economic Development Corporation, *Energizing Niagara's Wine Country Communities*, 4; http://www.winecountry niagara.com/index.php.

39 "Give the Parade Back to Those Who Support It," St. Catharines *Standard*, 28 September 2007, A14.

40 "Front-Row Parade Seat," St. Catharines *Standard*, 1 October 2007, A3.

41 "Rolling out a Tighter, Focused Festival"—Comments on this article, St. Catharines *Standard*, http://www.stcatharinesstandard.ca/ArticleDisplay.aspx?e=2758380.

42 "Festival Organizers Have Taken Grande Parade away from the People Who Support It," St. Catharines *Standard*, 5 October 2007, A14.

43 "Put the Grape back in the Wine Festival," St. Catharines *Standard*, 5 October 2005, A8.

44 "Why Were Changes Made to Festival?" St. Catharines *Standard*, 15 February 2003, A13.

45 *Niagara Grape and Wine Festival Souvenir Program* 1983, p. 4, Box 3, Folder 4, *Niagara Grape and Wine Festival Records, 1955, 1964–1991, n.d.—RG75-3*, Brock University Archives.

46 York Lions Steel Band, Home, http://www.yorklionssteelband.com/index.htm.

47 "Going to the Wine Festival Parade? Tell Us What You Think," St. Catharines *Standard: Facebook*, published 2010-09-24 18:58:00 GMT, http://www.facebook.com/stcatharinesstandard.

48 "Fall Celebration Hasn't Been the Same since the Grape Was Removed," St. Catharines *Standard*, 2 October 2007, A6.

49 "Floating New Parade Ideas," St. Catharines *Standard*, 23 June 2008, A1.

50 "Rolling out a Tighter, Focused Festival," St. Catharines *Standard*, 16 September 2010, A7.

51 "Grape and Wine Parade Entries to Be Screened," St. Catharines *Standard*, 15 January 2008, A4.

52 Rules and Regulations: All Units, *Niagara Wine Festival Grande Parade Application and Information* (pamphlet), 2009, St. Catharines, Niagara Wine Festival. Curiously, the Festival board allowed this application form to use the phrase "Niagara Grape & Wine Grande Parade."

53 "Wine Fest Changes with Times," St. Catharines *Standard*, 25 September 2007, A3.

54 "Grape and Wine Parade: Young Jedi Niagara/Force United Niagara"—Topic View, *Facebook.com*, http://www.facebook.com/topic.php?uid=3172355036&topic=2988.

55 "Drum Corps Has Sour Grapes with Wine Festival," *st.catharinesstandard.ca*, 28 September 2012, http://www.stcatharinesstandard.ca/2012/09/28/ drum-corps-has-sour-grapes-with-wine-festival.

56 "Some Still Gripe over Missing Grape," St. Catharines *Standard*, 15 September 2009, A1.

57 "Rolling Out a Tighter, Focused Festival," St. Catharines *Standard*, 16 September 2010, A7.

58 "Party on Weekend at the Wine Festival," St. Catharines *Standard*, 26 September 2011, A2; "Parade Will Be a Blast from the Past," St. Catharines *Standard*, 27 September 2012, A2.

59 "Attention Niagara Wine Festival Grande Parade attendees"—St. Catharines *Standard: Facebook*, published: 2010-09-25 18:18:00 GMT, http://www.facebook.com/stcatharinesstandard.

60 "Readers Rate Parade," St. Catharines *Standard*, 27 September 2010, A1.

61 "Wine Festival Draws Crowds," *st.catharinesstandard.ca*, http://www.stcatharines standard.ca/2012/09/30/wine-festival-draws-crowds; "Big Crowds, Soaring Spirits at Niagara Wine Festival Grande Parade," St. Catharines *Standard*, 25 September 2010, A1.

62 "Your Thoughts on the Grande Parade—Comments on this Article," St. Catharines *Standard*, http://www.stcatharinesstandard.ca/ArticleDisplay.aspx?e=2774032.

63 Doug Herod, "Let's Not Get Misty-Eyed about All Past Grande Parades," St. Catharines *Standard*, 1 October 2010, A3.

64 "Organizers Hail Success of the Wine Festival," St. Catharines *Standard*, 27 September 2010, A1; "Big Crowd Eats, Drinks Vendors Dry," St. Catharines *Standard*, 26 September 2011, A2.

65 Employment Insurance, Economic Region of Niagara, Human Resources and Skills Development Canada, http://srv129.services.gc.ca/rbin/eng/niagara.aspx?rates=1&period=289. The national unemployment rate was 7.3 percent as of August 2012 (latest release from the Labour Force Survey, Statistics Canada, http://www.statcan.gc.ca/daily-quotidien/120907/dq120907a-eng.htm).

66 Niagara Economic Development Corporation, *Energizing Niagara's Wine Country Communities*, 6.

Afterword

Tony Aspler

My first taste of Ontario wine was in London, England, on Canada Day 1975. It was called Dominion Day then, and as an employee of the CBC I had been invited to a celebratory lunch at Macdonald House. This magnificent building in Grosvenor Square is part of the High Commission of Canada in the British capital. At the lunch I was seated next to an elderly British diplomat. For the loyal toast to the Queen, we were served Château-Gai Champagne. I asked my neighbour what he thought of it. He sniffed, took a sip and replied, "Fine, dear boy, for launching enemy submarines."

Those were the bad old days. When you consider how the wine industry in Ontario has developed in thirty-five years, it's nothing short of miraculous. Today our vintners are making wines that can stand without a blush on the restaurant tables of the world alongside anything produced in the Old World or the New.

What has caused this sea change in Ontario wines? There are basically five factors, one of them accidental.

Free Trade

The watershed year for the transformation was 1988 and the introduction of free trade. The signing of the General Agreement on Tariffs and Trade meant a phasing out of tariffs on imported wines that had hitherto protected the local industry. Wineries in Ontario and British Columbia would have to compete with the wines of the world on an equal playing field. (The nascent wine regions of Quebec and Nova Scotia were not yet on the radar.) Up until this time, provincial governments had given guarantees to grape growers that their crops would be sold to wineries at negotiated prices; those grapes rejected by winemakers would be purchased by the government for distillation into industrial alcohol.

There was no incentive for the growers to move from hybrids to the noble *vinifera* varieties, although a few dedicated growers such as John Marynissen and Bill Lenko in Ontario had proven that you could grow and sustain Cabernet Sauvignon and Chardonnay in a marginal climate using the right viticultural practices.

Planting *vinifera* varieties

With the evaporation of trade barriers, the industry gave a collective shudder and convinced itself the sky was falling. Their response to this gloom-and-doom scenario—which turned out to be the right response—was to look at their vineyards and decide what they should get rid of. Ontario reduced its vineyard acreage from 24,000 to 17,000.

Banning *Labrusca*

In the same year as the Free Trade Agreement, the Wine Content Act in Ontario was revised and *labrusca* varieties were banned from table wines (they could still be used for "pop" wines, such as Baby Duck, and for faux sherry and port). If you have ever tasted Concord grape juice, you will be familiar with an aroma and taste of *labrusca* grapes. When fermented, these varieties, like Niagara and Isabella, give off a smell that the wine community calls "foxy." The marker for *labrusca*-based wines is a chemical called methyl anthranilate. It's a hard smell to define—I have never been close enough to a fox to confirm this descriptor. Thomas Pinney, in *A History of Wine in America: From the Beginnings to Prohibition* (1989), quotes the Russian-born American winemaker Alexander Brailow: "People have tried to compare the smell and taste to things that they know. In Russia, for instance, they say that the grape Isabella, which is grown extensively in Crimea for red wine, smells like bedbugs. It all depends on the association and personal taste."

Isabella, incidentally, was the variety most responsible for the spread of phylloxera through the vineyards of Europe in the 1860s. Cuttings of this vine contaminated with the phylloxera louse were shipped to the Rhône Valley for planting because of its resistance to powdery mildew, a fungus that had afflicted French vineyards from the 1850s. The scourge of phylloxera laid waste the vineyards of Europe. Its estimated damage: two and one half times the cost of the Franco-Prussian War. Ironically, the saviour of Europe's vines was the very cause of the initial disaster. North American *labrusca* varieties are immune to the depredations of the phylloxera louse. Now virtually all European vines are planted on North American rootstock.

Introduction of VQA

The pullout program in Ontario meant that future vineyards would be planted to *vinifera* varieties—which were those the consumer wanted. In the same year as free trade (1988), the Vintners Quality Alliance (VQA) was introduced in Ontario. This appellation system required that the wines be 100 percent grown in designated viticultural areas. The regulations also stipulated minimum sugar levels for grapes and the varieties that could be used for wine. And, importantly, all wines had to be blind tasted for typicity and quality by an independent review panel at the Liquor Control Board of Ontario.

Viticultural Know-How

From its earliest years, going back to the beginning of the nineteenth century, winemaking in Canada had been the preoccupation of passionate amateurs. At the urging of friends, farmers and basement hobbyists whose skills were sufficient to make decent wine opened commercial wineries. They may have taken courses in winemaking or followed the instructions of European relatives, but they had no experience of winemaking in other regions. Growers knew little of trellising techniques used around the world to maximize ripening and thought that dropping fruit to concentrate flavours was a crime against Mother Nature. With the advent of VQA, the more established wineries began to recruit winemakers from other parts of the world who could bring a different experience to bear on Ontario fruit. And they sent their own assistant winemakers to do harvests abroad. Today, the tally of winemakers in Ontario from Europe and other New World regions who bring their experience to Canadian wineries is impressive.

All in all, the future for the wine industries in Ontario is rosy. Consider that Ontario Icewine, in the space of a mere twenty years—when Inniskillin won the Prix d'honneur trophy for its Vidal Icewine 1989 at Vinexpo 1991—has become an icon wine around the world, even to the point that unscrupulous vintners are counterfeiting it for profit. But beyond Icewine, Ontario's Chardonnays and Rieslings, Pinot noirs and Cabernet blends are winning applause from the world's most discerning judges.

From a purely personal point of view—and as a writer who has followed the industry for over thirty-five years—I wish that British diplomat was around today to taste the sparkling wines now being made in the province. He would concur, I'm sure, that Ontario is entering its golden age of winemaking.

Contributors

Nick Baxter-Moore is Associate Professor in the Department of Communication, Popular Culture and Film at Brock University, where he teaches courses on media industries, theory and methods in popular culture, and popular music, among others. He is president of the Popular Culture Association of Canada. His current research interests include the politics of popular music, ethnographic approaches to re-enacting and "living history," and, as in this book, the study of local popular culture, with particular emphasis on the Niagara region.

Linda Bramble is the author of seven books on the wines and wine country of Ontario. In 2009 she won the Wine Writer of the Year Award for her book *Wine Visionaries of Niagara* (2009). Retired in 2005, she was on the faculty of Brock University's Cool Climate Oenology and Viticulture Institute and adjunct with the Faculty of Business. She continues to write about wine, does public speaking, is an Academic Coach for Athabasca University's M.B.A. program, and writes scripts for online educational business games.

Astrid Brummer holds the position of LCBO Product Manager, Ontario Wines. She joined the LCBO in 2001 and has had extensive experience sharing her passion for the world of wine, beer, and spirits. She is an expert on the provincial wine scene and thoroughly enjoys her role in helping consumers discover great local wines and ensuring that Ontario wineries find success at the LCBO.

Caroline Charest received a bachelor's degree in advertising from OCAD and a master's degree in Popular Culture from Brock University. Her graduate thesis examined the Ontario wine industry during its major transition during the 1980s. In particular, she explores industry leaders' efforts to use "authenticity" in marketing both wine regions and the wines themselves. She currently lives in Montreal, where she occupies a merchandising management position in the luxury industry.

Daryl F. Dagesse is Associate Professor of Geography at Brock University. His research centres primarily on physical changes that occur in soil during freezing and thawing through the winter. This work has been presented at international conferences and published in academic journals. He also investigates the inter-relationships between the postglacial landforms of the Niagara Peninsula and the soils developed upon them. His other interests include traditional martial arts and flying as a private pilot.

Dirk De Clercq is Professor of Management in the Goodman School of Business at Brock University. His research interests are in entrepreneurship, innovation, social exchange relationships, firm internationalization, and cross-country studies.

Victoria Fast completed her master's degree in Geography at Brock University in 2011 by blending her passion for maps and wine. Her research focuses on the application of geomatics technologies to characterize vineyard variability at Stratus Vineyards in Niagara-on-the-Lake. Victoria is currently a Ph.D. candidate in Environmental Applied Science and Management at Ryerson University (Toronto), where she is studying the application of Web-based mapping technologies to engage the public and policy-makers on global environmental change.

Christopher Fullerton is Associate Professor in the Department of Geography at Brock University. His research focuses largely on the dynamics and outcomes of rural land use planning and economic development initiatives, with a particular emphasis on grassroots-based projects. Prior to entering academia, he worked as a rural economic development practitioner in Manitoba and Saskatchewan.

Hugh Gayler is Professor Emeritus of Geography at Brock University. His research focuses on urban geography and urban planning, and he has published on various aspects of suburbanization and exurbanization in Western countries. Recent publications have examined issues relating to urban sprawl into areas of high resource value, and in particular how this sprawl influences the development of the wine industry in Niagara.

Atsuko Hashimoto is Associate Professor in the Department of Tourism and Environment at Brock University, where she is also an Associate Member of the Environmental Sustainability Research Centre. Her research focuses on socio-cultural, intercultural, and human aspects of tourism development and Green Tourism in Japan.

Alun Hughes graduated from Cambridge University and came to Brock in 1969 to teach geomatics—cartography, remote sensing, photogrammetry, geodesy, surveying, and geographic information systems. In 1988, as Cartographic Editor for *The Great Lakes: An Environmental Atlas and Resource Book*, he was presented with the British Cartographic Society Design Award. Since the late 1990s his research has moved from geomatics to local history, and he has studied and written on countless topics.

Debbie Inglis is Associate Professor of Biological Sciences and currently Director of CCOVI Director as well as a CCOVI Researcher. Her research areas include yeast biotechnology and biochemistry, icewine fermentation, yeast stress responses and their link to wine quality, etiology of grape sour rot, and wine remediation for taints. She was appointed the 2010 Niagara Grape King, an award recognizes excellence in vineyard management.

Ronald Jackson received his B.Sc. and M.Sc. from Queen's University and his Ph.D. from the University of Toronto. He was Professor and Chair, Botany Department, Brandon University, and technical advisor to the Manitoba Liquor Control Commission. He is the author of *Wine Science: Principles and Application, Wine Tasting: A Professional Handbook, Conserve Water Drink Wine*, and assorted book chapters and reviews, as well as editor of *Specialty Wines*. Officially retired, he remains active writing and editing, and is a Fellow, CCOVI, Brock University.

Russell Johnston is Associate Professor in the Department of Communication, Popular Culture and Film at Brock University. His research examines the early development of Canadian media industries, and with Michael Ripmeester he has explored the media's role in heritage work and popular memory. He is the author of *Selling Themselves: The Emergence of Canadian Advertising* (2001), as well as articles in *Media Culture & Society, Journalism & Mass Communication Quarterly, The International Journal of Heritage Studies, Canadian Geographer*, and *Canadian Historical Review*. His work in the history of advertising has been recognized by Advertising Standards Canada, the Institute of Communication Agencies (Canada), and the trade journal *Marketing*.

Marilyne Jollineau is Associate Professor in the Department of Geography at Brock University. Dr. Jollineau is primarily interested in using geospatial technologies to study a range of environmental issues, such as wetland protection and vineyard management. Recently, she has focused on the use of geospatial technologies for improved vineyard management in the Niagara Region. Her work has been published in quality journals, including the *International Journal of Remote Sensing*.

Phillip Gordon Mackintosh is Associate Professor and chair of the department of Geography at Brock University. His research in urban historical geography includes liberal conceptions of public space in Victorian, Edwardian, and inter-war Toronto.

Dan Malleck is Associate Professor in the Department of Community Health Sciences at Brock University. He researches medical history and is especially interested in the history of drug and alcohol regulation and prohibition. He is author of *Try to Control Yourself: The Regulation of Public Drinking in Post-Prohibition Ontario, 1927–44* and co-editor of *Consuming Modernity: Changing Gendered Behaviours and Consumerism, 1919–45*. He is editor-in-chief of the *Social History of Alcohol and Drugs: An Interdisciplinary Journal*.

Janet McLaughlin is Assistant Professor of Health Studies at Wilfrid Laurier University and a Research Associate with the International Migration Research Centre (IMRC). She has authored numerous book chapters and articles on health and human rights issues facing migrant agricultural workers in Canada. Dr. McLaughlin is also co-founder of the Migrant Worker Health Project <http://www.migrantworkerhealth.ca>, which aims to improve health care access for migrant workers in Ontario.

Narongsak (Tek) Thongpapanl is Associate Professor of Marketing and New Product Development at the Goodman School of Business. His main research interests include NPD, innovation management, and strategic marketing management in dynamic environments. His research examines the factors influencing developers' interaction with the marketplace as they advance innovations that potentially create both convergent and divergent values and interests for stakeholders. His research also investigates the exploration–exploitation tension that often presents problems in the implementation of both new-to-the-market and new-to-the-firm innovations, especially in highly dynamic settings, as is the case with the Ontario wine industry.

Gary Pickering is Professor of Biological Sciences and Psychology/Wine Science and a CCOVI Researcher. His areas of research include flavour science, wine quality, sensory evaluation, and the psychophysics of taste. He is also an inventor, international wine judge, and President of Picksen International Inc.

Andrew Reynolds is Professor of Viticulture, Brock University. His major research interests include canopy management; site, soil, and their impact upon flavor; irrigation and water relations; geomatics; and the use of GSP/GIS and remote sensing for studying terroir. He has won numerous awards, including the OIV Award for Best Book in Oenology (2011), the ASEV/ES Outstanding Achievement Award (2010), and Brock University Faculty Award for Excellence in Teaching (2009).

Michael Ripmeester is Professor of Geography at Brock University and a Research Fellow with CCOVI. His research focuses on the relations among landscape, power, and resistance. Recent work, with Russell Johnston, includes exploration of the place of wine in Niagara heritage narratives. His research has appeared in a number of journals, including *The Canadian Geographer*, *City, Culture and Society*, and the *International Journal of Heritage*, as well as in books focused on popular culture and popular memory.

Anthony B. Shaw is Professor in the Department of Geography and a Fellow of CCOVI (Cool Climate Enology and Viticulture Institute) at Brock University. His last major research projects include the demarcation of sub-appellations in Ontario's main wine regions for the Vintners' Quality Alliance, Ontario, and the assessment of new areas for wine production. His current major research area, funded by the Ontario Research Fund, is to determine evidence of climate change and assess the potential impacts on Ontario's main wine appellations and emerging regions.

David J. Telfer is Associate Professor in the Department of Tourism and Environment at Brock University. He is also an Associate Member of the Environmental Sustainability Research Centre at Brock. His research focuses on tourism and development theory, the linkages between tourism and agriculture, and Green Tourism in Japan.

Maxim Voronov (Ph.D., Columbia University) is an Associate Professor of Strategic Management at the Goodman School of Business at Brock University. His research centres on the ways people and organizations navigate and try to shape their socio-cultural environment. He has spent more than five years studying the Ontario wine industry, and his publications have appeared in such leading journals as *Academy of Management Review* and *Journal of Management Studies*.

Index